Design and Implemer
Inductive DC-DC Co

ANALOG CIRCUITS AND SIGNAL PROCESSING

Series Editors:
Mohammed Ismail. *The Ohio State University*
Mohamad Sawan. *École Polytechnique de Montréal*

For other titles published in this series, go to
www.springer.com/series/7381

Mike Wens · Michiel Steyaert

Design and Implementation of Fully-Integrated Inductive DC-DC Converters in Standard CMOS

Dr. Mike Wens
ESAT-MICAS
Dept. Elektrotechniek
K.U. Leuven
Room 91.22, Kasteelpark
Arenberg 10
Leuven B-3001
Belgium
Mike.Wens@esat.kuleuven.be

Prof. Dr. Michiel Steyaert
ESAT-MICAS
Dept. Elektrotechniek
K.U. Leuven
Kardinaal Mercierlaan 94
Heverlee B-3001
Belgium
michiel.steyaert@esat.kuleuven.ac.be

Series Editors:
Mohammed Ismail
205 Dreese Laboratory
Department of Electrical Engineering
The Ohio State University
2015 Neil Avenue
Columbus, OH 43210
USA

Mohamad Sawan
Electrical Engineering Department
École Polytechnique de Montréal
Montréal, QC
Canada

ISBN 978-94-007-1435-9   e-ISBN 978-94-007-1436-6
DOI 10.1007/978-94-007-1436-6
Springer Dordrecht Heidelberg London New York

Library of Congress Control Number: 2011928697

© Springer Science+Business Media B.V. 2011
No part of this work may be reproduced, stored in a retrieval system, or transmitted in any form or by any means, electronic, mechanical, photocopying, microfilming, recording or otherwise, without written permission from the Publisher, with the exception of any material supplied specifically for the purpose of being entered and executed on a computer system, for exclusive use by the purchaser of the work.

Cover design: VTeX UAB, Lithuania

Printed on acid-free paper

Springer is part of Springer Science+Business Media (www.springer.com)

To my wife Larissa and our daughter Anna

Preface

Technological progress in the semiconductor industry has led to a revolution towards new advanced, miniaturized, intelligent, battery-operated and wireless electronic applications. The base of this still ongoing revolution, commonly known as Moore's law, is the ability to manufacture ever decreasing transistor sizes onto a CMOS chip. In other words, the transistor density increases, leading to larger quantity of transistors which can be integrated onto the same single chip die area. As a consequence, more functionality can be integrated onto a single chip die, leading to Systems-on-Chip (SoC) and reducing the total system cost. Indeed, the cost of electronic applications depends in a inverse-proportional fashion on the degree of on-chip integration, which is the main drive for CMOS scaling.

A SoC requires both analog and digital circuitry to be combined in order for it to be able to interact with the analog world. Nevertheless, it is usually processed in a native digital CMOS technology. These CMOS technologies are optimized for the integration of large-scale digital circuits, using very small transistors and low power supply voltages to reduce the power consumption. Beside for the purpose of decreasing the (dynamic) power consumption, the power supply voltage of deep-submicron CMOS technologies is also limited due to the physically very thin gate-oxide of the transistors. This thin gate-oxide, of which the thickness may merely be a few atom layers, would otherwise suffer electrical breakdown. However, the analog circuitry generally needs higher power supply voltages, compared to the digital circuitry. For instance, a power amplifier needs a higher supply voltage to deliver sufficient power into the communication medium. Also, analog signal processing blocks require a higher supply voltage to achieve the desired Signal-to-Noise-Ratio (SNR).

Due to the trend towards electronic applications of portable and wireless nature, (rechargeable) batteries are mandatory to provide the required energy. Although also prone to innovation and improvement, the battery voltage does not scale with the CMOS technology power supply voltages. Obviously, this is due to their physical and chemical constraints. Moreover, their energy density remains limited, limiting the available power and/or the autonomy of the application. Therefore, it is clear that power-management on a SoC-scale is mandatory for ensuring the ongoing feasibility of these applications.

Matching the battery voltage to the required power supply voltage(s) of the SoC can essentially be done in two ways. The first method, which can only be used when the battery voltage is higher than the required power supply voltage(s), is the use of linear voltage converters. This method is very often applied in current state-of-the-art applications, due to the simplicity to integrated it onto the SoC and its low associated cost. However, the excess energy from the battery voltage is dissipated in the form of waste heat, negatively influencing the autonomy and/or physical size of the application. The second method, putting no constraints to the battery voltage, is the use of switched-mode Direct-Current to Direct-Current (DC-DC) voltage converters. These converters are able to increase or decrease the battery voltage in a power-efficient fashion, leading to potentially higher battery autonomies. As a drawback, these switched-mode DC-DC converters are more complex and difficult to integrate onto the SoC, which is why they still require off-chip electronic components, such as inductors and capacitors.

The focus of the presented work is to integrate the switched-mode DC-DC converters onto the SoC, thus reducing both the number of external components and the Printed Circuit Board (PCB) footprint area. However, the poor electrical properties (low Q-factors) of on-chip inductors and capacitors and their low associated values (nH, nF) poses many difficulties, potentially compromising the power conversion efficiency advantage. Combing both the concepts of monolithic SoC integration and achieving a maximal (overall) power conversion efficiency, is the key to success. Moreover, to minimize the costs, the power density of the fully-integrated DC-DC converter is to be maximized.

To achieve these goals a firm theoretical base on the matter of DC-DC conversion is provided, leading to the optimal inductive DC-DC converter topology choices. An extensive mathematical steady-state model is deduced, in order to accurately predict both the trade-offs and performance limits of the inductive DC-DC converters. A further increase the performance of DC-DC converters is achieved through the design of novel control techniques, which are particularly optimized for high-frequency monolithic inductive DC-DC converters. Finally, the theory and simulations are verified and validated through the realization of seven monolithic inductive CMOS DC-DC converters. As such, the highest power density and Efficiency Enhancement Factor (EEF) over a linear voltage converter are obtained, in addition to the feasibility proofing of various novel concepts.

The authors also wish to express their gratitude to all persons who have contributed to this scientific research and the resulting book. We would like to thank Prof. R. Puers and Prof. W. Dehaene for their useful comments. In addition we would like to thank the colleagues of the ESAT-MICAS laboratories of K.U. Leuven for both the direct and indirect contributions to the presented work. Finally, we thank our families for their unconditional support and patience.

Leuven   Mike Wens
Michiel Steyaert

Contents

1	**Introduction**		1
	1.1 The Origin of DC-DC Converters		2
		1.1.1 Basic Considerations	2
		1.1.2 Historical Notes	3
	1.2 Low Power DC-DC Converter Applications		9
		1.2.1 Mains-Operated	10
		1.2.2 Battery-Operated	11
	1.3 Monolithic DC-DC Converters: A Glimpse into the Future		14
		1.3.1 CMOS Technology	15
		1.3.2 The Challenges	20
	1.4 Structural Outline		23
	1.5 Conclusions		24
2	**Basic DC-DC Converter Theory**		27
	2.1 Linear Voltage Converters		27
		2.1.1 Series Converter	28
		2.1.2 Shunt Converter	29
	2.2 Charge-Pump DC-DC Converters		31
		2.2.1 On Capacitors	32
		2.2.2 Series-Parallel Step-Down Converter	34
		2.2.3 Series-Parallel Step-Up Converter	38
	2.3 Inductive Type DC-DC Converters		41
		2.3.1 On Inductors	41
		2.3.2 Inductors and Capacitors: The Combination	44
		2.3.3 Reflections on Steady-State Calculation Methods	49
	2.4 INTERMEZZO: The Efficiency Enhancement Factor		59
		2.4.1 The Concept	59
		2.4.2 Interpretations	61
	2.5 Conclusions		62
3	**Inductive DC-DC Converter Topologies**		65
	3.1 Step-Down Converters		65

ix

	3.1.1	Buck Converter	66
	3.1.2	Bridge Converter	72
	3.1.3	Three-Level Buck Converter	74
	3.1.4	$Buck^2$ Converter	77
	3.1.5	Watkins-Johnson Converter	79
	3.1.6	Step-Down Converter Summary	81
3.2	Step-Up Converters	82
	3.2.1	Boost Converter	84
	3.2.2	Current-Fed Bridge Converter	85
	3.2.3	Inverse Watkins-Johnson Converter	86
	3.2.4	Step-Up Converter Summary	88
3.3	Step-Up/Down Converters	90
	3.3.1	Buck-Boost Converter	91
	3.3.2	Non-inverting Buck-Boost Converter	92
	3.3.3	Ćuk Converter	93
	3.3.4	SEPIC Converter	94
	3.3.5	Zeta Converter	95
	3.3.6	Step-Up/Down Converter Summary	97
3.4	Other Types of Inductive DC-DC Converters	99
	3.4.1	Galvanic Separated Converters	99
	3.4.2	Resonant DC-DC Converters	104
3.5	Topology Variations	107
	3.5.1	Multi-phase DC-DC Converters	107
	3.5.2	Single-Inductor Multiple-Output DC-DC Converters	115
	3.5.3	On-Chip Topologies	118
3.6	Conclusions	121

4 A Mathematical Model: Boost and Buck Converter 123
4.1	Second-Order Model: Boost and Buck Converter		124
	4.1.1	Differential Equations: Boost Converter	124
	4.1.2	Calculating the Output Voltage: Boost Converter	126
	4.1.3	Differential Equations: Buck Converter	131
	4.1.4	Calculating the Output Voltage: Buck Converter	132
4.2	Non-ideal Converter Components Models		135
	4.2.1	Inductor	136
	4.2.2	Capacitor	142
	4.2.3	Switches	146
	4.2.4	Buffers	152
	4.2.5	Interconnect	154
4.3	Temperature Effects		158
	4.3.1	Inductor	159
	4.3.2	Switches	159
4.4	The Final Model Flow		160
	4.4.1	Inserting the Dynamic Losses	161
	4.4.2	Inserting the Temperature Effects	163
	4.4.3	Reflections on Design	164
4.5	Conclusions		167

Contents

5 Control Systems ... 169
  5.1 Inductive Type Converter Control Strategies ... 170
    5.1.1 Pulse Width Modulation ... 170
    5.1.2 Pulse Frequency Modulation ... 175
    5.1.3 Pulse Width Modulation vs. Pulse Frequency Modulation ... 176
  5.2 Constant On/Off-Time: COOT ... 181
    5.2.1 The COOT Concept ... 181
    5.2.2 Single-Phase, Single-Output Implementations ... 184
    5.2.3 Single-Phase, Two-Output SIMO Implementation ... 188
  5.3 Semi-Constant On/Off-Time: SCOOT ... 193
    5.3.1 The SCOOT Concept ... 193
    5.3.2 Multi-phase Implementations ... 195
  5.4 Feed-Forward Semi-Constant On/Off-Time: F^2-SCOOT ... 203
    5.4.1 The F^2-SCOOT Concept ... 203
    5.4.2 Single-Phase, Two-Output Implementation ... 205
  5.5 Start-up ... 209
    5.5.1 The Concept ... 210
    5.5.2 Implementations ... 210
  5.6 Conclusions ... 211

6 Implementations ... 213
  6.1 Monolithic Converter Components ... 214
    6.1.1 Inductor ... 214
    6.1.2 Capacitor ... 216
    6.1.3 Switches ... 220
  6.2 On Measuring DC-DC Converters ... 224
    6.2.1 Main Principles ... 224
    6.2.2 Practical Example ... 226
  6.3 Boost Converters ... 228
    6.3.1 Bondwire, Single-Phase, Single-Output ... 228
    6.3.2 Metal-Track, Single-Phase, Two-Output SIMO ... 232
  6.4 Buck Converters ... 235
    6.4.1 Bondwire, Single-Phase, Single-Output ... 236
    6.4.2 Metal-Track, Single-Phase, Single-Output ... 240
    6.4.3 Metal-Track, Four-Phase, Single Output ... 244
    6.4.4 Metal-Track, Four-Phase, Two-Output SMOC ... 248
    6.4.5 Bondwire, Single-Phase, Two-Output SMOC ... 250
  6.5 Comparison to Other Work ... 254
    6.5.1 Inductive Step-Up Converters ... 255
    6.5.2 Inductive Step-Down Converters ... 256
  6.6 Conclusions ... 259

7 General Conclusions ... 261
  7.1 Conclusions ... 261
  7.2 Remaining Challenges ... 263

References 265

Index 273

Abbreviations and Symbols

Abbreviations

AC	Alternating-Current
AC-AC	Alternating-Current to Alternating-Current
AC-DC	Alternating-Current to Direct-Current
ADC	Analog-to-Digital Converter
BCM	Boundary Conduction Mode
BiCMOS	Bipolar Complementary Metal-Oxide Semiconductor
BJT	Bipolar Junction Transistor
BW	BandWidth
CB	Conduction Boundary
CFL	Compact Fluorescent Lamps
CM	Conduction Mode
CCM	Continuous Conduction Mode
CMOS	Complementary Metal-Oxide Semiconductor
COOT	Constant On/Off-Time
CRT	Cathode Ray Tube
DAC	Digital-to-Analog Converter
DC	Direct-Current
DC-AC	Direct-Current to Alternating-Current
DC-DC	Direct-Current to Direct-Current
DCM	Discontinuous Conduction Mode
DIL	Dual In Line
EEF	Efficiency Enhancement Factor
EMI	Electro Magnetic Interference
FAIMS	High-Field Asymmetric waveform Ion Mobility Spectrometry
FET	Field-Effect Transistor
ESI	Electro-Spray Ionization
ESL	Electric Series inductance
ESR	Electric Series Resistance
FOX	Field Oxide
F^2SCOOT	Feed-Forward Semi-Constant On/Off-Time

GBW	Gain BandWidth
GND	GrouND
HF	High Frequency
IC	Integrated Circuit
IGBT	Insulated Gate Bipolar Transistor
LDO	Low Drop-Out
LIDAR	Laser Imaging Detection And Ranging
LiION	Lithium-ION
ME1	Metal-1
MIM	Metal-Insulator-Metal
MOM	Metal-Oxide-Metal
MOS	Metal-Oxide Semiconductor
MOSFET	Metal-Oxide Semiconductor Field-Effect Transistor
n-MOSFET	n-channel Metal-Oxide Semiconductor Field-Effect Transistor
MS	Mass Spectrometry
MUX	Multi-PleXer
NPN	n-type p-type n-type transition
OPAMP	OPerational AMPlifier
OTA	Operational Transconductance Amplifier
OX1	Oxide-1
PC	Personal Computer
PCB	Printed Circuit Board
PFC	Power Factor Correction
PFM	Pulse Frequency Modulation
p-MOSFET	p-channel Metal-Oxide Semiconductor Field-Effect Transistor
PNP	p-type n-type p-type transition
PSRR	Power Supply Rejection Ratio
PTC	Positive temperature coefficient
PWM	Pulse Width Modulation
Q.E.D.	Quod Erat Demonstrandum
RC	Resistor-Capacitor
RF	Radio-Frequency
RL	Resistor-Inductor
RLC	Resistor-Inductor-Capacitor
RMS	Root-Mean-Square
SCOOT	Semi-Constant On/Off-Time
SEPIC	Single-Ended Primary-Inductance Converter
SiGe	Silicon-Germanium
SIMO	Single-Inductor Multiple-Output
SMOC	Series Multiple-Output Converter
SMOS	Type of healthy sandwich
SMPS	Switched-Mode Power Supply
SoC	System-on-Chip
SPICE	Simulation Program with Integrated Circuit Emphasis

SRR	Supply Rejection Ratio
SW	Switch

Symbols and Quantities

A	Area
A_C	On-chip capacitor area
A_C^+	Positive charge balance area
A_C^-	Negative charge balance area
A_L	Perpendicular projected area of the inductor windings
A_L^+	Positive volt-second balance area
A_L^-	Negative volt-second balance area
arccos	Arc cosine
$A_\varnothing$	Perpendicular cross-sectional area
$A_{\varnothing\_eff}$	Effective $A_\varnothing$
C	Capacitance
C_{eq}	Equivalent capacitance
cos	Cosine
C_{db}	Parasitic drain-bulk capacitance
C_{dec}	Input decouple capacitance
C_{gb}	Parasitic gate-bulk capacitance
C_{gd}	Parasitic gate-drain capacitance
C_{gc}	Parasitic gate-source capacitance
C_{gg}	Parasitic gate capacitance
C_{g_min}	Parasitic gate of a minimal size inverter
C_{sb}	Parasitic source-bulk capacitance
C_{in}	Input capacitance
C_{out}	Output capacitance
C_{tot}	Total capacitance
C_{out_tot}	Total output capacitance
C_{out_1}	Capacitance of output 1
C_{pad}	Parasitic substrate capacitance of a bonding pad
C_{par}	Parasitic capacitance
C_{sub}	Parasitic substrate winding capacitance
C_1	Capacitance 1
d	Thickness
d	Pitch between two conductors
d_{ox}	Thickness of the oxide
e	Euler's constant: $2.718281828\ldots$
E_{C_in}	Energy stored in C
E_{C_out}	Energy delivered by C
$E_C(t)$	Energy stored in C as a function of t
E_{C_1}	Energy stored in C_1
E_{C_2}	Energy stored in C_2
$E_{C_1 C_2}$	Energy stored in C_1 and C_2
$E_{C_1 \to C_2}$	Transferred energy from C_1 to C_2
$E_{C_1 \to R}$	Transferred energy from C_1 to R

$E_{C_1 \to RC}$	Transferred energy from C_1 to RC
EEF	Efficiency Enhancement Factor
$\overline{EEF}$	Mean Efficiency Enhancement Factor
$\widetilde{EEF}$	Weighted Efficiency Enhancement Factor
$EEF(P_{out\_i})$	EEF as a function of the ith P_{out}
E_L	Magnetic energy stored in L
E_{L_in}	Energy stored in L
E_{L_out}	Energy delivered by L
$E_L(t)$	Magnetic energy stored in L as a function of t
$E_L(t)$	Energy stored in L as a function of t
$E_R(t)$	Energy dissipated in R as a function of t
E_{R_2}	Energy dissipated in R_2 in steady-state
$E_{R_2}(t)$	Energy dissipated in R_2 as a function of t
$E_{U_{in}}(t)$	Energy delivered by U_{in} as a function of t
$E_{U_{in} \to C_1}$	Transferred energy from U_{in} to C_1
$E_{U_{in} \to C_1 C_2}$	Transferred energy from U_{in} to C_1 and C_2
$E_{U_{in} C_1 \to C_2}$	Transferred energy from U_{in} and C_1 to C_2
$E_{U_{in} \to C}(t)$	Transferred energy from U_{in} to C as a function of t
$E_{U_{in} \to L}(t)$	Transferred energy from U_{in} to L as a function of t
$E_{U_{in} \to R}(t)$	Transferred energy from U_{in} to R as a function of t
$E_{U_{in} \to RC}(t)$	Transferred energy from U_{in} to R and C as a function of t
$E_{U_{in} \to RL}(t)$	Transferred energy from U_{in} to R and L as a function of t
$E_{U_{in} \to RLC}(t)$	Transferred energy from U_{in} to R, L and C as a function of t
f	Frequency
F	Global effective fan-out
f_{scale}	Scaling factor
f_{SW}	Switching frequency
f_0	Resonance frequency
$g_1\{\}$	Function g_1
$H(f_{SW})$	Transfer function as a function of f_{SW}
I	Current
I_{ak}	Anode-cathode current
I_b	Base current
I_c	Collector current
i_{C_charge}	Charge current through C
$i_{C\_discharge}$	Discharge current through C
I_{C_leak}	Leakage current through C
I_{cs}	Control system supply current
$i_C(t)$	Current through C as a function of t
i_{C1}	Current 1 through C
i_{C2}	Current 2 through C
I_{ds}	Drain-source current
I_e	Emitter current
I_{in}	Input current
I'_{in_max}	Maximum input current

I'_{in_min}	Minimum input current
I_{in_RMS}	RMS input current
$\overline{I_L}$	Mean current through L
I_{L_max}	Maximal current through L
I_{L_min}	Minimal current through L
$i_L(t)$	Current through L as a function of t
$i_L(0)$	Initial current through L
I_{out}	Output current
$\overline{I_{out}}$	Mean output current
I_{out_RMS}	RMS output current
$I_{out}(t)$	Output current as a function of t
i_{prim}	Current through primary winding
$i_{prim}(s)$	Current through primary winding, in the Laplace-domain
$i_{Rb}(t)$	Current through R_b as a function of t
$i_{Rc}(t)$	Current through R_c as a function of t
i_{sec}	Current through secondary winding
$i_{sec}(s)$	Current through secondary winding, in the Laplace-domain
i_{SW}	Current through SW
I_{SW1_RMS}	RMS current through $SW1$
$\overline{i_{SW2}}$	Mean current through $SW2$
$i_{SW2}(t)$	Current through $SW2$ as a function of t
$i(t)$	Current as a function of t
k	Voltage conversion ratio
K	Form-factor fitting parameter
$k(f_{SW})$	Voltage conversion ratio as a function of f_{SW}
k_{lin}	Voltage conversion ratio of a linear voltage converter
k_{lin_max}	Maximal voltage conversion ratio of a linear voltage converter
k_M	Magnetic coupling factor
k_{SW}	Voltage conversion ratio of a switched-mode voltage converter
$k(\delta)$	Voltage conversion ratio as a function of δ
ℓ	Length of a conductor
L	Inductance
L_{Cs}	Parasitic series inductance of C
lim	Limit
L_{line}	Metal line length
L_M	Magnetizing inductance
ln	Natural logarithm
L_n	Gate-length of an nMOSFET
L_{p_buff}	L_p of a buffer
$\ell_{overlap}$	Overlapping length of two conductors
L_p	Gate-length of an pMOSFET
L_{prim}	Primary winding inductance
L_{sec}	Secondary winding inductance
L_{self}	Self inductance
L_{tot}	Total inductance

L_{track}	Length of a metal track
L_1	Inductance 1
$\mathcal{L}^{-1}$	Inverse Laplace-transform
M	Mutual inductance
M^+	Positive mutual inductance
M^-	Negative mutual inductance
n	Number of stages/phases
N_a	Doping concentration
n_{prim}	Number of turns in the primary winding
n_{sec}	Number of turns in the secondary winding
n_{Tr}	Winding turn ratio
n_1	Number of turns of winding 1
P_{buff_cpar}	Power loss in parasitic capacitances in buffers
P_{buff_short}	Power loss due to short-circuit current in buffers
P_C	Power for charging a capacitor
P_{Df}	Diode forward conduction power loss
P_{diss}	Dissipated power
P_{in}	Input power
P_{in_lin}	Input power of a linear DC-DC voltage converter
P_{in_SW}	Input power of switched-mode DC-DC voltage converter
P_{L_Csub}	Parasitic substrate capacitance power loss of an inductor
P_{out}	Output power
P'_{out}	Real output power
P_{out_lin}	Output power of a linear DC-DC voltage converter
P_{out_max}	Maximal output power
P_{out_SW}	Output power of a switched-mode DC-DC voltage converter
P_{Rcs}	Parasitic series resistance power loss
P_{Rcp}	Parasitic parallel resistance power loss
P_{Rin}	Power loss in R_{in}
P_{Ron}	Power loss in R_{on}
P_{Rout}	Power loss in R_{out}
P_{Rsw1}	Power loss in R_{SW1}
P_{tf_SW1}	Fall-time power loss of $SW1$
P_{tr_SW1}	Rise-time power loss of $SW1$
Q	Q-factor
Q_d	Charge in the drain
Q_g	Charge in the gate
Q_s	Charge in the source
Q_{in}	Stored charge
Q_{out}	Delivered charge
r	Perpendicular cross-section radius a round conductor
R	Resistance
R_a	Equivalent resistance
R_b	Equivalent resistance
$R_{bondwire}$	Parasitic series resistance of a bondwire

R_c	Equivalent resistance
$R_{channel\square}$	Square-resistance of the induced channel
R_{Cdec}	Parasitic series resistance of C_{dec}
R_{cont_f}	Parasitic series resistance of gate contacts
R_{cont_ds}	Parasitic series resistance of drain/source contacts
R_{Cp}	Parasitic parallel resistance of C
R_{Cs}	Parasitic series resistance of C
R_e	Equivalent load resistance
R_{eq}	Equivalent resistance
R_{in}	Input resistance
R_{in}	Parasitic series resistance of U_{in}
R_L	Load resistance
R'_L	Real load resistance
R_{left}	Conductor series resistance, seen from the left
R_{Ls}	Parasitic series resistance of L
$R_{Ls@T}$	R_{Ls} at temperature T
$R_{Ls@T+\Delta T}$	R_{Ls} at temperature $T + \Delta T$
R_{line}	Line resistance
R_{loss}	Additional loss resistance
$R_{n+\square}$	Square-resistance of n^+-region
R_{on}	On-resistance
$R_{on@T}$	R_{on} at temperature T
$R_{on@T+\Delta T}$	R_{on} at temperature $T + \Delta T$
R_{on_n}	On-resistance of an n-MOSFET
R_{on_p}	On-resistance of an p-MOSFET
R_{out}	Parasitic output resistance
$R_{ploy\square}$	Square-resistance of poly-silicon
R_{right}	Conductor series resistance, seen from the right
R_{sen}	Sense resistance
R_{series}	Variable series resistance of a series voltage converter
R_{shunt}	Variable shunt resistance of a shunt voltage converter
R_{SW1}	Parasitic series resistance of SW_1
R_{track}	Parasitic series resistance of a metal track
R_{via}	Parasitic via series resistance
R_{via_tot}	Total parasitic via series resistance
R_0	Output resistance at f_0
$R_\square$	Square-resistance
s	Laplace-transform operator
sin	Sine
t	Time
T	Period
T	Temperature
$t_{a \to b}$	Time from point a to point b
$t_{a \to c}$	Time from point a to point c
$t_{b \to c}$	Time from point b to point c

t_d	Dead-time
t_f	Fall-time
t_{f_SW1}	Fall-time SW_1
t_{flank}	Mean rise/fall-time
t_{on}	On-time
t_{off}	Off-time
t_{off_real}	Real off-time
t_{ox}	MOSFET gate-oxide thickness
t_r	Rise-time
$t_{r/f}$	Rise/fall-time
t_{r_SW1}	Rise-time SW_1
Tr	Transformer
t_{SW}	Switching/Charging time
t_{zero1}	Intersect time 1 with the X-axis
t_1	Time 1
U	Voltage
U_{be}	Base-emitter voltage
U_{ce}	Collector-emitter voltage
U_{C_max}	Maximal voltage over C
U_{C_min}	Minimal voltage over C
$u_C(t)$	Voltage over C as a function of t
$U_C(T)$	Voltage over C at the end of T
$U_C(0)$	Initial voltage over C
U_{dd}	Nominal technology supply voltage
U_{Df}	Diode forward voltage drop
U_{ds}	Drain-source voltage
U_{dsatp}	Drain-source saturation voltage of a p-MOSFET
U_{dsn}	Drain-source voltage of an n-MOSFET
U_{err}	Error-voltage
U_{gb}	Gate-bulk voltage
U_{g_od}	Gate-overdrive voltage
U_{gs}	Gate-source voltage
U_{gsn}	Gate-source voltage of an n-MOSFET
U_{in}	Input voltage
U'_{in_max}	Maximum input voltage
U'_{in_min}	Minimum input voltage
U_{in_peak}	Peak value of U_{in}
$u_L(t)$	Voltage over L as a function of t
U_{L1}	Voltage 1 over L
U_{offset}	Offset voltage
U_{out}	Output voltage
$\overline{U_{out}}$	Mean output voltage
U_{out_max}	Maximal output voltage
U_{out_min}	Minimal output voltage
U_{out_RMS}	RMS output voltage

U'_{out_RMS}	Real RMS output voltage
$u_{out}(t)$	Output voltage as a function of t
$u_{out}(x)$	Output voltage as a function of x
$\hat{u}_{out}(x)$	Output voltage amplitude as a function of x
$\hat{u}_{out}(\theta)$	Output voltage amplitude as a function of θ
U_{prim}	Voltage over the primary winding
U_{out_ptp}	Peak-to-peak output voltage
$u_{Ra}(t)$	Voltage over R_a as a function of t
$u_{Rb}(t)$	Voltage over R_b as a function of t
$u_{Rc}(t)$	Voltage over R_c as a function of t
$u_{RCp}(t)$	Voltage over R_{Cp} as a function of t
$u_{RCs}(t)$	Voltage over R_{Cs} as a function of t
U_{ref}	Reference voltage
$u_R(t)$	Voltage over R as a function of t
U_{sb}	Source-bulk voltage
U_{sen}	Sense voltage
U_{sec}	Voltage over the secondary winding
u_{SW}	Voltage over SW
U_{SW_3}	Voltage over SW_3
U_{tria}	Triangular waveform voltage
V_t	Threshold voltage
W_{drain}	Drain-width
W_n	Gate-width of an nMOSFET
W_p	Gate-width of an pMOSFET
W_{p_buff}	W_p of a buffer
W_{source}	Source-width
W_{track}	Width of a metal track
x	ΔU_{out} approximation variable
Z_{in}	Input impedance
Z_k	Impedance ratio
Z_{out}	Output impedance
Z_1	Impedance 1
α	Resistance temperature coefficient
$\alpha(P_{out})$	Power activity probability distribution
δ	Duty-cycle
δ_{skin}	Skin-depth
ΔI_{in}	Input current ripple
ΔI_L	Current ripple through L
ΔI_{L_tot}	Total current ripple through L
ΔI_{L_1}	Current ripple through L_1
ΔP_{in}	Input power difference
ΔT	Temperature difference
ΔU	Voltage difference
ΔU_C	Voltage swing over a capacitor
ΔU_{in}	Input voltage ripple

ΔU_L	Voltage swing over an inductor
ΔU_{out}	Output voltage ripple
$\Delta U_{out}(\delta)$	Output voltage ripple as a function of δ
ΔQ	Charge difference
ΔQ_{SW}	Transferred charge in one switch cycle
$\Delta \eta$	Power conversion efficiency difference
ϵ	Dielectric permittivity
ϵ_0	Permittivity of vacuum
ϵ_{r_ox}	Relative permittivity of an oxide
η	Power conversion efficiency
η_{C_charge}	Energy charging efficiency of C
$\eta_{C\_charge}(t)$	η_{C_charge} as a function of t
η_{L_charge}	Energy charging efficiency of L
$\eta_{L\_charge}(t)$	η_{L_charge} as a function of t
$\eta_{RLC\_charge}(t)$	η_{C_charge} in an RLC-circuit as a function of t
η_{lin}	Power conversion efficiency of a linear DC-DC voltage converter
η_{sp_down}	Power conversion efficiency of a step-down charge-pump
η_{sp_up}	Power conversion efficiency of a step-up charge-pump
η_{SW}	Power conversion efficiency of a switched-mode DC-DC converter
η_{SW_max}	Maximal η_{SW}
η_{Tr}	Power conversion efficiency of an ideal transformer
$\eta_{Tr}(t)$	η_{Tr} as a function of t
η_{Φ_1}	Energy conversion efficiency of Φ_1
γ	Thermal resistance
Φ_1	Phase 1
κ	CMOS technology scaling factor
λ_p	Early voltage of a p-MOSFET
μ_n	n-carrier mobility
μ_p	p-carrier mobility
μ	Magnetic Permeability
μ_r	Relative permeability
π	Circumference/diameter ratio of a circle: $3.141592654\ldots$
ρ	Resistivity
τ_C	Time constant of an RC-circuit
τ_L	Time constant of an RL-circuit
τ_{LC}	Time constant of an RLC-circuit
τ_{Tr}	Time constant of the primary winding of a transformer
θ	Phase difference
ω_{LC}	Angular frequency of an RLC-circuit
Υ	ΔU_{out} approximation function
#fingers	Number of gate fingers of a MOS capacitor
#C_{out_1}	Total required C to implement C_{out_1}
#C_{out_tot}	Total required C to implement C_{out_tot}
#seg	Number of segments
#via	Number of vias

∞	Infinite
■	Q.E.D.
✔	A benefit
✘	A drawback

List of Figures

Fig. 1.1	A black-box representation of a DC-DC converter	2
Fig. 1.2	The power-balance of a DC-DC converter	2
Fig. 1.3	The principle of ideal switching	3
Fig. 1.4	(a) Faraday's original 1831 *induction ring* [Ins10] and (b) the schematic representation of the induction ring experiment	4
Fig. 1.5	A mechanical rotary DC-AC step-up converter, for powering gas-discharge lamps [Ran34]	4
Fig. 1.6	(a) A Cockcroft-Walton voltage multiplier build in the year 1937, which was used for an early particle accelerator [Wik10]. (b) A half-wave, two-stage Cockcroft-Walton voltage multiplier .	5
Fig. 1.7	A mechanical vibratory DC-DC step-up converter [Sta34]	6
Fig. 1.8	A DC-DC step-up converter, using an inverted vacuum-tube triode as primary switch and secondary rectifier [Haz40]	6
Fig. 1.9	(a) The first commercial vacuum-tube triode: *the audion*. (b) The schematic symbol of a direct-heated triode and its simplified construction principle	7
Fig. 1.10	A transistorized DC-DC step-up converter, for powering a vacuum-tube pentode audio amplifier for a hearing aid device [Phi53]	8
Fig. 1.11	A two-phase DC-DC step-up converter [Wes67]	8
Fig. 1.12	(a) The first contact bipolar junction transistor [Rio10b] and (b) a schematic cross section of an NPN BJT	9
Fig. 1.13	The block-diagram representation of a mains-operated application, using step-down AC-DC converter	10
Fig. 1.14	(a) The block diagram of a battery-operated application using a DC-DC step-down converter with external components. (b) The same system implemented as a SoC, with a monolithic DC-DC converter	12
Fig. 1.15	(a) The block diagram of a battery-operated application using a DC-DC step-up converter with external components. (b) The same system implemented as a SoC, with a monolithic DC-DC converter	14

Fig. 1.16	(a) The first integrated circuit [Lee10b] and (b) the schematic circuit representation [Lee10c]	15
Fig. 1.17	(a) The schematic symbol of an n-MOSFET and (b) its schematic perspective cross-section view	16
Fig. 1.18	The qualitative behavior for an n-MOSFET of I_{ds} as a function of U_{ds}, for different U_{gs}	17
Fig. 1.19	An n-MOSFET (*left*) and a p-MOSFET (*right*) in a six-mask CMOS process. The *upper* drawings show the lay-out view and the *lower* ones a cross-section of the according physical devices	18
Fig. 1.20	The minimum feature size and the transistor count per chip as a function of time, for Intel CMOS technologies [Boh09]	19
Fig. 1.21	A cross-sectional view of the interconnect of the 32 nm Intel CMOS process [Boh10]	20
Fig. 1.22	A perspective microphotograph of a DC-DC step-up converter, using a bondwire inductor [Wen07]	22
Fig. 1.23	The graphical representation of the structural outline of the dissertation	24
Fig. 2.1	(a) The principle of a linear series voltage converter and (b) a simple practical implementation	28
Fig. 2.2	(a) The power conversion efficiency η_{lin} as a function of the output power P_{out} for a linear series voltage converter, at a constant voltage conversion ratio k_{lin}. The *black curve* is valid for a zero control system supply current I_{cs} and the *gray curve* is valid for a non-zero I_{cs}. (b) The power conversion efficiency η_{lin} as a function of the voltage conversion ratio k_{lin} for a linear series voltage converter, at a constant output power P_{out}. The *black curve* is valid for a zero control system supply current I_{cs} and the *gray curve* is valid for a non-zero I_{cs}	29
Fig. 2.3	(a) The principle of a linear shunt voltage converter and (b) a simple practical implementation	30
Fig. 2.4	(a) The power conversion efficiency η_{lin} as a function of the output power P_{out} for a linear shunt voltage converter, at a constant voltage conversion ratio k_{lin}. The *black curve* is valid for a zero control system supply current I_{cs} and the *gray curve* is valid for a non-zero I_{cs}. (b) The power conversion efficiency η_{lin} as a function of the voltage conversion ratio k_{lin} for a linear shunt voltage converter, for a constant value of $P_{out} = P_{out\_max}$. The *black curve* is valid for a zero control system supply current I_{cs} and the *gray curve* is valid for a non-zero I_{cs}	31
Fig. 2.5	(a) The circuit for charging a capacitor C with a series resistor R by means of a voltage source U_{in}. (b) The voltage $u_C(t)$ over C and the current $i_C(t)$ through C, as a function of time. (c) The energy $E_{U_{in} \to RC}(t)$ delivered by U_{in}, the energy $E_{U_{in} \to C}(t)$ stored in C and the energy $E_{U_{in} \to R}(t)$ dissipated in R, as a function of time	32

List of Figures

Fig. 2.6	The energy charging efficiency η_{C_charge} of a capacitor charged by means of a voltage source U_{in} as a function of the initial voltage $U_C(0)$ over the capacitor, for three different charge times t	33
Fig. 2.7	(a) The circuit for charging C_b with C_a. (b) The $E_{C_a \to RC_b}$, $E_{C_a \to C_b}$ and $E_{C_a \to R}$ as a function of $\Delta U = U_{C_a}(0) - U_{C_b}(0)$, for $C_a \gg C_b$ and (c) for $C_b \gg C_a$	35
Fig. 2.8	(a) The circuit of an ideal series-parallel charge-pump step-down DC-DC converter, together with (b) its equivalent charge circuit and (c) its equivalent discharge circuit	36
Fig. 2.9	The *black curves* show the power conversion efficiency η_{sp_down} of an ideal series-parallel charge-pump step-down DC-DC converter, as a function of the voltage conversion ratio k_{SW}, for three different cases of the values of C_1 and C_2. The *gray curve* shows the power conversion efficiency of a linear series converter, as a function of k_{SW}	38
Fig. 2.10	(a) The circuit of an ideal series-parallel charge-pump step-up DC-DC converter, together with (b) its equivalent charge circuit and (c) its equivalent discharge circuit	39
Fig. 2.11	The *black curves* show the power conversion efficiency η_{sp_up} of an ideal series-parallel charge-pump step-up DC-DC converter, as a function of the voltage conversion ratio k_{SW}, for three different cases of the values of C_1 and C_2. The *gray curve* shows the power conversion efficiency η_{lin} of a linear series converter, as a function of k_{SW}	40
Fig. 2.12	(a) The circuit for charging an inductor L in series with a resistor R by means of a voltage source U_{in}. (b) The voltage $u_L(t)$ over and the current $i_L(t)$ through L, as a function of time. (c) The energy $E_{U_{in} \to RL}(t)$ delivered by U_{in}, the energy $E_{U_{in} \to L}(t)$ stored in L and the energy $E_{U_{in} \to R}(t)$ dissipated in R, as a function of time	42
Fig. 2.13	The energy charging efficiency η_{L_charge} of an inductor with a series resistance charged by a voltage source U_{in}, as a function of the initial current $I_C(0)$ through the inductor, for three different charge time t	43
Fig. 2.14	The circuit for charging a series inductor L and a series capacitor C with a series resistor R, by means of a voltage source U_{in}	44
Fig. 2.15	(a) The current $i(t)$, the voltage $u_L(t)$ over L, the voltage $u_R(t)$ over R and the voltage $u_C(t)$ over C as a function of time, for in ideal ($R = 0$) and (b) a non-ideal ($R \neq 0$) series *RLC*-circuit .	46
Fig. 2.16	(a) The energy $E_{U_{in} \to RLC}(t)$ delivered by U_{in} the energy $E_L(t)$ stored in L, the energy $E_R(t)$ dissipated in R and the energy $E_C(t)$ stored in C as a function of time, for an ideal ($R = 0$) and (b) a non-ideal ($R \neq 0$) series *RLC*-circuit	47

Fig. 2.17 (a) The energy charging efficiency $\eta_{RLC\_charge}(t)$ of a capacitor in a periodically-damped series RLC-circuit as a function of time, for different values of the initial voltages $U_C(0)$ over C and (b) for different initial currents $I_L(0)$ through L 48

Fig. 2.18 (a) The circuit of an ideal boost DC-DC converter. (b) The equivalent circuit of the inductor charge phase and (c) the inductor discharge phase 50

Fig. 2.19 (a) The convention of the voltage over and the current through a capacitor C and an inductor L. (b) The current $i_C(t)$ through C and (c) the voltage $u_L(t)$ over L, both in energetic equilibrium . 51

Fig. 2.20 The linearized current $i_L(t)$ through L, the linearized voltage $u_L(t)$ over L, the linearized current $i_C(t)$ through C and the linearized output voltage $u_{out}(t)$ as a function of time, for an ideal boost converter in CCM 52

Fig. 2.21 The voltage conversion ratio $k(\delta)$ as a function of the duty-cycle δ, for an ideal boost converter in CCM 53

Fig. 2.22 The linearized current $i_L(t)$ through L, the linearized voltage $u_L(t)$ over L, the linearized current $i_C(t)$ through C and the linearized output voltage $u_{out}(t)$ as a function of time, for an ideal boost converter in DCM 54

Fig. 2.23 The *upper graph* shows the voltage conversion ratio $k(\delta)$ as a function of the duty-cycle δ, where the *black curve* is valid for CCM and the *gray curves* for DCM. In the *lower graph* the *black curve* shows the boundary between the two CMs and the *gray curves* illustrate three numerical examples. These graphs are valid for an ideal DC-DC boost converter 58

Fig. 2.24 The *upper graph* shows the power conversion efficiencies η_{SW} and η_{lin} of a switched-mode DC-DC converter and a linear series voltage converter having the same voltage conversion ratio $k_{lin} = k_{SW}$, as a function of the output power P_{out}. The *lower graph* shows the corresponding EEF and $\overline{EEF}$, as a function of P_{out} 61

Fig. 3.1 (a) The circuit of an ideal buck DC-DC converter. (b) The equivalent circuit of the inductor charge phase and (c) the inductor discharge phase 67

Fig. 3.2 The linearized $i_L(t)$, the linearized $u_L(t)$, the linearized $i_C(t)$ and the linearized $u_{out}(t)$ as a function of time, for an ideal buck DC-DC converter in CCM 68

Fig. 3.3 The linearized $i_L(t)$, the linearized $u_L(t)$, the linearized $i_C(t)$ and the linearized $u_{out}(t)$ as a function of time, for an ideal buck DC-DC converter in DCM 69

Fig. 3.4	The *upper graph* shows $k(\delta)$ as a function of δ, where the *black curve* is valid for CCM and the *gray curves* for DCM. In the *lower graph* the *black curve* shows the boundary between the two CMs and the *gray curves* illustrate three numerical examples. These graphs are valid for an ideal DC-DC buck converter	70
Fig. 3.5	The circuit of an ideal bridge DC-DC converter	73
Fig. 3.6	The voltage conversion ratio $k(\delta)$ as a function of the duty-cycle δ, for a bridge converter in CCM	73
Fig. 3.7	The circuit of an ideal three-level buck DC-DC converter	75
Fig. 3.8	(a) The timing of the four switches of an ideal three-level buck DC-DC converter in CCM, for $\delta < 0.5$ and (b) for $\delta > 0.5$	75
Fig. 3.9	The circuit of an ideal buck^2 DC-DC converter	77
Fig. 3.10	The voltage conversion ratio $k(\delta)$ as a function of the duty-cycle δ, for an ideal buck^2 converter in CCM	78
Fig. 3.11	(a) The circuit of an ideal Watkins-Johnson DC-DC converter, using an inductor and (b) using two coupled inductors	80
Fig. 3.12	The voltage conversion ratio $k(\delta)$ as a function of the duty-cycle δ, for a Watkins-Johnson converter in CCM	80
Fig. 3.13	The total required capacitance C_{tot} of five step-down DC-DC converter topologies as a function of the output power P_{out}. These values are obtained by means of SPICE-simulations, such that the five converters meet with the specifications of Table 3.1	83
Fig. 3.14	The circuit of an ideal current-fed bridge DC-DC converter	85
Fig. 3.15	The voltage conversion ratio $k(\delta)$ as a function of the duty-cycle δ, for a current-fed bridge converter in CCM	85
Fig. 3.16	(a) The circuit of an ideal inverse Watkins-Johnson DC-DC converter, using an inductor and (b) using two coupled inductors	87
Fig. 3.17	The voltage conversion ratio $k(\delta)$ as a function of the duty-cycle δ, for an inverse Watkins-Johnson converter in CCM	87
Fig. 3.18	The total required capacitance C_{tot} of three DC-DC step-up converter topologies as a function of the output power P_{out}. These values are obtained by means of SPICE-simulations, such that the three converters meet with the specifications of Table 3.8	90
Fig. 3.19	The circuit of an ideal buck-boost DC-DC converter	91
Fig. 3.20	The voltage conversion ratio $k(\delta)$ as a function of the duty-cycle δ, for an ideal buck-boost converter in CCM	92
Fig. 3.21	The circuit of an ideal non-inverting buck-boost DC-DC converter	92
Fig. 3.22	The voltage conversion ratio $k(\delta)$ as a function of the duty-cycle δ, for an ideal non-inverting buck-boost converter in CCM	92
Fig. 3.23	The circuit of an ideal Ćuk DC-DC converter	93
Fig. 3.24	The circuit of an ideal SEPIC DC-DC converter	94
Fig. 3.25	The circuit of an ideal zeta DC-DC converter	96

Fig. 3.26	The total required capacitance C_{tot} of three DC-DC step-up/down converter topologies as a function of the output power P_{out}. These values are obtained by means of SPICE-simulations, such that the three converters meet with the specifications of Table 3.13	99
Fig. 3.27	The model of an ideal transformer Tr, together with its magnetizing inductance L_M	100
Fig. 3.28	(a) The circuit for calculating the energy transfer of Tr and (b) the equivalent T-circuit, both in the Laplace-domain	101
Fig. 3.29	(a) $i_{prim}(t)$ and $u_{out}(t)$, (b) $E_{U_{in}}(t)$ and $E_{R_2}(t)$ and (c) $\eta_{Tr}(t)$ as a function of time t, different values of coupling factor k_M	102
Fig. 3.30	The circuit of an ideal forward DC-DC converter	102
Fig. 3.31	The circuit of an ideal full-bridge buck DC-DC converter	103
Fig. 3.32	The circuit of an ideal push-pull boost DC-DC converter	103
Fig. 3.33	The circuit of an ideal flyback DC-DC converter	104
Fig. 3.34	The circuit of an ideal series resonant DC-DC converter	105
Fig. 3.35	The voltage conversion ratio $k(f_{SW})$ as a function of the switching frequency f_{SW}, for a series resonance DC-DC converter	105
Fig. 3.36	(a) The circuit of a halve-bridge galvanic separated series resonance DC-AC high-voltage converter for the FAIMS setup. (b) U_{out} of the DC-AC converter as a function of t. (c) A photograph of the realization of the DC-AC converter	106
Fig. 3.37	The concept of multi-phase DC-DC converters	107
Fig. 3.38	The example of how a two-phase DC-DC converter can achieve a higher power conversion efficiency η_{SW} than a single-phase DC-DC converter, at the same output power P_{out}	108
Fig. 3.39	(a) The timing signals of a two-phase converter and (b) the equivalent representation with sine waves, assuming that the converter is operating in CCM	109
Fig. 3.40	The circuit of an ideal n-phase boost DC-DC converter	110
Fig. 3.41	(a) The current $i_C(t)$ through the output capacitor C of a 2-phase boost converter for $\delta < 50\%$ and (b) for $\delta > 50\%$, both valid for CCM. $i_C(t)$ is divided into the respective parts from the first (*black curve*) and second converter (*gray curve*)	111
Fig. 3.42	The output voltage ripple ΔU_{out} as a function of the duty-cycle δ for an ideal 1-phase, 2-phase and 4-phase boost DC-DC converter. For the 1-phase and 2-phase boost converter both the exact and approximated functions are plotted. For the 2-phase and 4-phase boost converter the approximated functions are plotted	112
Fig. 3.43	The circuit of an ideal n-phase buck DC-DC converter	113
Fig. 3.44	(a) The respective currents $i_{L1}(t)$ and $i_{L2}(t)$ through inductors L_1 and L_2 of a 2-phase buck converter for $\delta < 50\%$ and (b) for $\delta > 50\%$, both valid for CCM	114

Fig. 3.45	The output voltage ripple ΔU_{out} as a function of the duty-cycle δ for an ideal 1-phase, 2-phase and 4-phase buck DC-DC converter. For the 1-phase and 2-phase buck converter both the exact functions are plotted. For the 2-phase and 4-phase buck converter the approximated functions are plotted	115
Fig. 3.46	The concept of Single-Inductor Multiple-Output (SIMO) DC-DC converters	116
Fig. 3.47	The circuit of an ideal SIMO boost DC-DC converter with n outputs	117
Fig. 3.48	The circuit of an ideal SIMO buck DC-DC converter with n outputs	117
Fig. 3.49	The circuit of an ideal DC-DC boost Series Multiple Output Converter (SMOC) with n outputs	118
Fig. 3.50	The circuit of an ideal DC-DC buck Series Multiple Output Converter (SMOC) with n outputs	120
Fig. 4.1	The circuit of a boost DC-DC converter with all its resistive losses	125
Fig. 4.2	(a) The equivalent circuit of the charge phase and (b) discharge phase of the inductor L for a boost DC-DC converter with all its resistive losses	125
Fig. 4.3	The current $i_{SW2}(t)$ through SW_2 as a function of time t for a boost converter in steady-state DCM is shown by the *gray curve*, the *black curve* shows its linear approximation	128
Fig. 4.4	The voltage $u_C(t)$ over C as a function of time t for a boost converter in steady-state DCM is shown by the *gray curve*, the *black curve* shows its piecewise linear approximation	128
Fig. 4.5	The circuit of a buck DC-DC converter with all its resistive losses	131
Fig. 4.6	(a) The equivalent circuit of the charge phase and (b) discharge phase of the inductor L for a buck DC-DC converter with all its resistive losses	132
Fig. 4.7	The current $i_L(t)$ through L as a function of time t for a buck converter in steady-state DCM is shown by the *gray curve*, the *black curve* shows its linear approximation	133
Fig. 4.8	The voltage $u_C(t)$ over C as a function of time t for a buck converter in steady-state DCM is shown by the *gray curve*, the *black curve* shows its piecewise linear approximation	134
Fig. 4.9	The lumped model for a metal-track or bondwire inductor, taking both the parasitic series resistance R_{Ls} and parasitic substrate capacitance C_{sub} into account	136
Fig. 4.10	(a) The top-view of a planar square spiral inductor above a conductive substrate and (b) the cross-sectional view with indication of the most significant mutual inductances	137
Fig. 4.11	The perpendicular cross-sectional view of a conductor which is prone to the skin-effect	139

Fig. 4.12	The *black curve* shows the series resistance $R_{bondwire}$ per millimeter of length ℓ for a gold bondwire with $r = 12.5$ μm, as a function of frequency f and the *gray curve* denotes the DC value	140
Fig. 4.13	The model for a capacitor, taking the parasitic series resistance R_{Cs}, the parasitic parallel resistance R_{Cp} and the parasitic series inductance L_{Cs} into account	142
Fig. 4.14	The output voltage $u_{out}(t)$ of a boost converter in (**a**) DCM and (**b**) CCM, as a function of time t. The *gray curves* are valid for $R_{Cs} = 0$ and the *black curves* for a finite value of R_{Cs}	143
Fig. 4.15	The output voltage $u_{out}(t)$ of a buck converter in (**a**) DCM and (**b**) CCM, as a function of time t. The *gray curves* are valid for $R_{Cs} = 0$ and the *black curves* for a finite value of R_{Cs}	144
Fig. 4.16	The parallel circuit of two capacitors C_1 and C_2, with their respective parasitic series resistances R_1 and R_2, and the equivalent circuit with one capacitor $C_{eq}(f)$ and resistor $R_{eq}(f)$	145
Fig. 4.17	The *upper graph* shows the equivalent capacitance $C_{eq}(f)$ and the *lower graph* shows the equivalent resistance $R_{eq}(f)$, both as a function of frequency f	146
Fig. 4.18	The power loss P_{Df} of forward voltage drop of a diode (*black curve*) and the power loss P_{Ron} of the on-resistance of a MOSFET (*gray curve*), both as a function of the current I	147
Fig. 4.19	The parasitic capacitances in an n-MOSFET	148
Fig. 4.20	The currents $i_{SW1}(t)$ and $i_{SW2}(t)$ through SW_1 and SW_2 and the voltages $u_{SW1}(t)$ and $u_{SW2}(t)$ over SW_1 and SW_2 for a boost converter in (**a**) DCM and (**b**) CCM	150
Fig. 4.21	The currents $i_{SW1}(t)$ and $i_{SW2}(t)$ through SW_1 and SW_2 and the voltages $u_{SW1}(t)$ and $u_{SW2}(t)$ over SW_1 and SW_2 for a buck converter in (**a**) DCM and (**b**) CCM	151
Fig. 4.22	The physical cross-sections (**a**) of the freewheeling p-MOSFET in a boost converter and (**b**) the freewheeling n-MOSFET in a buck converter. In both cross-sections the bulk current, which occurs at the transition between the charge and discharge phase, is shown	153
Fig. 4.23	The circuit of a digital tapered CMOS buffer with n-stages ...	153
Fig. 4.24	A perspective view of a square metal-track conductor, with the definition of its width W_{track}, its length L_{track} and its thickness d	155
Fig. 4.25	(**a**) The model for the parasitic input resistance R_{in} and inductance L_{in}, with an on-chip decouple capacitor C_{dec} and its parasitic series resistance R_{Cdec}. (**b**) The equivalent impedance circuit of this model	156
Fig. 4.26	The on-chip input voltage ripple ΔU_{in} as a function of the capacitance of the decouple capacitor C_{dec}, for three different values of the parasitic series resistance R_{Cdec} of the decouple capacitor. The parameters for which this plot is valid are given in Table 4.1	157

Fig. 4.27	The flow-chart of the model flow for the boost and the buck converter, starting from the differential equations and taking all the significant resistive and dynamic losses into account, except for the temperature effects	162
Fig. 4.28	The flow-chart showing the additional flow to take the temperature and self-heating effects into account for the model of the boost and the buck converter	163
Fig. 4.29	The qualitative design trade-offs for monolithic DC-DC converters: (a) f_{SW} as a function of L for different values of C, (b) η_{SW} as a function of L for different values of C, (c) I_{L_max} and I_{L_min} as a function of L for different values of C, (d) A_L as a function of L, for different values of R_{Ls}, (e) η_{SW} as a function of A_L for different values of C, (f) f_{SW} as a function of C for different values of R_{Cs}, (g) ΔU_{out} as a function of C for different values of R_{Cs} and (h) η_{SW} as a function of $C \sim A_C$, for different values of R_{Cs}	166
Fig. 5.1	The concept of a control system for an inductive DC-DC converter	170
Fig. 5.2	The concept of Pulse Width Modulation (PWM) signal Φ_1 generation by means of comparing a triangular waveform U_{tria} to an error-voltage U_{err}	171
Fig. 5.3	The basic principle of subharmonic oscillations in a DC-DC converter with a PWM control loop	172
Fig. 5.4	The block diagram of the PWM control system implementation of a fully-integrated boost converter [Wen07]	173
Fig. 5.5	The circuit of a symmetrical cascoded OTA with a current-loaded common emitter output stage	174
Fig. 5.6	The circuit of a comparator	174
Fig. 5.7	The circuit of a time-delay	175
Fig. 5.8	The circuit of a level-shifter [Ser05]	175
Fig. 5.9	The concept of Pulse Frequency Modulation (PFM), with a constant on-time t_{on}. The *upper graph* shows the timing for low load, low frequency operation and the *lower graph* shows the timing for high load, high frequency operation	176
Fig. 5.10	The power conversion efficiencies η_{SW_PFM} and η_{SW_PWM} of a PFM (constant t_{on}) and a PWM controlled DC-DC (*gray curve*) converter, as a function of the output power P_{out}. The *solid black curve* and the *dashed black curve* denote η_{SW_PFM} for equal switching frequencies $f_{SW\_PFM} = f_{SW\_PWM}$ at the maximal output power P_{out_max} and at the minimal output power P_{out_max}, respectively	177

Fig. 5.11　The boundaries of the output voltages U_{out_PFM} and U_{out_PWM} of a PFM and a PWM (*gray curve*) controlled DC-DC converter as a function of the output power P_{out}, **(a)** for a boost converter and **(b)** for a buck converter in DCM. The *solid black curve* and the *dashed black curve* denote the boundary of U_{out_PFM} for equal switching frequencies $f_{SW\_PFM} = f_{SW\_PWM}$ at the maximal output power P_{out_max} and at the minimal output power P_{out_max}, respectively 178

Fig. 5.12　The boundaries of the output voltages U_{out} (*gray curves*) of **(a)** a PWM controlled boost converter, **(b)** a PWM controlled buck converter and **(c)** a PFM controlled boost or boost converter. All the graphs are valid for DCM. The *black curves* denote the mean output voltage $\overline{U_{out}}$ 180

Fig. 5.13　The basic concept of a Constant On/Off-Time (COOT) control system, illustrated by means of the current $i_L(t)$ through the inductor as a function of time t. The *upper* and *lower graphs* show the respective timing for low and high load operation ... 183

Fig. 5.14　The current $i_L(t)$ trough the inductor as a function of time t, for a single switching cycle of a COOT controlled DC-DC converter. The *solid black curve* denotes the nominal timing, whereas the *dashed black line* and the *solid gray curve* denote the respective case where bulk-conduction and a short-circuit of $i_L(t)$ occur 183

Fig. 5.15　The block diagram of the COOT control system implementation for a single-phase, single-output, fully-integrated buck DC-DC converter, using a bondwire inductor [Wen08b] 185

Fig. 5.16　The block diagram of the COOT control system implementation for a single-phase, single-output, fully-integrated buck DC-DC converter, using a metal-track inductor [Wen08a] 186

Fig. 5.17　The circuit of a time-delay with external reset functionality ... 187

Fig. 5.18　The block diagram of the COOT control system implementation for a single-phase, two-output, SIMO, fully-integrated buck DC-DC converter, using a metal-track inductor 189

Fig. 5.19　**(a)** The circuit of a level-shifter, which shifts the input from U_{dd}–GND to $3 \cdot U_{dd}$–$2 \cdot U_{dd}$ and **(b)** the circuit of a modified inverter, of which the in- and output may vary between $3 \cdot U_{dd}$ and U_{dd} 191

Fig. 5.20　The current $i_L(t)$ through the inductor as a function of time t, for different two-output SIMO/SMOC DC-DC converter switching schemes in DCM: a dedicated switching cycle scheme for **(a)** a boost and **(b)** a buck converter, a shared switching scheme for **(c)** a boost and **(d)** a buck converter 192

List of Figures

Fig. 5.21 (a) The power conversion efficiency η_{SW} as a function of the output power P_{out} for a PWM (*gray curve*) and two COOT controlled DC-DC converters (*solid* and *dashed black curves*). (b) The power conversion efficiency η_{SW} as a function of the output power P_{out} for PWM (*gray curve*) and a SCOOT controlled DC-DC converter 193

Fig. 5.22 The basic concept of a four-phase Semi-Constant On/Off-Time (SCOOT) control system, illustrated by means of the currents $i_L(t)$ through the respective inductors as a function of time t. The *upper* and *lower graphs* show the respective timing for low and high load operation 194

Fig. 5.23 The block diagram of the SCOOT control system implementation for a four-phase, single-output, fully-integrated buck DC-DC converter, using metal-track inductors [Wen09b] .. 196

Fig. 5.24 The circuit of the selectable time-delay 198

Fig. 5.25 The circuit of the low-pass RC filter 198

Fig. 5.26 The circuit of the schmitt-trigger 199

Fig. 5.27 The block diagram of the SCOOT control system implementation for a four-phase, two-output SMOC, fully-integrated buck DC-DC converter, using metal-track inductors 200

Fig. 5.28 Two circuit implementations of a level-shifter: (a) to shift the input from U_{dd}–GND to $2 \cdot U_{dd}$–U_{dd} and (b) to shift the input from U_{dd}–GND to $3 \cdot U_{dd}$–$2 \cdot U_{dd}$ 202

Fig. 5.29 (a) The power conversion efficiency η_{SW} as a function of the input voltage U_{in}, at constant output power P_{out} for a buck converter. The *gray curve* denotes a constant on-time control scheme, the *black curve* denotes an F^{2}SCOOT control scheme. (b) The on-time t_{on} and the real off-time t_{off_real} of an F^{2}SCOOT control scheme, as a function of the input voltage U_{in} 204

Fig. 5.30 The block diagram of the F^{2}SCOOT control system implementation of the feed-forward control loop 206

Fig. 5.31 The circuit of a 4-bit binary Digital to Analog Converter (DAC), using binary weighted current sources 207

Fig. 5.32 The circuit of a high voltage ratio level-shifter 207

Fig. 5.33 The circuit of a rail-shifter for generating a fixed offset voltage U_{rail}, which is referred to U_{in}, together with its start-up circuit .. 208

Fig. 5.34 The circuit of a basic symmetrical Operational Transconductance Amplifier (OTA) 209

Fig. 5.35 The start-up method concept, suited for DC-DC step-down converters 210

Fig. 5.36 The start-up circuit, used in various practical realizations in this work 211

Fig. 6.1	A schematic perspective view of (**a**) a hollow-spiral rectangular bondwire inductor with a patterned capacitor underneath and (**b**) an integrated hollow-spiral octagonal metal-track inductor, in top metal above the silicon substrate 215
Fig. 6.2	(**a**) A schematic perspective view of a MIM capacitor and (**b**) a MOM capacitor in a parallel, interleaved wire configuration ... 217
Fig. 6.3	The resistance of a conductive plate as a function of its length L, when the plate is connected from the left, the right and both sides 218
Fig. 6.4	(**a**) The parasitic series resistance R_{C_S} and (**b**) the capacitance density C/A of a MOS capacitor, both as a function of the width W and the length L of the individual fingers 219
Fig. 6.5	(**a**) The considered CMOS circuit. (**b**) The physical cross-section of the CMOS circuit, with the parasitic thyristor structure. (**c**) The equivalent BJT circuit of a thyristor 221
Fig. 6.6	The concept of stacked MOSFETs, for a dual-stack n-MOSFET example, (**a**) in the off-state and (**b**) in the on-state 221
Fig. 6.7	The lay-out of a MOSFET using (**a**) a linear finger structure and (**b**) using a waffle-shaped structure 222
Fig. 6.8	The alternative lay-out of a waffle-shaped MOSFET, modified for large current handling. The figure on the *left* shows the detail of the waffle-shaped structure and the *right-hand* figure shows the entire transistor 222
Fig. 6.9	(**a**) The micro-photograph of the driver chip. (**b**) The laser-diode together with the driver chip, mounted on a PCB. (**c**) The circuit of the driver chip 223
Fig. 6.10	The circuit for measuring monolithic DC-DC converters 225
Fig. 6.11	The circuit of (**a**) an electronic controlled load and (**b**) an electronic controlled voltage source 226
Fig. 6.12	The schematic representation of the measurement setup, containing: a DC-DC converter chip mounted on a substrate, a PCB with various measurement and biasing circuits, and laboratory measurement equipment 227
Fig. 6.13	The circuit of the implementation of the bondwire, single-phase, single-output boost DC-DC converter, with a PWM control system 229
Fig. 6.14	The micro-photograph of the naked chip die of the bondwire, single-phase, single-output boost DC-DC converter, with the indication of the building blocks 230
Fig. 6.15	The power conversion efficiency η_{SW} as a function of the output power P_{out}, of the bondwire, single-phase, single-output boost DC-DC converter implementation 230
Fig. 6.16	The load regulation, measured for P_{out} varying between 25 mW and 150 mW, at a frequency of 1 kHz 231

Fig. 6.17	The micro-photograph of the chip die of the bondwire, single-phase, single-output DC-DC boost converter, with the bondwire inductor added	232
Fig. 6.18	The circuit of the implementation of the metal-track, single-phase, two-output SIMO boost DC-DC converter, with a COOT control system	233
Fig. 6.19	The micro-photograph of the naked chip die of the metal-track, single-phase, two-output SIMO boost DC-DC converter, with the indication of the building blocks	234
Fig. 6.20	The circuit of the implementation of the bondwire, single-phase, single-output buck DC-DC converter, with a COOT control system	236
Fig. 6.21	The micro-photograph of the naked chip die of the bondwire, single-phase, single-output buck DC-DC converter, with the indication of the building blocks	238
Fig. 6.22	The power conversion efficiency η_{SW} as a function of the output power P_{out}, of the bondwire, single-phase, single-output buck DC-DC converter implementation	238
Fig. 6.23	The load regulation, measured for P_{out} varying between 30 mW and 300 mW, at a frequency of 1 kHz	239
Fig. 6.24	The micro-photograph of the chip die of the bondwire, single-phase, single-output buck DC-DC converter, with the bondwire inductor added	240
Fig. 6.25	The micro-photograph of the naked chip die of the metal-track, single-phase, single-output buck DC-DC converter, with the indication of the building blocks	242
Fig. 6.26	The power conversion efficiency η_{SW} as a function of the output power P_{out}, of the metal-track, single-phase, single-output buck DC-DC converter implementation	242
Fig. 6.27	The load regulation, measured for P_{out} varying between 5 mW and 180 mW, at a frequency of 100 kHz	243
Fig. 6.28	The circuit of the implementation of the metal-track, four-phase, single-output buck DC-DC converter, with a SCOOT control system	244
Fig. 6.29	The micro-photograph of the naked chip die of the metal-track, four-phase, single-output buck DC-DC converter, with the indication of the building blocks	246
Fig. 6.30	The power conversion efficiency η_{SW} as a function of the output power P_{out}, of the metal-track, four-phase, single-output buck DC-DC converter implementation	247
Fig. 6.31	The load regulation, measured for P_{out} varying between 0 mW and 720 mW, at a frequency of 10 kHz	247
Fig. 6.32	The circuit of the implementation of the metal-track, four-phase, two-output SMOC buck DC-DC converter, with a SCOOT control system	249

Fig. 6.33	The micro-photograph of the naked chip die of the metal-track, four-phase, two-output SMOC buck DC-DC converter, with the indication of the building blocks	251
Fig. 6.34	The circuit of the implementation of the bondwire, single-phase, two-output SMOC DC-DC buck converter, with a F^2SCOOT control system	252
Fig. 6.35	The micro-photograph of the naked chip die of the bondwire, single-phase, two-output SMOC DC-DC buck converter, with the indication of the building blocks	254
Fig. 6.36	The micro-photograph of the chip die of the bondwire, single-phase, two-output SMOC DC-DC buck converter, with the bondwire inductor added	255

List of Tables

Table 1.1 Classical CMOS scaling laws for circuit and interconnect performance, anno 1974 [Den74] 19

Table 2.1 A comparison to clarify the concept of the *EEF* figure of merit for step-down DC-DC converters. Each comparison is made between a linear series voltage converter and a switched DC-DC step-down voltage converter, having the same voltage conversion ratio $k_{lin} = k_{SW}$ 60

Table 3.1 The input and output parameters, together with their values, used to compare different DC-DC step-down converter topologies 67

Table 3.2 The SPICE-simulations results for the required capacitance C, to comply with the specifications of Table 3.1, of an ideal DC-DC buck converter, for four different output powers P_{out}. The required duty-cycle δ and the CM are also provided 72

Table 3.3 The SPICE-simulations results for the required capacitance C, to comply with the specifications of Table 3.1, of an ideal DC-DC bridge converter, for four different output powers P_{out}. The required duty-cycle δ and the CM are also provided 74

Table 3.4 The SPICE-simulations results for the required capacitances C_1 and C_2, to comply with the specifications of Table 3.1, of an ideal three-level buck DC-DC converter, for four different output powers P_{out}. The required duty-cycle δ and the CM are also provided 76

Table 3.5 The SPICE-simulations results for the required capacitances C_1 and C_2, to comply with the specifications of Table 3.1, of an ideal buck^2 DC-DC converter, for four different output powers P_{out}. The required duty-cycle δ and the CM are also provided . . 78

Table 3.6 The SPICE-simulations results for the required capacitance C, to comply with the specifications of Table 3.1, of an ideal DC-DC Watkins-Johnson converter, for four different output powers P_{out}. The required duty-cycle δ and the CM are also provided 81

Table 3.7	The comparison of key properties and parameters of five types of step-down DC-DC converters, with respect to monolithic integration (✔ = yes, ✘ = no)	82
Table 3.8	The input and output parameters, together with their values, used to compare different DC-DC step-up converter topologies	84
Table 3.9	The SPICE simulations results for the required capacitance C, to comply with the specifications of Table 3.8, of an ideal DC-DC boost converter, for four different output powers P_{out}. The required duty-cycle δ and the CM are also provided	84
Table 3.10	The SPICE simulations results for the required capacitance C, to comply with the specifications of Table 3.8, of an ideal DC-DC current-fed bridge converter, for four different output powers P_{out}. The required duty-cycle δ and the CM are also provided	86
Table 3.11	The SPICE simulations results for the required capacitance C, to comply with the specifications of Table 3.8, of an ideal DC-DC inverse Watkins-Johnson converter, for four different output powers P_{out}. The required duty-cycle δ and the CM are also provided	88
Table 3.12	The comparison of key properties and parameters of three types of DC-DC step-up converters, with respect to monolithic integration (✔ = yes, ✘ = no)	89
Table 3.13	The input and output parameters, together with their values, used to compare different DC-DC step-up/down converter topologies	91
Table 3.14	The SPICE simulations results for the required capacitance C, to comply with the specifications of Table 3.13, of an ideal DC-DC non-inverting buck-boost converter, for four different output powers P_{out}. The required duty-cycle δ and the CM are also provided	93
Table 3.15	The SPICE simulations results for the required capacitances C_1 and C_2, to comply with the specifications of Table 3.13, of an ideal SEPIC DC-DC converter, for four different output powers P_{out}. The required duty-cycle δ and the CM are also provided	95
Table 3.16	The SPICE simulations results for the required capacitances C_1 and C_2, to comply with the specifications of Table 3.13, of an ideal zeta DC-DC converter, for four different output powers P_{out}. The required duty-cycle δ and the CM are also provided	97
Table 3.17	The comparison of key properties and parameters of five types of step-down DC-DC converters, with respect to monolithic integration (✔ = yes, ✘ = no)	98
Table 3.18	The comparison of the required total output capacitance for a two-output boost SIMO and SMOC converter in CCM	119
Table 3.19	The comparison of the required total output capacitance for a two-output buck SIMO and SMOC converter in CCM	120

Table 4.1	The parameters used for the calculation example of the input decouple capacitor of a DC-DC converter, shown in Fig. 4.26	156
Table 6.1	The circuit parameters of the bondwire, single-phase, single-output boost DC-DC converter implementation	229
Table 6.2	The main measured parameters of the bondwire, single-phase, single-output DC-DC boost converter implementation	231
Table 6.3	The circuit parameters of the metal-track, single-phase, two-output boost DC-DC converter implementation	234
Table 6.4	The main expected parameters of the metal-track, single-phase, two-output SIMO boost DC-DC converter implementation	235
Table 6.5	The circuit parameters of the bondwire, single-phase, single-output buck DC-DC converter implementation	237
Table 6.6	The main measured parameters of the bondwire, single-phase, single-output buck DC-DC converter implementation	239
Table 6.7	The circuit parameters of the metal-track, single-phase, single-output buck DC-DC converter implementation	241
Table 6.8	The main measured parameters of the metal-track, single-phase, single-output buck DC-DC converter implementation	243
Table 6.9	The circuit parameters of the metal-track, four-phase, single-output buck DC-DC converter implementation	245
Table 6.10	The main measured parameters of the metal-track, four-phase, single-output buck DC-DC converter implementation	248
Table 6.11	The circuit parameters of the metal-track, four-phase, two-output SMOC buck DC-DC converter implementation	250
Table 6.12	The main expected parameters of the metal-track, four-phase, two-output SMOC buck DC-DC converter implementation	251
Table 6.13	The circuit parameters of the bondwire, single-phase, two-output SMOC buck DC-DC converter implementation	253
Table 6.14	The main simulated parameters of the bondwire, single-phase, two-output SMOC buck DC-DC converter implementation	254
Table 6.15	The comparison of the most important measured parameters of the monolithic inductive DC-DC step-up converter, presented in this work, to the other known work	256
Table 6.16	The comparison of the most important measured parameters of the monolithic inductive DC-DC step-down converters, presented in this work, to the other known work	258

Chapter 1
Introduction

This work aims to provide a comprehensive dissertation on the matter of monolithic inductive Direct-Current to Direct-Current (DC-DC) converters. For this purpose six chapters are defined which will allow the designer to gain specific knowledge on the design and implementation of monolithic inductive DC-DC converters, starting from the very basics.

DC-DC have been around since the use of electricity became common practice. Over the years many technological developments have led to a wide variety of different types and applications for DC-DC converters. In the recent years a trend has emerged towards very compact low-power (100 mW–1 W) and low-voltage (1 V–80 V) DC-DC converters, for main use in battery-operated applications. The two key specifications for this recent breed of DC-DC converters are power conversion efficiency and power density. The first specification determines the battery autonomy of the target application and the second specification determines the required space of the converter. DC-DC converters featuring a high power conversion efficiency, only requiring a limited number of off-chip (passive) components are considered the established state-of-the-art. The next technological step is to integrate the remaining off-chip components of the DC-DC converter on-chip, causing both the required area and the costs to decrease. The technology of choice to achieve this ongoing on-chip integration is CMOS, as it is by far the most widely used and thus potentially the most economical chip technology. The target of this work is to determine the feasibility of monolithic integration of inductive DC-DC converters in CMOS, in addition with the fundamental limits that apply.

This chapter provides some basic considerations and a few historical notes, in Sect. 1.1. Examples of low power applications for DC-DC converters situate the relevance of the work, in Sect. 1.2. The challenges of creating monolithic inductive DC-DC are highlighted in Sect. 1.3. The outline of this dissertation is provided in Sect. 1.4. The conclusions of this chapter are given in Sect. 1.5.

Fig. 1.1 A black-box representation of a DC-DC converter

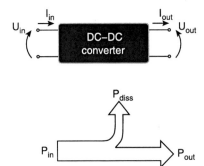

Fig. 1.2 The power-balance of a DC-DC converter

1.1 The Origin of DC-DC Converters

In order to make the reader familiar with the topic of DC-DC converters, some basic considerations are discussed in Sect. 1.1.1. A broader perspective is provided through a brief number of historical notes, which are provided in Sect. 1.1.2.

1.1.1 Basic Considerations

A DC-DC converter is a device that converts a Direct-Current (DC) voltage U_{in} into a lower or higher DC voltage U_{out}, as illustrated by the black-box representation in Fig. 1.1. The ratio of U_{out} over U_{in} is defined as the voltage conversion ratio k, which is given by (1.1). When k is larger than 1 the DC-DC converter is denoted as a step-up converter and when k is lower than 1 the term step-down converter is used.

$$k = \frac{U_{out}}{U_{in}} \quad (1.1)$$

The output current I_{out} can in turn become lower or higher than the input current I_{in}. Therefore, a DC-DC converter can be considered as a DC impedance transformer. It can be proven that the impedance ratio Z_k of an ideal impedance transformer is given by (1.2). For an ideal DC-DC converter Z_k will be lower than 1 if U_{out} is lower than U_{in} and vice versa.

$$Z_k = \frac{Z_{out}}{Z_{in}} = \left. \frac{\frac{U_{out}}{I_{out}}}{\frac{U_{in}}{I_{in}}} \right|_{U_{in}I_{in}=U_{out}I_{out}} = \left(\frac{U_{out}}{U_{in}}\right)^2 = \left(\frac{I_{in}}{I_{out}}\right)^2 \quad (1.2)$$

As such, a certain input power P_{in} is demanded by the DC-DC converter and a certain output power P_{out} is delivered by it. For obvious reasons P_{out} can never exceed P_{in}. The difference between them, the dissipated power P_{diss}, is transferred into heat, which is undesired. This power-balance is illustrated by Fig. 1.2. Note that for an ideal DC-DC converter P_{diss} equals zero.

Again, it is intuitively clear that an ideal DC-DC converter cannot be realized, as any given implementation will be associated with losses that cause unwanted power

1.1 The Origin of DC-DC Converters

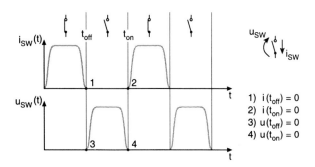

Fig. 1.3 The principle of ideal switching

1) $i(t_{off}) = 0$
2) $i(t_{on}) = 0$
3) $u(t_{off}) = 0$
4) $u(t_{on}) = 0$

dissipation. Thus, the goal of any DC-DC converter realization is to minimize P_{diss}. In order to benchmark this, the power conversion efficiency η is defined by (1.3). However, for step-down converters η alone is insufficient for comparing converters with a different k. A solution for this problem is proposed in Sect. 2.4.

$$\eta = \frac{I_{out}U_{out}}{I_{in}U_{in}} = \frac{P_{in} - P_{diss}}{P_{in}} = \frac{P_{out}}{P_{out} + P_{diss}} = \frac{P_{out}}{P_{in}} \quad (1.3)$$

For the sake of completeness it is noted that for any given DC-DC converter the value of η will be lower than 100% and Z_{in} will be lower than for the ideal case, causing Z_k to increase.

As explained previously, η is an important parameter that needs to be maximized. To accomplish this task, the presented work will focus on switched-mode DC-DC converters. Indeed, switched-mode DC-DC converters can achieve the highest possible η, since switching can theoretically be lossless. This is not the case for linear voltage converters (see Sect. 2.1), which are used to decrease the input voltage through dissipation.

The principle of ideal, lossless switching is illustrated in Fig. 1.3, where u_{SW} is the voltage over the switch and i_{SW} the current through it. It can be seen that there are four requirements to establish ideal switching, as denoted in Fig. 1.3. These requirements are summarized as: Never at the same moment in time should there exist a voltage u_{SW} over a switch and a current i_{SW} through it.

As will be explained in Sect. 4.2.3, real-world switches are associated with a number of losses. For this reason ideal switching can only be attempted for, but never achieved.

1.1.2 Historical Notes

For the reader's interest, a few historical notes are provided in this section. They indicate that DC-DC converters are a far from recent invention and that their application is required since the use of electricity became common practice. These notes should not be regarded as being complete, they rather highlight some interesting historical facts related to the topic.

⚘FARADAY'S INDUCTION RING

The electrical transformer is invented by Michael Faraday in the year 1831 [Roy10]. In those days it was recently discovered that an electrical current produces a magnetic field. Scientists were searching for a way to prove the opposite, which would lead to mechanical electricity production. For this purpose, Faraday developed the *induction ring* (a toroidal transformer), shown in Fig. 1.4(a). The transformer consisted of both a secondary and primary winding of copper wire on an iron toroidal core, insulated from each other by means of cotton.

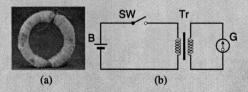

Fig. 1.4 (a) Faraday's original 1831 *induction ring* [Ins10] and (b) the schematic representation of the induction ring experiment

The circuit used by Faraday for this experiment is illustrated in Fig. 1.4(b). The primary winding of the transformer Tr is connected to a battery B, through a switch SW and the secondary winding of Tr is connected to a galvano-meter G. The needle of G will swing back and forth upon the closing and opening of SW. The reason for this is that the current through the primary winding will induce a magnetic field in the core, which is picked-up by the secondary winding and in turn produces an electrical current in it, despite the insulation. This phenomenon is known as mutual inductance. In essence one can regard this as the first DC-AC converter.

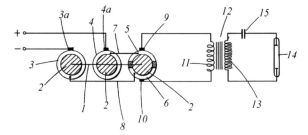

Fig. 1.5 A mechanical rotary DC-AC step-up converter, for powering gas-discharge lamps [Ran34]

Rotary DC-DC Converters

Inductive DC-DC converters are in essence Direct-Current to Alternating-Current (DC-AC) converters with an output rectifier. The introduction of DC-AC converters initiated the realization of DC-DC converters, by simply rectifying the AC output voltage.

The first DC-AC converters where mechanical rotary converters [Ran34], as illustrated in Fig. 1.5. This example has the purpose of feeding electrical gas-discharge tubes. In rotary DC-AC converters an electrically split commutator in

1.1 The Origin of DC-DC Converters

> ⊛THE COCKCROFT-WALTON VOLTAGE MULTIPLIER
>
> Although capacitive type voltage converters are beyond the scope of this dissertation, an important milestone for this type of converter is noted here. The Cockcroft-Walton converter is probably the first capacitive type AC-DC converter, which is invented by Heinrich Greinacher in the year 1919 [Wik10]. It became well known by the two physicists, namely John D. Cockcroft and Ernest T.S. Walton, who used it to perform the first artificial nuclear disintegration in history, in the year 1932.
>
>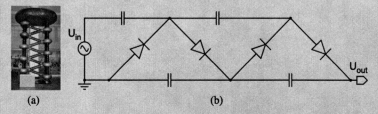
>
> **Fig. 1.6** (a) A Cockcroft-Walton voltage multiplier build in the year 1937, which was used for an early particle accelerator [Wik10]. (b) A half-wave, two-stage Cockcroft-Walton voltage multiplier
>
> The circuit of a half-wave, two-stage Cockcroft-Walton voltage multiplier is illustrated in Fig. 1.6. During a positive cycle of the AC input voltage U_{in}, the lower capacitors are charged to the voltage $2 \cdot U_{in\_peak}$ and during the negative cycle of U_{in} the upper capacitors are charged to the voltage U_{in_peak}. The ideal output voltage U_{out} in steady-state equals $2 \cdot n U_{in\_peak}$, with n the number of stages. The advantage of the Cockcroft-Walton voltage multiplier is that the maximum voltage over the components is $2 \cdot U_{in\_peak}$, reducing the electrical stress on the individual components and making it suitable for generating high voltages (MV-range).

combination with (three) brushes is used to generate an AC voltage, which is fed to the primary winding of a transformer. The commutator itself is driven by a small electrical motor (not shown on the figure). These kind of DC-DC converters are not compact, produce audible noise, have a low efficiency at low output power, are prone to mechanical wear, etc.

Rotary converters can also be used for DC-DC conversion, by adding an extra commutator on the secondary side of the transformer [Gra29]. This commutator will then serve the purpose of a mechanical rectifier.

Vibratory DC-DC Converters

A variant on the mechanical rotary DC-DC converter is the mechanical vibratory DC-DC converter [Sta34], which basically consists of a self-oscillating mechanical relay. The circuit of a vibratory DC-DC converter is shown in Fig. 1.7. A first set of contacts in the oscillating beam is used to produce an AC voltage out of the DC voltage source. This AC voltage is fed to a transformer for the step-up function,

Fig. 1.7 A mechanical vibratory DC-DC step-up converter [Sta34]

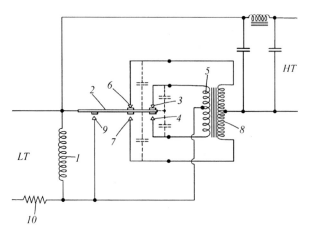

Fig. 1.8 A DC-DC step-up converter, using an inverted vacuum-tube triode as primary switch and secondary rectifier [Haz40]

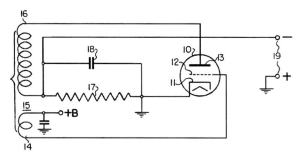

whereafter it is rectified by a second set of contacts on the same oscillating beam. The mechanical nature of this type of DC-DC converters implies that it has the same disadvantages of rotary DC-DC converters.

Vacuum-Tube DC-DC Converters

The invention of the vacuum-tubes triggered a revolution in the field of electronics. For DC-DC converters this meant that the mechanical primary switch could be replaced by a vacuum-tube. The first use of this technology is reported in [Ter28], where an inverted[1] vacuum-tube triode is acquired as a primary switch for a DC-AC converter. The reported switching frequency of this converter is 300 Hz to 2000 Hz.

In a later stage, a similar inverted vacuum-tube triode functions both as primary switch and as secondary rectifier, forming a vacuum-tube DC-DC step-up converter [Haz40]. The circuit of this converter is shown in Fig. 1.8. Vacuum-tube DC-DC converters do not possess the mechanical drawbacks as do rotary and vibratory DC-DC converters, as explained previously. However, they are not suitable to deliver a

---

[1] An inverted vacuum-tube triode is an ordinary triode where the grid is biased with a positive voltage, thereby drawing current from the cathode.

1.1 The Origin of DC-DC Converters

⓸The Vacuum-Tube Triode

By adding a grid to the vacuum-tube diode of John A. Fleming, Lee De Forest created the vacuum-tube triode, *the audion*, in the year 1906 [Moe08]. The first commercial vacuum-tube triode is shown in Fig. 1.9(a).

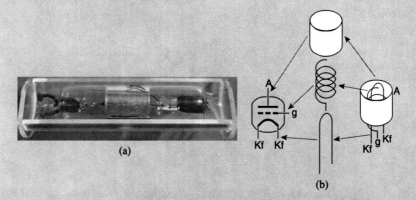

Fig. 1.9 (a) The first commercial vacuum-tube triode: *the audion*. (b) The schematic symbol of a direct-heated triode and its simplified construction principle

The schematic symbol of a direct-heated vacuum-tube triode and the simplified construction principle are illustrated in Fig. 1.9(b). The negative biased cathode Kf serves as the heating filament, thereby emitting electrons to the positive biased anode A. The grid g voltage is more negative than the cathode, hence some of the emitted electrons towards the anode are repelled by the grid. This reduces the anode-to-cathode current. Thus, the anode-to-cathode current can be adjusted by altering the grid-to-cathode voltage, similar as a depletion MOSFET, causing transconductance amplification. The construction is sealed in a glass envelope (not shown), in vacuum. The term direct-heated is due to the fact that the filament physically acts as the cathode.

large output power and also have poor low load efficiency, because the filament of the vacuum-tube triode needs to be heated.

Transistorized DC-DC Converters

Modern DC-DC converters use a transistor as switching device, which can be a Metal-Oxide Semiconductor Field-Effect Transistor (MOSFET), a Bipolar Junction Transistor (BJT), or other specialized semiconductor switch devices.[2] This kind of transistorized DC-DC converters where first reported in the early 1950's. One of the

---

[2]Where MOSFETs and BJTs are found in virtually every low to modest power DC-DC converters, high power/voltage DC-DC converters use Insulated Gate Bipolar Transistors (IGBT) and other exotic electronic switch devices.

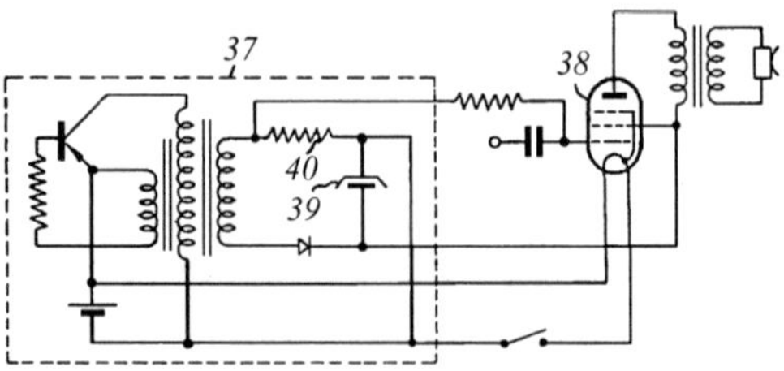

Fig. 1.10 A transistorized DC-DC step-up converter, for powering a vacuum-tube pentode audio amplifier for a hearing aid device [Phi53]

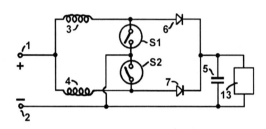

Fig. 1.11 A two-phase DC-DC step-up converter [Wes67]

first reported implementations [Phi53] is shown in Fig. 1.10, which uses a PNP BJT. The step-up DC-DC converter (shown in the box drawing) of this example is used to generate the high-tension anode voltage for a vacuum-tube pentode audio amplifier (shown on the left), used in early portable electronic hearing aids. Its switching frequency is about 15 kHz to 20 kHz.

Transistorized DC-DC converters obviously have an intrinsic advantage over mechanical and vacuum-tube DC-DC converters. They are not prone to mechanical wear and do not produce audible noise, as no moving parts are present. Also, in contrast to vacuum-tube DC-DC converters, their low-load efficiency can be drastically improved. Furthermore, the switching frequency of transistors can be chosen higher, avoiding audible noise/humming and requiring smaller transformers and passive filters for the DC-DC converters. Last but not least they are much more compact than their predecessors. Therefore, ways where sought to replace the older types DC-DC converters with transistorized versions [Kre57].

Multi-phase DC-DC Converters

A final historical note is made upon the invention of multi-phase DC-DC converters [Wes67]. The concept of multi-phase DC-DC embodies the idea that two or more DC-DC converters are used to power the same load, possibly interleaved in the time-domain. This concept can provide a number of intrinsic advantages, such as:

- Higher output power.
- Higher power conversion efficiency.
- Lower output voltage ripple.
- Lower input current ripple.
- Higher output power per area/volume.

⊛ THE BIPOLAR JUNCTION TRANSISTOR

In mid-December of the year 1947 physicists at the Bell Telephone Laboratories, John Bardeen and Walter Brattain, discovered the Bipolar Junction Transistor (BJT), as they were trying to make William Shockley's Field-Effect Transistor (FET) work [Rio10a]. A photograph of this experimental first BJT, based on the semiconductor germanium, is shown in Fig. 1.12(a). For this achievement, which initiated the second revolution in electronics (after the vacuum-tube triode), the three shared the 1956 Nobel Prize in physics.

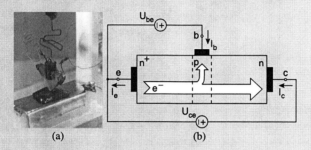

Fig. 1.12 (a) The first contact bipolar junction transistor [Rio10b] and (b) a schematic cross section of an NPN BJT

Figure 1.12(b) illustrates a schematic cross-section of an NPN BJT. An NPN BJT consists of a semiconductor with three different doped regions: a high n^+-doped emitter e, a p-doped base b and an n-doped collector c. Thus, two anti-series diode pn-junctions are formed, being the respective base-collector and base-emitter diodes. In the forward, linear operating region the base-collector diode is reverse biased and the base-emitter diode is forward biased. This causes electrons to diffuse from the emitter to the base. If the width of the base is smaller than the diffusion length of these electrons, they will not be able to recombine in the base and consequentially reach the collector. In reality a certain amount of recombination in the base will occur. The effect of these mechanisms is that a small base current I_b will cause a much higher collector current I_c. In other words, I_c is the result of the amplified I_b.

Note that, apart from the NPN BJT, also the PNP BJT exists. This device functions in an analogue, yet complementary, fashion an NPN BJT.

Needless to say that this invention is massively used in many DC-DC converter implementations. Figure 1.11 illustrates this concept by means of a two-phase DC-DC step-up converter. For a more technical discussion of this topic the reader is referred to Sect. 3.5.1.

1.2 Low Power DC-DC Converter Applications

DC-DC converters are present in virtually every piece of electronic equipment. As such they are responsible for providing the adequate voltage or current to the application. The source of which this electrical power is derived can either be the AC

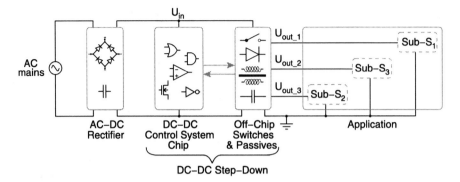

Fig. 1.13 The block-diagram representation of a mains-operated application, using step-down AC-DC converter

mains or an autonomic DC source, such as (rechargeable) batteries.[3] Depending on the power source different requirements are set for the DC-DC converter, which can be in terms of output power, voltage conversion ratio, power conversion efficiency, power density, volume, weight, etc.

This section discusses the main system-level principles of DC-DC conversion, used in modern electronic (consumer) equipment. Section 1.2.1 describes the principles involved in systems that use mains-operated DC-DC converters, whereas in Sect. 1.2.2 the principles of DC-DC converters used in battery-operated systems are discussed. The latter section also explains the use and possible advantages of monolithic DC-DC converters.

1.2.1 Mains-Operated

The major part of mains-operated electronic (consumer) equipment has an on-board power supply, providing the adequate voltage(s) and/or current(s) for the inner electronic and electric circuits. This power supply usually consists of an AC-DC step-down converter, which can be either a 50/60 Hz transformer or a Switched-Mode Power Supply (SMPS). Only the latter is considered, as 50/60 Hz transformers are becoming obsolete due to their intrinsic disadvantages.[4] Also, the few applications that require step-up converters, such as Cathode Ray Tubes (CRT) monitors, are not considered in this discussion.

Figure 1.13 illustrates the system-level block-diagram of a mains-operated application, that uses a SMPS AC-DC converter. As can be seen, the SMPS in the AC-DC converter is a DC-DC converter. Indeed, the AC-DC conversion is performed by a

---

[3]Other examples are solar cells and fuel cells.

[4]They have a low power density, require vast copper windings (expensive) and they have a poor low load efficiency.

passive bridge-rectifier, followed by a decouple capacitor. The resulting DC voltage U_{in} is fed to the DC-DC converter,[5] which serves two functions. The first function is to lower the input voltage to the desired level and the second function is to provide galvanic[6] separation between the mains voltage and the output voltage(s). To achieve these purposes the topology of the DC-DC converter can be either a fly-back or a forward converter, which are discussed Sect. 3.4.1.

The DC-DC step-down converter can be regarded as two separate parts. The first part is a DC-DC control system, which is most commonly integrated on a chip. This control system serves the purpose of controlling the switch(es) of the DC-DC converter in such a way that the output voltage(s) remain constant to the desired level, under varying load and line conditions. The second part consists of the power switch(es), the transformer and output capacitor (= the passives) and the output rectifier. These are the components that perform the actual power conversion. Up to this time the power conversion components are all placed on the system's Printed Circuit Board (PCB) and thus off-chip. An exception is sometimes observed for the switch(es), which can also be on-chip if the required output power is fairly low (order of magnitude of 10 W).

Mains-operated SMPS do also often provide multiple output voltages, as required by the sub-systems of the application. An example of such a system is a Personal Computer (PC) which typically needs five or more different output voltages. The output power of mains-operated SMPSs can vary from the W-range to the kW-range, which in many cases exceeds that of battery-operated DC-DC converters.

The consequence of both the requirement of galvanic separation and the high output power is that monolithic mains-operated SMPSs are not yet feasible with current technology. Therefore, they will not be considered for full-integration in this dissertation. The presented work rather focusses on the monolithic integration of battery-operated, low voltage (1 V–80 V), low power (100 mW–1 W) inductive DC-DC converters.

1.2.2 Battery-Operated

Many modern electronic applications make use of advanced Integrated Circuits (IC), which enables these applications to provide ever more functionality and become small enough to be portable, battery-operated and wireless. This is achieved by placing an increasing amount of the building blocks of the application on one single chip die, leading to so called Systems-on-Chip (SoC). These SoCs contain mostly mixed-signal circuits, digital circuits, analog circuits and in some case sensor devices. Due to the different nature and power requirements of this variety of circuits

---

[5] In SMPSs with Power Factor Correction (PFC), the DC voltage is first fed into a DC-DC buck or boost converter, before it is fed to the actual DC-DC step-down converter.

[6] No electrical connection between input and output terminals exists, which is mostly required in mains-operated SMPSs as safety precaution, whereby the reference from the output voltage(s) to the physical earth connection is eliminated.

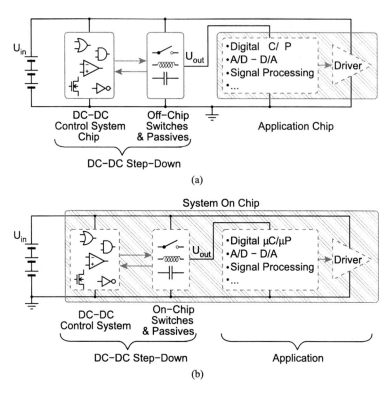

Fig. 1.14 (a) The block diagram of a battery-operated application using a DC-DC step-down converter with external components. (b) The same system implemented as a SoC, with a monolithic DC-DC converter

and devices, many SoCs require multiple supply voltages. However, due to space, weight and cost constraints, most applications only contain one battery. Therefore, the need for multiple on-chip supply voltages emerges.

The following sections describe the general implementation of DC-DC converters in battery-operated applications. Also the evolution from conventional DC-DC converters, using off-chip components, towards monolithic DC-DC converters is discussed, for both step-up and step-down DC-DC converters.

Step-Down Converters

In many applications the battery voltage is too high[7] for at least a part of the circuitry. It therefore needs to be decreased, which is done by means of an off-chip DC-DC step-down converter. This concept is illustrated in Fig. 1.14(a). Consider the

---

[7]This is usually the case for Lithium-ION (LiION) batteries, which have a typical cell voltage of 3.7 V.

application chip to be a complex SoC, which requires two different supply voltages. The part that contains for instance the digital circuits, such as the analog-to-digital converters (ADC), digital-to-analog converters (DAC), signal processing circuits, etc. typically needs a lower supply voltage as this part needs to consume minimal power. Whereas the circuitry used to transmit data into the communication medium, which can be a line-driver or a power amplifier, typically needs a higher supply voltage as adequate power is required to communicate over a specified distance.

As illustrated in Fig. 1.14(a), the high voltage might be directly derived from the battery voltage U_{in}. The lower voltage U_{out} needs to be converted by an off-chip DC-DC step-down converter. This converter in turn consists of a control system chip and some external passive components and power switch(es), similar as the system described in Sect. 1.2.1. Notice that current commercial DC-DC control system ICs can have an integrated power switch(es). Clearly, the separate application chip, in combination with the DC-DC chip, introduces an overhead in terms of packaging, mounting and PCB area. Moreover, the additional external components of the DC-DC converter are costly and also require extra PCB area.

By placing the DC-DC step-down converter and its external components on the application SoC, the PCB area and the system's number of external components can be decreased. This concept of a monolithic DC-DC step-down converter is shown in Fig. 1.14(b). In contrast to mains-operated applications, the full-integration of DC-DC converters in battery-operated applications is possible with current technology. Indeed, no galvanic separation is required and the output power levels are in most cases limited.

However, it is understood that the complete SoC, including on-board monolithic DC-DC converter, requires more chip area. This (costly!) area should be minimized, which implies that the power density of the converter is to be maximized. Finally, the battery-autonomy should not be compromised, requiring maximal power conversion efficiency. A more detailed discussion on the needs and challenges that are involved with the realization of monolithic DC-DC converters is provided in Sect. 1.3.2.

Step-Up Converters

As opposed to the situation discussed in the previous section, there are also many cases where the battery voltage of the application is insufficient[8] for at least a part of the circuitry. In this case a DC-DC step-up converter is required to provide the required higher voltage.

The classical system-level solution for this problem is illustrated in Fig. 1.15(a). In this case the part of the circuitry that needs a low voltage can be directly connected to the battery voltage U_{in}, whereas the circuits that require a higher voltage U_{out} are supplied by the output of the off-chip DC-DC step-up converter. This implementation has similar disadvantages as the example from the previous section.

---

[8]This is usually the case for Nickel Metal-Hydride (NiMH) batteries, which have a typical cell voltage of 1.2 V.

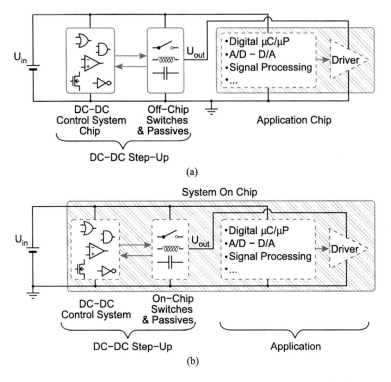

Fig. 1.15 (a) The block diagram of a battery-operated application using a DC-DC step-up converter with external components. (b) The same system implemented as a SoC, with a monolithic DC-DC converter

In this case the integration of the DC-DC converter control system and the external passives and switch(es) can also be integrated on the same chip die of the application itself. This complete SoC implementation yields the same advantages and poses similar problems, as discussed earlier for a DC-DC step-down converter, which will be more elaborated upon in Sect. 1.3.2.

1.3 Monolithic DC-DC Converters: A Glimpse into the Future

The hint towards monolithic integration of DC-DC converters for battery-operated applications has been given in Sect. 1.2.2. However, many questions on the details of these SoC implementations still remain unresolved. The answer to the fundamental question of which technology to use is already given in this section: CMOS. Answers to other important questions are provided throughout the following chapters. This section will therefore also give an idea of which problems are to be solved, in order to achieve the realization of monolithic inductive DC-DC converters.

Section 1.3.1 clarifies the basics of CMOS technology and scaling, as these concepts will be used to explain the implementations in Chap. 6. A selection of the most stringent problems that are to be solved, is given in Sect. 1.3.2.

⊛ THE INTEGRATED CIRCUIT

In the year 1958 Jack Kilby realized the first IC at the Texas Instruments laboratories [Lee10a], for which he received the Nobel Prize in physics in the year 2000. He successfully demonstrated an integrated 1.3 MHz BJT Resistor-Capacitor (RC) oscillator, which is shown in Fig. 1.16(a). The IC is based on the semiconductor germanium and used bondwires for the on-chip interconnect. The production method and materials where not yet practical for mass-production, but nevertheless proved the feasibility of ICs.

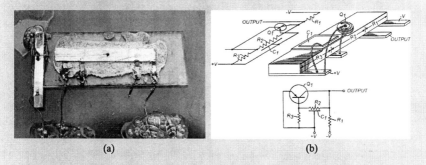

Fig. 1.16 (a) The first integrated circuit [Lee10b] and (b) the schematic circuit representation [Lee10c]

The schematic lay-out and the equivalent circuit of Kilby's IC is shown in Fig. 1.16(b). The intrinsic electrical resistance of the germanium substrate is used to create the resistors, the capacitor is implemented as a plate capacitor and the transistor is a PNP BJT.

1.3.1 CMOS Technology

The technology choice for the design and implementation of monolithic inductive DC-DC converters in this work is CMOS. This silicon based IC-technology is the most common and widely used in the semiconductor industry. As a consequence, CMOS technologies are less expensive than other, more exotic, technologies.[9] The main drivers of CMOS IC-technology fabrication and refinement are digital ICs, which are mostly computer processors and digital memories. This implies that CMOS technology is optimized for high speed and low power digital circuits rather than for analog and other circuitry, which is be a crucial part of any given SoC. Thus, CMOS technology introduces a major challenge for the successful realization of monolithic inductive DC-DC converters. Moreover, the trend towards smaller feature sizes, called scaling, in CMOS technologies and the associated supply voltage decrease, poses even more problems for other than digital circuit implementations.

---

[9]Such as: Bipolar technologies, mixed Bipolar-CMOS (Bi-CMOS) technologies and Silicon-Germanium (SiGe) technologies.

Fig. 1.17 (a) The schematic symbol of an n-MOSFET and (b) its schematic perspective cross-section view

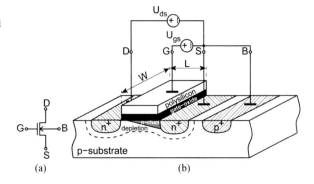

The MOSFET

Figure 1.17(a) shows the schematic symbol and Fig. 1.17(b) the schematic perspective cross-section view of an n-MOSFET, which has four terminals: the Drain (D), the Gate (G), the Source (S) and the Bulk (B). An n-MOSFET is created on a p-type silicon substrate, which is physically connected to its bulk terminal. The gate consists of a polysilicon layer on top of a very thin oxide, of which the physical length L_n and width W_n are the two design parameters. The drain and source are n^+-type regions alongside the gate.

For the basic operation[10] of an enhancement MOSFET, two cases[11] for the gate-source voltage U_{gs} are considered: $U_{gs} = 0$ and $U_{gs} > V_t$, with V_t the threshold voltage. For both cases a positive drain-source voltage U_{ds} and a zero source-bulk voltage U_{sb} are assumed. In the first case, when $U_{gs} = 0$, the electrical path between drain and source consists of two anti-series connected pn-junction diodes. Thus, no drain-source current I_{ds} is able to flow. In the second case, when $U_{gs} < V_t$, the positive biased gate will repel the free holes (which are positively charged) from the p-substrate under the gate and push them towards the depletion region. It will also attract free electrons from the source and the drain regions. This causes the formation of an inducted n-channel, also called inversion layer, between the drain and the source regions. This n-channel allows for a current to flow, by means of free electrons. The basic operation of p-MOSFET is dual to the operation of an n-MOSFET and is therefore omitted here.

The shape of the inducted n-channel can be pinched off before the drain, as drawn in Fig. 1.17(b). This situation occurs if $U_{ds} > U_{gs} - V_t$ and this operation region is called *saturation*. In this case I_{ds} saturates because U_{ds} no longer affects the channel.[12] For this operation region the behavior of the n-MOSFET is described by (1.4), where C_{ox} is the capacitance per unit of area of the parasitic gate capaci-

---

[10] For more information the reader is referred to [Sed98].

[11] A third case exits when $U_{gs} < V_t$. This operating region is called sub-threshold and will not be considered here.

[12] In realty I_{ds} further increases with U_{ds}, due to channel-length modulation.

Fig. 1.18 The qualitative behavior for an n-MOSFET of I_{ds} as a function of U_{ds}, for different U_{gs}

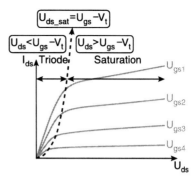

tance, formed between the gate electrode and the induced channel, μ_n is the electron mobility and λ is the channel-length modulation.

$$I_{ds} = \frac{\mu_n C_{ox}}{2} \frac{W}{L} (U_{gs} - V_t)^2 (1 + \lambda U_{ds}) \tag{1.4}$$

If $U_{ds} < U_{gs} - V_t$ then the induced n-channel extends from the source to the drain and the n-MOSFET operates in the *triode* region (also called *linear* region). In this operation region the n-MOSFET shows a resistive behavior, which is described by (1.5).

$$I_{ds} = \mu_n C_{ox} \frac{W}{L} \left((U_{gs} - V_t) U_{ds} - \frac{U_{ds}^2}{2} \right) \tag{1.5}$$

Finally, Fig. 1.18 illustrates for an n-MOSFET I_{ds} as a function of U_{ds}, for different U_{gs}. The previously explained respective operation regions, saturation and triode, are also shown.

Scaling Towards Nanometer CMOS

From the invention of the IC onwards the minimal processable physical gate length L_n and L_p, also known as minimum feature size, of transistors decreased. This allows for more transistors to be placed on one chip. Gordon Moore predicted in the year 1965 that the number of transistors per chip would double every 12 months [Moo65]. This was roughly the case until the mid 1970's. In the year 1975 Moore predicted that this factor two increase would be maintained for every two years, rather than every year [Moo75]. In reality this last statement is still true until the present day and is commonly known as Moore's law. Figure 1.20(a) illustrates this law for Intel CMOS technologies [Boh09]. The law also implies that, in order to maintain this increase of transistor count per chip, the minimum feature size needs to scale with the scaling factor $\kappa = 1.4$ every two years. This is also illustrated in Fig. 1.20(a), for Intel CMOS technologies [Boh09]. The current (Q2 2010) commercial CMOS technology node is 32 nm.

The scaling of CMOS technologies has a number of implications on the performance of the ICs produced with them. These implications are known as the scaling

⊕THE BASIC SIX-MASK CMOS PROCESS LAY-OUT

The CMOS technology in its most basic form requires six masks for processing. This process (for a *p*-type substrate) enables the following native devices: n-MOSFETs, p-MOSFETs, lateral PNP-BJTs, substrate *pn*-junction diodes, n-well *pn*-junction diodes, diffusion resistors, pinched-diffusion resistors, n-well resistors, polysilicon resistors, metal-1 resistors, MOS-capacitors and MOM-capacitors (polysilicon, metal-1 and their combination). A lay-out view of an n-MOSFET and a p-MOSFET, together with the cross-section of the according physical devices is shown in Fig. 1.19.

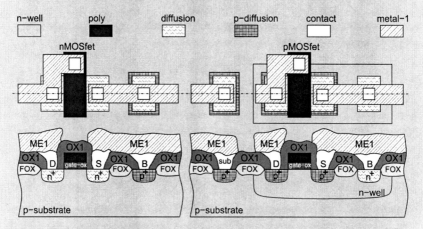

Fig. 1.19 An n-MOSFET (*left*) and a p-MOSFET (*right*) in a six-mask CMOS process. The *upper* drawings show the lay-out view and the *lower* ones a cross-section of the according physical devices

The following list denotes the six masks, of a basic CMOS process:

1. *n-well*: The regions in the *p*-type substrate where n-wells will occur.
2. *polysilicon*: The regions where polysilicon will occur. This mask defines the length L_n and L_p of the gates and also, in combination with the overlapping *active* mask, where these gates will occur.
3. *active*: Defines where *n*-type and *p*-type diffusion will occur, by leaving apertures in the Field Oxide (FOX). Active regions that are not surrounded by the *p-diffusion* mask will be *n*-type. Also defines the width W_n and W_p of the gates and, in combination with the overlapping *polysilicon* mask, where these gates will occur.
4. *p-diffusion*: The regions of the substrate, kept clear of FOX by the *active* mask, which will be *p*-type.
5. *contact*: Defines apertures in the first oxide (OX1), which are filled with metal (tungsten), contacting active areas and polysilicon.
6. *metal-1*: Defines the regions where metal-1 (ME1) (aluminum) tracks will occur, for connecting contacts defined by the *contact* mask.

1.3 Monolithic DC-DC Converters: A Glimpse into the Future

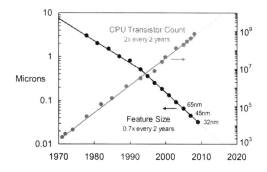

Fig. 1.20 The minimum feature size and the transistor count per chip as a function of time, for Intel CMOS technologies [Boh09]

laws. Table 1.1 describes the classical scaling laws for the device, the circuit and the interconnect parameters of CMOS technologies. From the early twenty-first century onwards, some of these classical scaling laws are no longer valid [Boh09]:

- *Gate-oxide thickness*: t_{ox} ceased decreasing, because of the excess gate-leakage. The use of novel high dielectric permittivity insulating materials, known as high-k material, instead of silicon dioxide (SiO_2) made this possible.
- *Supply voltage*: Is kept constant at about 1 V to 1.2 V, since the 130 nm technology node. Further decrease is ceased due to limitations towards minimal gate-overdrive voltage $U_{g\_od} = U_{gs} - V_t$ and the associated V_t variability. In addition, the ceased scaling of the gate-oxide thickness also prevents the further decrease of the supply voltage.
- *Power dissipation*: Down-scaling slope has decreased, as the supply voltage ceased scaling.
- *Power density*: Increases, as the power dissipation decreases slower.

Table 1.1 Classical CMOS scaling laws for circuit and interconnect performance, anno 1974 [Den74]

Device or circuit parameter	Scaling factor
Device dimensions t_{ox}, L_n, L_p, W_n, W_p	$1/\kappa$
Doping concentration N_a	κ
Voltage U	$1/\kappa$
Current I	$1/\kappa$
Capacitance $\epsilon A/d$	$1/\kappa$
Delay time/circuit UC/I	$1/\kappa$
Power dissipation/circuit UI	$1/\kappa^2$
Power density UI/A	1
Interconnect parameter	Scaling factor
Line resistance $R_{line} = \rho L_{line}/Wd$	κ
Normalized voltage drop $I R_{line}/U$	κ
Line response time $R_{line} C$	1
Line current density I/A	κ

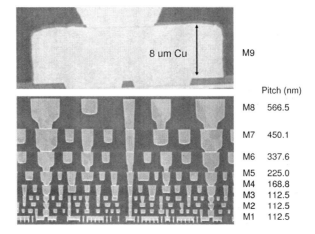

Fig. 1.21 A cross-sectional view of the interconnect of the 32 nm Intel CMOS process [Boh10]

- *Line resistance*: The overall line-resistance increase is slowed down for three reasons: 1) The use of copper interconnect together, or instead of, aluminum, as copper has a lower resistivity ρ. 2) The number of available metal-layers is increased for every new technology node. 3) The use of thick(-top) metal layers. These improvements are illustrated in Fig. 1.21, which shows a cross-sectional view of the available metal layers in the 32 nm Intel CMOS process.
- *Normalized line voltage drop*: Increases slower due to the improvements in the interconnect line resistance.
- *Line response time*: Decreases as the line resistance is improved and also because the parasitic line capacitance decreases due to the use of low dielectric permittivity insulating materials, known as low-k materials.
- *Line current density*: Ceased to increase as the limits of electromigration are met.

The fact that some of the classical scaling laws are no longer valid for nanometer CMOS processes will prove to be advantageous for the purpose of monolithic inductive DC-DC converters. Indeed, the availability of an increasing number of metal layers will improve the characteristics of on-chip passives and power-routing, as explained in Chap. 6. Also the decreasing gate lengths L_n and L_p will enable faster power switches, needed for high-frequency switching, which is discussed in Chap. 4.

1.3.2 The Challenges

As discussed in Sect. 1.1.2 DC-DC converters are not a recent invention. Moreover, they are widely used in many modern electronic applications. Therefore, their operation is well understood and described in literature. In contrast, the feasibility of monolithic DC-DC converters has only been proved quite recently [Ric04] and by the authors knowledge no commercial ICs are available until now (Q4 2010). This is a clear indication that the established design and implementation techniques of

1.3 Monolithic DC-DC Converters: A Glimpse into the Future

DC-DC converters are not suitable for the purpose of monolithic integration. Therefore, this work aims to provide new insights and develop novel techniques that can lead to the practical use of monolithic inductive DC-DC converters.

To achieve this goal, a clear understanding of the differences, with respect to DC-DC converters with external components, is required. Also the problems that will be encountered upon monolithic integration are to be pinpointed. Hence, the following sections will reveal the some significant issues that are to be solved, including the integration of passives on-chip, the required high switching frequencies and the high currents, voltages and powers which need to be dealt with.

Integrating Passives

When performing energy conversion using switched-mode DC-DC converter techniques a minimum of two energy-storing elements is required. These elements can either be two capacitors, yielding a capacitive converter, or an inductor and a capacitor, yielding an inductive converter. Integrating these passive components on-chip in a standard CMOS IC technology poses a number of problems since these IC technologies are optimized for digital circuits, which do not require native passive components (see Sect. 1.3.1).

The realization of on-chip capacitors can be achieved using native CMOS Metal-Oxide-Metal (MOM) capacitors, native CMOS MOS capacitors or Metal-Insulator-Metal (MIM) capacitors, of which the latter are only available in Radio-Frequency (RF) optimized CMOS technologies. Each of these three implementations has its specific advantages and disadvantages, on which will be elaborated in Sect. 6.1.2. The two fundamental bottlenecks are the limited capacitance density and the relatively high parasitic series resistance. The costly and thus limited available chip area together with the low capacitance density implies that the on-chip capacitors will have a low capacitance. More concretely, on-chip standard CMOS[13] capacitors have a capacitance that is roughly three orders of magnitude lower than their off-chip equivalents (from µF-range off-chip to nF-range on-chip). The relatively high parasitic series resistance of on-chip capacitors leads to a low Q-factor and associated losses.

The integration of inductors is commonly achieved using metal-track inductors. A number of the implementations in this work are achieved using the hollow-spiral variant of metal-track inductors [Wen08a, Wen09b]. As with on-chip capacitors, on-chip inductors introduce some difficulties for the use in DC-DC converters. These difficulties are mainly the result of a low inductance density and a high parasitic series resistance. The inductance density of on-chip inductors yields small inductances (nH-range), compared to their off-chip equivalents (µH-range), and the high parasitic series resistance causes Joule-losses. The hollow-spiral bondwire inductor, of which an implementation is shown in Fig. 1.22, can improve some of

---

[13]Through specialized processing, capacitance densities of two orders of magnitude larger, compared to standard CMOS, can be realized [Cha10].

Fig. 1.22 A perspective microphotograph of a DC-DC step-up converter, using a bondwire inductor [Wen07]

these issues [Wen07, Wen08b]. A discussion on integrated inductors is provided in Sect. 6.1.1.

Towards Higher Frequencies

The low absolute values of the inductor and capacitor, combined with their high parasitic series resistance, explained previously, has an implication on the switching frequency of the DC-DC converter.

First, the consequence of the high parasitic series resistance of the inductor is that the maximum value of the inductor current $i_L(t)$ is limited because of the associated Joule-losses. Moreover, the amount of magnetic energy E_L, determined by (1.6), that can be stored in the inductor is limited by both the limited $i_L(t)$ and also by the low inductance L value. The result of this limited E_L is that more switching cycles per unit of time will be needed to achieve a certain output power, thus requiring a higher switching frequency.

$$E_L(t) = \frac{L i_L(t)^2}{2} \tag{1.6}$$

Secondly, the low capacitances and the high parasitic series resistances of on-chip capacitors cause an increased output voltage ripple. The primary solution for this problem is again increasing the switching frequency, as this limits the peak capacitor current for a given output power.

It is understood that the switching frequency of monolithic inductive DC-DC converters will be much higher (100 MHz-range), compared to DC-DC converters with external components (100 kHz-range). This will in-turn lead to an increased stress on the power switches and other building blocks, requiring special care in their design, as explained in Sect. 6.1.3. Moreover, the increased switching frequency will lead to higher dynamic power losses and therefore lower power conversion efficiencies η. This problem can be partly overcome by using new timing-schemes [Wen08a, Wen08b, Wen09b] in the control systems of the converters. These timing-schemes and their practical implementation are discussed in Chap. 5.

1.4 Structural Outline

Coping with Current, Voltage and Power

The implemented designs in this work use input and/or output voltages that exceed the nominal technology supply voltage. One of the features enabling this is the use of stacked transistors circuit topologies. Another option would be to use special (optional) thick-oxide transistors, which can withstand higher voltages, however this would not harmonize with the standard-CMOS philosophy. The design and layout of the switches is explained in Sect. 6.1.3. Another way to cope with the high voltages is to use novel adopted DC-DC converter topologies, which are optimized for monolithic integration, which is discussed in Sect. 3.5.3. Note that high on-chip peak currents, even in the order of 10 A, are feasible and compatible with nanosecond switching in CMOS technologies [Wen09a]. Therefore, these high currents will not pose many problems, providing the interconnect is properly designed, as explained in Sect. 4.2.5.

Non-ideal DC-DC converters introduce power conversion losses and the associated heat dissipation. This leads to increased die temperatures, which can affect the performance of the DC-DC converter. For this purpose the effects of temperature, and their correlation with the efficiency, on the crucial DC-DC converter components are modeled. This is discussed in Sect. 4.3.

Designing for the Limits

The aim of this work, being the monolithic integration of inductive DC-DC converters, is expanded by the search for the limits of these converters. This will be done in two phases. First, adequate inductive DC-DC converter topologies are sought in Chap. 3. Second, the main specifications, for which these limits are sought, are optimized. These specifications are namely: maximal power conversion efficiency, maximal output power, maximal power density and minimal output voltage ripple.

These requirements will counteract on one another, requiring optimizations and justified design choices. This process becomes quite complicated as there is a large number of design parameters (>10). From this point of view it is clear that accurate models of the used DC-DC converter topologies are necessary. For this purpose, mathematical design models, for both boost and buck converters, are deduced in Chap. 4. These models are especially optimized for the specific design of monolithic converters and will enable the deployment of a straightforward design strategy.

1.4 Structural Outline

The structural outline of this dissertation is graphically illustrated by Fig. 1.23. Chapter 1 discusses the basics of DC-DC conversion and the trend towards monolithic DC-DC converters. Chapter 2 gives an overview of DC-DC converter types, in addition with basic DC-DC converter theory. A comparison and a selective overview of inductive DC-DC converter topologies, with regard to monolithic integration, is

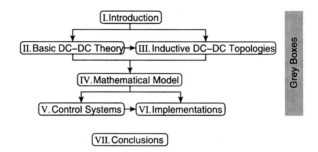

Fig. 1.23 The graphical representation of the structural outline of the dissertation

provided in Chap. 3. For the purpose of the design of inductive DC-DC converters, a mathematical model for both a monolithic boost and buck converter is derived in Chap. 4. New control strategies and systems, which are deployed in this work, are discussed in Chap. 5. The actual implementations, including notions of design guidelines and the measurement setups, of the monolithic inductive DC-DC converters are discussed in Chap. 6. The conclusion of the presented work is formulated in Chap. 7. Finally, it is noted that additional side-information is provided throughout the text, emphasized by the gray boxes.

1.5 Conclusions

This chapter introduces the reader into the domain of DC-DC converters, with a strong affiliation towards monolithic integration.

In Sect. 1.1 a black-box DC-DC voltage converter is introduced, along with its basic properties. These properties are the voltage conversion ratio k, the impedance ratio Z_k and the power conversion efficiency η. Afterwards, the extension towards the principle of ideal switching is discussed. The section is concluded with historical notes on the origin of DC-DC converters, starting in the early 20th century. It is noted that the first DC-DC converters where of mechanical nature, using either electromechanically driven commutators or vibrators as switch devices. Later on these mechanical components where replaced by vacuum-tube triodes and eventually by solid-state transistor switches. The important invention of the concept of multi-phase converters is also noted.

Section 1.2 clarifies the practical use of DC-DC converters in modern low power electronic applications. This is done by distinguishing three types of DC-DC converters:

1. *Mains-operated step-down converters*: Are usually required to provide galvanic separation between the in- and output. For this purpose they require the use of transformers, which are rather difficult to integrate with current standard CMOS technologies. Therefore, they will not be considered for integration in this work.
2. *Battery-operated step-down converters*: Since no galvanic separation is required, the use of transformers can be omitted. Also the output power levels of battery-operated applications are limited to about 1 W. Thus, this type of converter will

1.5 Conclusions

be considered for monolithic integration. Typical applications are SoCs that require multiple supply voltages, which are equal or lower than the battery voltage.
3. *Battery-operated step-up converters*: This type of DC-DC converter will also be considered for monolithic integration, for similar reasons as the previously described step-down converters. Typical applications are SoCs that require multiple supply voltages, which are equal or higher than the battery voltage.

Section 1.3 concludes this first chapter with the technological considerations of monolithic inductive DC-DC CMOS converters. For this purpose, a short introduction to CMOS technology and the MOSFET is given. The prospects on nanometer CMOS technologies gives the reader an idea of what CMOS technology scaling is basically about and how this is related to the topic of this work. Furthermore, the challenges associated with the monolithic integration of DC-DC converters are cited, such as: the integration of the passive components, dealing with high switching frequencies and coping with high on-chip currents, voltages and power. Finally, it is noted that this work will aim for the limits in the realization of monolithic inductive DC-DC converters in CMOS.

Chapter 2
Basic DC-DC Converter Theory

Several methods exist to achieve DC-DC voltage conversion. Each of these methods has its specific benefits and disadvantages, depending on a number of operating conditions and specifications. Examples of such specifications are the voltage conversion ratio range, the maximal output power, power conversion efficiency, number of components, power density, galvanic separation of in- and output, etc. When designing fully-integrated DC-DC converters these specifications generally remain relevant, nevertheless some of them will gain weight, as more restrictions emerge. For instance the used IC technology, the IC technology options and the available chip area will be dominant for the production cost, limiting the value and quality factor of the passive components. These limited values will in-turn have a significant impact upon the choice of the conversion method.

In order for the designer to obtain a clear view of the DC-DC voltage conversion methods and their individual advantages and disadvantages, with respect to monolithic integration, the three fundamental methods are discussed in this chapter. The first and oldest method of performing DC-DC voltage conversion is by means of linear voltage converters (resistive dividers), which are explained in Sect. 2.1. The second method, which also has an interesting potential for the purpose of monolithic voltage conversion, is by means of capacitor charge-pumps, as explained in Sect. 2.2. The latter two methods are explained more briefly as this work will mainly concentrate on inductive type DC-DC converters, which are discussed in Sect. 2.3.

Power conversion efficiency is in most cases a primary specification for any given energy converter. Therefore, a formal method for the fair comparison of DC-DC step-down voltage converters, in terms of power conversion efficiency, is introduced in Sect. 2.4. This method is referred to as the Efficiency Enhancement Factor (EEF). The chapter is concluded with the conclusions in Sect. 2.5.

2.1 Linear Voltage Converters

The most elementary DC-DC converters are linear voltage converters. They achieve DC-DC voltage conversion by dissipating the excess power into a resistor, making

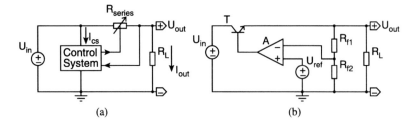

Fig. 2.1 (a) The principle of a linear series voltage converter and (b) a simple practical implementation

them resistive dividers. Clearly, this is not quite ideal for the power conversion efficiency η_{lin}. Another implication of their operating principle is the fact that they can only convert a certain input voltage U_{in} into a lower output voltage U_{out}, having the same polarity. In other words, the value of their voltage conversion ratio k_{lin}, given by (1.1), is always between zero and one.

The advantage of linear voltage converters is that they are fairly simple to implement. Moreover, they generally do not need large, and space consuming, inductors or capacitors, making them an attractive option for monolithic integration [Rin98]. Therefore, the two types of linear voltage converters, namely the series and the shunt regulator, are discussed in Sects. 2.1.1 and 2.1.2.

2.1.1 Series Converter

The operating principle of a linear series voltage converter is shown in Fig. 2.1(a). A variable resistor R_{series} is placed in series with the load R_L, lowering U_{in} to U_{out}. The resistance of R_{series} is controlled by the control system, which keeps U_{out} constant under varying values of U_{in} and R_L, by measuring U_{out}. The control system also consumes power, which is illustrated by its supply current I_{cs}. In this case the control system uses U_{in} as supply voltage, which can also be provided by U_{out}. However, the latter case will require a start-up circuit, as U_{out} is initially zero.

A practical implementation example for a linear series voltage converter is shown in Fig. 2.1(b). In this example R_{series} is implemented as an NPN BJT and the control system as an OPerational AMPlifier (OPAMP), which performs the task of an error amplifier. By doing so, U_{out} is determined by (2.1).

$$U_{out} = U_{ref} \frac{R_{f1} + R_{f2}}{R_{f2}} \qquad (2.1)$$

By examining the operating principle of Fig. 2.1(a), η_{lin} can be calculated through (2.2). When I_{cs} is neglected and assumed to be zero, η_{lin} is equal to k_{lin} and thus independent of R_L. This is graphically illustrated by the black curve in Fig. 2.2(a). The gray curve illustrates the more realistic situation, where I_{cs} has a finite positive value. It can be seen that η_{lin} will tend to decrease when P_{out} decreases.

2.1 Linear Voltage Converters

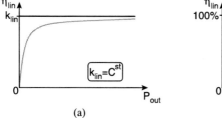
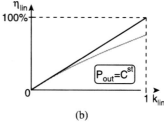

(a)   (b)

Fig. 2.2 (a) The power conversion efficiency η_{lin} as a function of the output power P_{out} for a linear series voltage converter, at a constant voltage conversion ratio k_{lin}. The *black curve* is valid for a zero control system supply current I_{cs} and the *gray curve* is valid for a non-zero I_{cs}. (b) The power conversion efficiency η_{lin} as a function of the voltage conversion ratio k_{lin} for a linear series voltage converter, at a constant output power P_{out}. The *black curve* is valid for a zero control system supply current I_{cs} and the *gray curve* is valid for a non-zero I_{cs}

Clearly, linear series voltage converters have an intrinsic advantage, in terms of power conversion efficiency, at high voltage conversion ratios.[1] This is illustrated by Fig. 2.2(b), where the black curve is valid for $I_{cs} = 0$ and the gray curve for a finite, non-zero I_{cs}. The gray curve shows that the impact of the power consumption of the control system on η_{lin} becomes more dominant when k_{lin} approaches unity.

$$\eta_{lin} = \frac{P_{out}}{P_{in}} = \frac{U_{out} I_{out}}{U_{in}(I_{out} + I_{cs})}\bigg|_{I_{cs}=0} = \frac{U_{out}}{U_{in}} = k_{lin} \qquad (2.2)$$

It is already mentioned that this type of converter is well suited for monolithic integration, due to its simple nature and lack of large passives. However, as the excess power is dissipated as a Joule-loss in R_{series}, the maximal P_{out} is limited by the allowed on-chip power dissipation. This limitation becomes more dominant for low values of k_{lin}.

Note that variants of this type of voltage converter are used in several designs in this work for start-up and rail-shifting. More details can be found in Chap. 6.

2.1.2 Shunt Converter

The alternative for a linear series voltage converter is a linear shunt voltage converter. The principle of operation for this type of DC-DC converter is shown in Fig. 2.3(a). U_{in} is lowered to U_{out} by means of the resistive division between the fixed input resistor R_{in} and both the load R_L and the variable shunt resistor R_{shunt}, where U_{out} is calculated through (2.3). R_{in} can either be the intrinsic output resistance of U_{in}, an added resistor or the combination of both. U_{out} is kept constant

---

[1] Linear series voltage converters are often used in applications which require a small voltage difference $U_{in} - U_{out}$, as they achieve a high η_{lin} in such cases. Therefore, they are commonly denoted as Low Drop-Out (LDO) regulators, which is not necessarily the case.

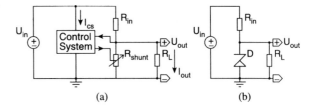

Fig. 2.3 (a) The principle of a linear shunt voltage converter and (b) a simple practical implementation

under varying R_L and U_{in} conditions by adapting the value of R_{shunt}. This operation can be performed by a control system, providing feedback from U_{out}. The control system consumes a certain amount of power by drawing a current I_{cs} from U_{in} or U_{out}. The fact that U_{out} can also be directly used to supply the control systems is due to the self-starting nature of this circuit, as opposed to the linear series voltage converter. The voltage used for supplying the control system will depend on whether U_{out} has a sufficiently large value. In the following analysis it is assumed that U_{in} is used for this purpose, for which the results merely differ little from the other possibility.

$$U_{out} = U_{in} \frac{R_L // R_{shunt}}{R_{in} + R_L // R_{shunt}} \qquad (2.3)$$

Feedback of U_{out} is however not always required, as illustrated by the simple practical implementation of Fig. 2.3(b). For this implementation the shunt resistor is replaced by a reverse-biased zener diode D. In this way a quasi constant U_{out} can be achieved, if the current through D is kept large enough for it to operate in the zener-region.

For a shunt converter η_{lin} is calculated through (1.3), yielding (2.4).

$$\eta_{lin} = \frac{P_{out}}{P_{in}} = \frac{\frac{U_{out}^2}{R_L}}{\frac{U_{in}^2}{R_{in}+R_L//R_{shunt}} + U_{in}I_{cs}} \qquad (2.4)$$

The explicit notation of R_{shunt} from (2.3) substituted into (2.4), yields (2.5). This expression for η_{lin} is not dependent on R_{shunt}.

$$\eta_{lin} = \frac{U_{out}^2}{R_L} \frac{R_{in}}{I_{cs}R_{in}U_{in} + U_{in}^2 - U_{in}U_{out}} \bigg|_{I_{cs}=0} = \frac{U_{out}^2}{R_L} \frac{R_{in}}{U_{in}^2 - U_{in}U_{out}} \qquad (2.5)$$

Figure 2.4(a) graphically illustrates η_{lin} as a function of the output power P_{out}, for a constant voltage conversion ratio k_{lin}. The black curve is valid for the ideal case where I_{cs} is zero and the gray curve is valid for a finite non-zero I_{cs}. As opposed to a linear series converter, η_{lin} is intrinsically linear dependent on P_{out}. It can be seen that η_{lin} is zero for $P_{out} = 0$ and that it has a maximal value equal to k_{lin}, occurring at the maximal output power P_{out_max} which is given by (2.6). For a given U_{in} and U_{out}, P_{out_max} is determined by the inverse of the value of R_{in}.

$$P_{out\_max} = \frac{U_{out}(U_{in} - U_{out})}{R_{in}} \qquad (2.6)$$

2.2 Charge-Pump DC-DC Converters

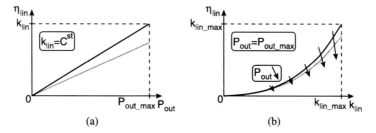

Fig. 2.4 (a) The power conversion efficiency η_{lin} as a function of the output power P_{out} for a linear shunt voltage converter, at a constant voltage conversion ratio k_{lin}. The *black curve* is valid for a zero control system supply current I_{cs} and the *gray curve* is valid for a non-zero I_{cs}. (b) The power conversion efficiency η_{lin} as a function of the voltage conversion ratio k_{lin} for a linear shunt voltage converter, for a constant value of $P_{out} = P_{out\_max}$. The *black curve* is valid for a zero control system supply current I_{cs} and the *gray curve* is valid for a non-zero I_{cs}

The dependency of η_{lin} on k_{lin} is illustrated in Fig. 2.4(b), for a constant $P_{out} = P_{out\_max}$. The black curve is valid when I_{cs} is zero and the gray curve is valid for a finite, non-zero value of I_{cs}. As explained for Fig. 2.4(a), η_{lin} is maximal for P_{out_max}. Therefore, the curves will become lower upon decreasing P_{out}, eventually congregating with the X-axis. The maximal achievable voltage conversion ratio k_{lin_max} is calculated by (2.7) and is for a given U_{out} inversely proportional to P_{out} and R_{in}.

$$k_{lin\_max} = \frac{U_{out}^2}{U_{out}^2 + P_{out\_max} R_{in}} \tag{2.7}$$

Unlike a linear series converter, where η_{lin} is ideally independent of P_{out}, a linear shunt converter only achieves its maximal η_{lin} at P_{out_max}. This behavior makes a linear shunt converter inferior compared to a series converter, in terms of η_{lin}. However, its simple practical implementation makes it suitable for applications that require a small and quasi constant P_{out}. Furthermore, a linear shunt converter can prove to be more practical than a linear series converter in applications that have a low value for k_{lin} and P_{out}. In such a case the voltage difference $U_{in} - U_{out}$ will only be present over the passive resistor R_{in} rather than over an active device, of which the maximal voltage is limited. The simple nature of a linear shunt voltage converter, and its lack of large passives, makes it suitable for monolithic integration in non-critical applications. Obviously, the problem of on-chip power dissipation remains and becomes more limiting than for linear series voltage converters.

Note that this type of voltage converter is also used in several designs in this work, for the purpose of start-up. More details can be found in Chap. 6.

2.2 Charge-Pump DC-DC Converters

Rather than linear converter methods to perform DC-DC voltage conversion, which are described in Sect. 2.1, switched-mode DC-DC converters acquire energy-storing

Fig. 2.5 (a) The circuit for charging a capacitor C with a series resistor R by means of a voltage source U_{in}. (b) The voltage $u_C(t)$ over C and the current $i_C(t)$ through C, as a function of time. (c) The energy $E_{U_{in} \to RC}(t)$ delivered by U_{in}, the energy $E_{U_{in} \to C}(t)$ stored in C and the energy $E_{U_{in} \to R}(t)$ dissipated in R, as a function of time

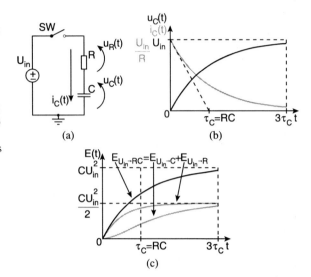

passives and switches to alter the connections between them. Ideally, this switching can be performed lossless, as explained in Sect. 1.1.1, which is preliminary assumed in this section. The losses associated with switching in inductive DC-DC converters are discussed in Chap. 4, of which the main principles remain valid for other types of converters. This section discusses the basics of switched-mode charge-pump DC-DC converters, which only use capacitors as energy-storing elements.

In Sect. 2.2.1 the principles for using capacitors as energy reservoirs is explained and some basic calculations are conducted. These calculations will be used to investigate two ideal charge-pump DC-DC converters. In Sect. 2.2.2 a series-parallel step-down DC-DC converter is discussed, followed by a series-parallel step-up DC-DC converter in Sect. 2.2.3.

2.2.1 On Capacitors

The charging of capacitors is a process that is intrinsically lossy, even when all the components in the charge-circuit are considered to be ideal and lossless. This is explained by means of the circuit for the charging of a capacitor C, with a series resistor R and out of a voltage source U_{in}. This circuit is shown in Fig. 2.5(a). The voltage $u_C(t)$ over C and the current $i_C(t)$ through C, as a function of time, are shown in Fig. 2.5(b), where it is assumed that the initial voltage $U_C(0)$ over C is zero. $u_C(t)$ and $i_C(t)$ are calculated by means of (2.8) and (2.9) respectively. At the time equal to the time constant $t = \tau_C$ the values of $u_C(t)$ and $i_C(t)$ are 63% of the steady-state values, where $t = \infty$, and at $t = 3\tau_C$ their values are 95% of the steady-state values.

2.2 Charge-Pump DC-DC Converters

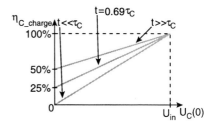

Fig. 2.6 The energy charging efficiency η_{C_charge} of a capacitor charged by means of a voltage source U_{in} as a function of the initial voltage $U_C(0)$ over the capacitor, for three different charge times t

$$u_C(t) = U_{in} + (U_C(0) - U_{in})e^{-\frac{t}{RC}} \quad (2.8)$$

$$i_C(t) = \frac{U_{in} - U_C(0)}{R} e^{-\frac{t}{RC}} \quad (2.9)$$

The corresponding energy $E_{U_{in} \to RC}(t)$ that is delivered by U_{in}, the energy $E_{U_{in} \to R}(t)$ that is dissipated in R and the energy $E_{U_{in} \to C}(t)$ that is stored in C, as a function of time, are calculated through (2.10), (2.11) and (2.12), respectively. These energies are plotted as a function of time in Fig. 2.5(c). Analogue to Fig. 2.5(b), $U_C(0)$ is assumed to be zero. It can be seen that for this case the steady-state values of $E_{U_{in} \to R}(t)$ and $E_{U_{in} \to C}(t)$ are equal. However, in the transient region $E_{U_{in} \to R}(t)$ is higher than $E_{U_{in} \to C}(t)$, making this region undesired for the charging of C. Therefore, it is understood that, from a energy efficiency point of view, when charging a capacitor by means of a voltage source, the steady-state region should be approached sufficiently close. It is noted that it this conclusion can also be proven for the charging of a capacitor by means of another capacitor, which is omitted here.

$$E_{U_{in} \to RC}(t) = \int_0^t U_{in} i_C(t) \, dt = C U_{in}(U_C(0) + U_{in})\left(1 - e^{-\frac{t}{RC}}\right) \quad (2.10)$$

$$E_{U_{in} \to R}(t) = \int_0^t u_R(t) i_C(t) \, dt = \frac{C(U_C(0) - U_{in})^2}{2}\left(1 - e^{-\frac{2t}{RC}}\right) \quad (2.11)$$

$$E_{U_{in} \to C}(t) = E_{U_{in} \to RC} - E_{U_{in} \to R} \quad (2.12)$$

In the case of Fig. 2.5(c), where the value of $U_C(0)$ is assumed zero, half of the energy in steady-state is dissipated in R. This is obviously not the case when $U_C(0) > 0$, as otherwise a charge-pump would never reach a power conversion efficiency higher than 50% · 50% = 25%. In order to comprehend the relation between the charging time t, $U_C(0)$ and the energy charging efficiency $\eta_{C\_charge}(t)$ of a capacitor by means of a voltage source, $\eta_{C\_charge}(t)$ is calculated accordingly to (2.13).

$$\eta_{C\_charge}(t) = \frac{E_{U_{in} \to C}(t)}{E_{U_{in} \to RC}(t)} = \frac{(2.12)}{(2.10)} = \frac{U_C(0) + U_{in} + (U_C(0) - U_{in})e^{-\frac{t}{RC}}}{2U_{in}} \quad (2.13)$$

This relation is graphically represented in Fig. 2.6, where η_{C_charge} is plotted as a function of $U_C(0)$, for three different values of t: $t \ll \tau_C$, $t = 0.69 \cdot \tau_C$ and $t \gg \tau_C$. It is observed that $\eta_{C\_charge}(t)$ increases upon an increasing t, for a certain constant value of $U_C(0)$, meaning that steady-state is approached more closely. This is already concluded earlier in this section, for the case where $U_C(0) = 0$. It can also be seen that $\eta_{C\_charge}(t)$ increases when $U_C(0)$ increases, for a certain constant t. Moreover, $\eta_{C\_charge}(t)$ approaches 100% for the case when $U_C(0) = U_{in}$. Using this knowledge allows for the following conclusions to be made, for maximizing $\eta_{C\_charge}(t)$:

- Sufficient settling time $t \gg \tau_C$ is to be provided, however the effect of settling on $\eta_{C\_charge}(t)$ becomes less dominant if $U_C(0)$ approaches U_{in}.
- The value of $U_C(0)$ should be as close as possible to U_{in}. However, the more $U_C(0)$ approaches to U_{in}, the less energy is transferred to C, which becomes zero for the following condition: $U_C(0) = U_{in}$.
- These conclusions are also valid for the charging of capacitors with one another.

Because of the fact that switching can be ideally lossless and because the charging of capacitors can be performed with $\eta_{C\_charge}(t)$ approaching a value of 100%, charge-pump DC-DC converters have a promising prospect on the achievable conversion efficiency. Indeed, they could theoretically achieve higher power conversion efficiencies than do linear voltage converters, which are explained in Sect. 2.1.

2.2.2 Series-Parallel Step-Down Converter

The circuit topology of an ideal series-parallel charge-pump step-down DC-DC converter is shown in Fig. 2.8(a). Its basic operation consists of two phases:

1. *The flying capacitor charge phase* Φ_1: During Φ_1 both the flying-capacitor C_1 and the output capacitor C_2 are charged, in series with each other, through the voltage source U_{in}. This is done by closing the switches SW_1–SW_4 and opening SW_2–SW_3, yielding the equivalent charge circuit shown in Fig. 2.8(b). In Φ_1 a part of the charge current through C_1 also flows through the load R_L.
2. *The flying capacitor discharge phase* Φ_2: During Φ_2 C_2 is charged by C_1, by placing C_1 parallel with C_2. This is achieved by opening SW_1–SW_4 and closing SW_2–SW_3, yielding the equivalent discharge circuit shown in Fig. 2.8(c). Similar as in Φ_1 a part of the charge current from C_1 also flows through R_L. After this second phase the first phase is started again. The frequency at which this is done is denoted as the switching frequency f_{SW}.

From this operation principle it follows that U_{out} can never exceed $U_{in}/2$, as this would cause reversing of the energy flow from the output to the input. Obviously, this is physically not possible. By means of these principles a simple expression for U_{out} can be formulated, given by (2.23). In this equation I_{out} is the current through the load, ΔQ_{SW} is the amount of charge being transferred to the output in each switching cycle and ΔU_{C_1} is the voltage difference over C_1, between Φ_1 and Φ_2. The equation readily shows that, for an ideal converter, U_{out} is proportional to R_L, f_{SW} and C_1 and that U_{out} will asymptotically reach the value of $U_{in}/2$.

2.2 Charge-Pump DC-DC Converters

⊛CHARGED CAPACITORS IN EQUILIBRIUM

The circuit for charging a capacitor C_b with another capacitor C_a is shown in Fig. 2.7(a). It is proved that for $t \gg \tau_C$ the energy transfer is independent of the value of the series resistor R. This is done by calculating the limit, where $t/\tau_C \to \infty$, for the energy equations $E_{C_a \to RC_b}$ and $E_{C_a \to R}$, yielding (2.14) and (2.15). $E_{C_a \to C_b}$ is subsequently calculated by (2.16).

$$E_{C_a \to RC_b} = \lim_{\frac{t(C_a+C_b)}{C_a C_b R} \to \infty} \left[\int_0^t U_{C_a}(t) I(t) \, dt \right]$$

$$= -\frac{C_a C_b^2}{2(C_a + C_b)^2} \Delta U^2 + \frac{U_{C_a}(0) C_a C_b}{C_a + C_b} \Delta U \quad (2.14)$$

$$E_{C_a \to R} = \lim_{\frac{t(C_a+C_b)}{C_a C_b R} \to \infty} \left[\int_0^t U_R(t) I(t) \, dt \right] = \frac{C_a C_b}{2(C_a + C_b)} \Delta U^2 \quad (2.15)$$

$$E_{C_a \to C_b} = E_{C_a \to RC_b} - E_{C_a \to R}$$

$$= -\frac{C_a C_b (C_a + 2C_b)}{2(C_a + C_b)^2} \Delta U^2 + \frac{U_{C_a}(0) C_a C_b}{C_a + C_b} \Delta U \quad (2.16)$$

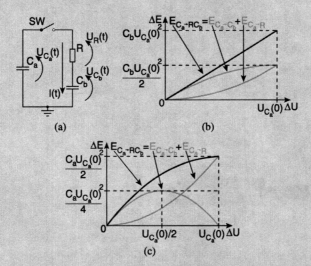

Fig. 2.7 (a) The circuit for charging C_b with C_a. (b) The $E_{C_a \to RC_b}$, $E_{C_a \to C_b}$ and $E_{C_a \to R}$ as a function of $\Delta U = U_{C_a}(0) - U_{C_b}(0)$, for $C_a \gg C_b$ and (c) for $C_b \gg C_a$

The obtained expressions for the energy transfers can be subdivided for two special cases. The first case, where $C_a \gg C_b$, results in (2.17), (2.18) and (2.19). This is illustrated in Fig. 2.7(b), where $E_{C_a \to RC_b}$, $E_{C_a \to C_b}$ and $E_{C_a \to R}$ are plotted normalized as a function of time. The second case, where $C_b \gg C_a$, results in (2.20), (2.21) and (2.22). This is illustrated in Fig. 2.7(c), similar as in Fig. 2.7(b).

⊛CHARGED CAPACITORS IN EQUILIBRIUM (CONTINUED)

$$C_a \gg C_b \implies E_{C_a \to RC_b} \simeq C_b U_{C_a}(0) \Delta U \qquad (2.17)$$

$$\implies E_{C_a \to R} \simeq \frac{C_b}{2} \Delta U^2 \qquad (2.18)$$

$$\implies E_{C_a \to C_b} \simeq -\frac{C_b}{2} \Delta U^2 + C_b U_{C_a}(0) \Delta U \qquad (2.19)$$

$$C_b \gg C_a \implies E_{C_a \to RC_b} \simeq -\frac{C_a}{2} \Delta U^2 + C_a U_{C_a}(0) \Delta U \qquad (2.20)$$

$$\implies E_{C_a \to R} \simeq \frac{C_a}{2} \Delta U^2 \qquad (2.21)$$

$$\implies E_{C_a \to C_b} \simeq -C_a \Delta U^2 + C_a U_{C_a}(0) \Delta U \qquad (2.22)$$

Please note that for the case where $C_a \gg C_b$ the obtained results are also valid for the circuit of Fig. 2.5(a), with $U_C(0) = U_{in}$.

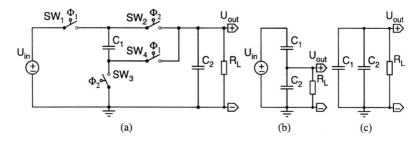

Fig. 2.8 (a) The circuit of an ideal series-parallel charge-pump step-down DC-DC converter, together with (b) its equivalent charge circuit and (c) its equivalent discharge circuit

$$U_{out} = R_L I_{out} = R_L f_{SW} \Delta Q_{SW} = R_L f_{SW} C_1 \Delta U_{C_1} = R_L f_{SW} C_1 (U_{in} - 2U_{out})$$

$$\implies U_{out} = \frac{R_L f_{SW} C_1 U_{in}}{1 + 2R_L f_{SW} C_1} \blacksquare \qquad (2.23)$$

The process of charging capacitors by means of voltage sources, as well as other capacitors, is intrinsically prone to losses, as discussed earlier in this section. Hence, by making the assumption that the series-parallel charge-pump consists of ideal and lossless components, the ideal power conversion efficiency η_{sp_down} as a function of the voltage ratio $k_{SW} = U_{in}/U_{out}$ can be calculated. For this ideal converter it is assumed that the output voltage ripple ΔU_{out} is infinitesimal, implying that f_{SW} is infinitely large and that there is no dependency on R_L, which follows from (2.23). Furthermore, as the switches are assumed ideal there are no switch losses, therefore η_{sp_down} is independent of f_{SW}. The calculation of η_{sp_down} is performed in three steps. In the first step the energy conversion efficiency η_{Φ_1} of the charge phase is determined. The circuit of this phase, where C_1 and C_2 are charged through U_{in}, is similar to that of Fig. 2.7(a), for the case where $C_a \gg C_b$. As a result, η_{Φ_1} is given by (2.24).

2.2 Charge-Pump DC-DC Converters

$$\frac{E_{C_a \to C_b}}{E_{C_a \to RC_b}} = \frac{(2.19)}{(2.17)} \quad \text{with} \quad \begin{cases} \Delta U = U_{in} - 2U_{out} \\ C_b = \dfrac{C_1 C_2}{C_1 + C_2} \\ U_{C_a}(0) = U_{in} \end{cases}$$

$$\implies \eta_{\Phi_1} = \frac{E_{C_1 C_2}}{E_{U_{in} \to C_1 C_2}} = \frac{U_{in} + 2U_{out}}{2U_{in}} \qquad (2.24)$$

In the second step the energy conversion efficiency η_{Φ_2} of the discharge phase is determined. In this phase C_1 is discharged through C_2, as illustrated by Fig. 2.7(a). Accordingly, η_{Φ_2} can be calculated analogue to (2.24). However, for this phase C_1 and C_2 are assumed to obtain arbitrary values, yielding (2.25).

$$\frac{E_{C_a \to C_b}}{E_{C_a \to RC_b}} = \frac{(2.16)}{(2.14)} \quad \text{with} \quad \begin{cases} \Delta U = U_{in} - 2U_{out} \\ U_{C_a}(0) = U_{in} - U_{out} \end{cases}$$

$$\implies \eta_{\Phi_2} = \frac{E_{C_2}}{E_{C_1 \to C_2}} = \frac{C_1 U_{in} + 2C_2 U_{out}}{U_{in}(2C_1 + C_2) - 2C_1 U_{out}} \qquad (2.25)$$

Finally, for the third step η_{sp_down} can be calculated, resulting in (2.26).

$$\eta_{sp\_down} = \eta_{\Phi_1} \eta_{\Phi_2} = \frac{(C_1 U_{in} + 2C_2 U_{out})\left(\frac{U_{in} + 2U_{out}}{2U_{in}}\right)}{C_2 U_{in} + 2C_1(U_{in} - U_{out})} \blacksquare \qquad (2.26)$$

In addition, for the purpose of graphically representing this information, two special cases of (2.26) are considered: The case where the flying capacitor C_1 is significantly larger than the output capacitor C_2 and vice versa. The first case is given by (2.27) and the second case by (2.28).

$$C_1 \gg C_2 \implies \eta_{sp\_down} = \frac{U_{in} + 2U_{out}}{4U_{in} - 4U_{out}} \qquad (2.27)$$

$$C_2 \gg C_1 \implies \eta_{sp\_down} = \frac{U_{in} U_{out} + 2U_{out}^2}{U_{in}^2} \qquad (2.28)$$

The graphical representation of (2.26), (2.27) and (2.28), illustrated in Fig. 2.9 by the black curves, shows η_{sp_down} as a function of k_{SW}. For the case of (2.26) the value of C_1 and C_2 are assumed to be equal. The gray curve shows η_{lin} of an ideal linear series converter, for comparison (refer to Sect. 2.1.1). It is observed that for all three cases of the values of C_1 and C_2, a higher power conversion efficiency than achievable with a linear series converter is obtained. Also, for all three cases η_{sp_down} reaches its maximum value of 100% at $k_{SW} = 0.5$ and decreases upon decreasing values of k_{SW}. This can be understood by reconsidering Fig. 2.6, where it can be seen that η_{C_charge} decreases when $U_{in} - U_C(0)$ increases. Similar behavior can be observed when k_{SW} decreases. For the charge phase this implies that $U_{in} - (U_{C_1}(0) + U_{C_2}(0))$ increases and for the discharge phase $U_{C_1}(0) - U_{C_2}(0)$ increases. Finally, it can be seen that the highest overall η_{sp_down} is obtained for the case where $C_1 \gg C_2$. This follows from the fact that the charging of capacitors with one another is more efficient if the charging capacitor is larger than the capacitor being charged, as can be seen by comparing the graphical representations from Fig. 2.7(a) and Fig. 2.7(b).

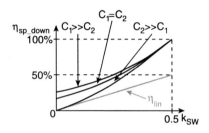

Fig. 2.9 The *black curves* show the power conversion efficiency η_{sp_down} of an ideal series-parallel charge-pump step-down DC-DC converter, as a function of the voltage conversion ratio k_{SW}, for three different cases of the values of C_1 and C_2. The *gray curve* shows the power conversion efficiency of a linear series converter, as a function of k_{SW}

This kind of converter topologies is quite promising, seen from the perspective towards monolithic integration. The advantage over linear voltage converters, which are explained in Sect. 2.1, in terms of η_{sp_down}, is clearly illustrated in Fig. 2.9. Practical monolithic realizations in standard CMOS technologies of charge-pump step-down converters show that high values for η_{sp_down} are realistic [Ram10]. However, care must be taken when comparing η_{sp_down}, as it should be normalized conform to k_{SW}. This is elaborated upon in Sect. 2.4. There are also a few potential drawbacks. First, this topology is limited in a sense that the maximal k_{SW} can never exceed the value of 0.5. Obviously, this can be overcome by using a different topology, which is however beyond the scope of this work. Secondly, two capacitors for storing the energy are necessary. Therefore, this type of switched converter will inevitable require more chip area, compared to a linear converter with a similar maximal output power. Thirdly, although η_{sp_down} can theoretically reach a value of 100%, this case for which $k_{SW} = 0.5$ will not always prove to be practical in a real-world application. This can also be partially solved by using different, more complex, gear-box topologies [Mak95], of which a discussion is also omitted in this work. Finally, four switches are required, where other switched converter topologies can provide similar functionality with only two switches, as explained in Sect. 3.1.

2.2.3 Series-Parallel Step-Up Converter

Unlike linear voltage converters switched-mode voltage converters are capable of converting a given input voltage to a higher output voltage. Such converters are commonly denoted as step-up converters. A straightforward example of a charge-pump step-up converter is the series-parallel converter, of which the circuit topology with ideal components is shown in Fig. 2.10. The basic operation of this converter consists of two phases:

1. *The flying capacitor charge phase* Φ_1: During Φ_1 the flying-capacitor C_1 is charged through the voltage source U_{in}. This is done by closing the switches SW_1–SW_3 and opening SW_2–SW_4, yielding the equivalent charge circuit shown

2.2 Charge-Pump DC-DC Converters

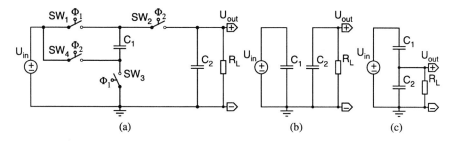

Fig. 2.10 (a) The circuit of an ideal series-parallel charge-pump step-up DC-DC converter, together with (b) its equivalent charge circuit and (c) its equivalent discharge circuit

in Fig. 2.10(b). Also, in this phase the output capacitor C_2 is discharged through the load R_L.

2. *The flying capacitor discharge phase* Φ_2: During Φ_2 C_2 is charged by C_1 and U_{in}, by placing C_1 in series with U_{in}. This is achieved by opening SW_1–SW_3 and closing SW_2–SW_4, yielding the equivalent charge circuit shown in Fig. 2.10(c). A part of the charge current from C_1 and U_{in} also flows through R_L. After this second phase the first phase is started again. The frequency at which this is done is denoted as the switching frequency f_{SW}.

The conclusions and calculations made in the following discussion are similar to those of the series-parallel step-down charge-pump of Sect. 2.2.2, hence the following explanation will be more briefly. U_{out} of the series-parallel step-up converter is limited to $2U_{in}$ and is calculated by means of (2.29). This leads to the conclusion that $0 < U_{out} < 2U_{in}$, where the value of $2U_{in}$ is asymptotically reached.

$$U_{out} = R_L I_{out} = R_L f_{SW} \Delta Q_{SW} = R_L f_{SW} C_1 \Delta U_{C_1} = R_L f_{SW} C_1 (2U_{in} - U_{out})$$
$$\implies U_{out} = \frac{2R_L f_{SW} C_1 U_{in}}{1 + R_L f_{SW} C_1} \blacksquare \qquad (2.29)$$

For this converter topology the ideal power conversion efficiency η_{sp_up} will also depend on k_{SW}. In order to calculate this dependency, ΔU_{out} is assumed to be zero, implying that f_{SW} is infinitely large. Thus, it follows from (2.29) that there is no dependency on R_L. The calculation itself is performed in three steps. First, the energy conversion efficiency η_{Φ_1} of the charge phase is determined, which is done analogue to the calculation of (2.24). In this case only C_1 is charged, yielding (2.30).

$$\frac{E_{C_a \to C_b}}{E_{C_a \to RC_b}} = \frac{(2.19)}{(2.17)} \quad \text{with} \quad \begin{cases} \Delta U = 2U_{in} - U_{out} \\ C_b = C_1 \\ U_{C_a}(0) = U_{in} \end{cases}$$

$$\implies \eta_{\Phi_1} = \frac{E_{C_1}}{E_{U_{in} \to C_1}} = \frac{U_{out}}{2U_{in}} \qquad (2.30)$$

Secondly, the energy conversion efficiency η_{Φ_1} of the discharge phase is calculated. The calculation method is analogue to (2.25), resulting in (2.31).

Fig. 2.11 The *black curves* show the power conversion efficiency η_{sp_up} of an ideal series-parallel charge-pump step-up DC-DC converter, as a function of the voltage conversion ratio k_{SW}, for three different cases of the values of C_1 and C_2. The *gray curve* shows the power conversion efficiency η_{lin} of a linear series converter, as a function of k_{SW}

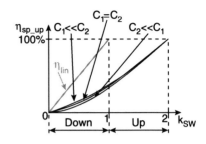

$$\frac{E_{C_a \to C_b}}{E_{C_a \to RC_b}} = \frac{(2.16)}{(2.14)} \quad \text{with} \quad \begin{cases} \Delta U = 2U_{in} - U_{out} \\ U_{C_a}(0) = 2U_{in} \end{cases}$$

$$\implies \eta_{\Phi_2} = \frac{E_{C_2}}{E_{U_{in}C_1 \to C_2}} = \frac{2C_1 U_{in} + C_1 U_{out} + 2C_2 U_{out}}{4C_1 U_{in} + 2C_2 U_{in} + C_2 U_{out}} \quad (2.31)$$

Thirdly, the resulting η_{sp_up} is calculated by means of (2.32).

$$\eta_{sp\_up} = \eta_{\Phi_1} \eta_{\Phi_2} = \frac{U_{out}(2C_2 U_{out} + C_1(2U_{in} + U_{out}))}{2U_{in}(4C_1 U_{in} + C_2(2U_{in} + U_{out}))} \blacksquare \quad (2.32)$$

Two special cases for (2.32): 1) the flying capacitor C_1 is significantly larger than the output capacitor C_2 and 2) vice versa, are also considered. The results for these two cases are respectively given by (2.33) and (2.34).

$$C_1 \gg C_2 \implies \eta_{sp\_up} = \frac{2U_{in}U_{out} + U_{out}^2}{8U_{in}^2} \quad (2.33)$$

$$C_2 \gg C_1 \implies \eta_{sp\_up} = \frac{U_{out}^2}{2U_{in}^2 + U_{in}U_{out}} \quad (2.34)$$

The graphical representation of (2.32), (2.33) and (2.34) is illustrated in Fig. 2.11, where the black curves show η_{sp_up} as a function of k_{SW}. For (2.32), the values of C_1 and C_2 are chosen equally large. The gray curve shows η_{lin} of an ideal linear series converter, for comparison (refer to Sect. 2.1.1). Although this converter is designated for step-up voltage conversion, it is also capable to perform a step-down function. This is indicated on the X-axis. For the step-down operation region, the achievable η_{sp_up} is however significantly lower than for a series-parallel step-down converter, as can be seen in Fig. 2.9, and it is also significantly lower that for a linear series regulator. Thus, the step-down operation region of this converter has no practical use. The step-up operation region yields theoretical values for η_{sp_up} ranging from about 35%, for $k_{SW} = 1$, to 100%, for $k_{SW} = 2$. The reason for η_{sp_up} to decrease upon decreasing values for k_{SW} is due to the fact that the charging of capacitors becomes less efficient when $U_{in} - U_C(0)$ increases, as is illustrated in Fig. 2.6. Translated for this converter, this means that upon decreasing values for k_{SW}, $2U_{in} - U_{C_1}(0)$ increases for the charge phase and $(U_{in} + U_{C_1}(0)) - U_{C_2}(0)$ increases for the discharge phase. Finally, it is observed that the dependency of

η_{sp_up} upon the difference in values for C_1 and C_2 is far less significant than for the step-down variant of this converter. This follows from the fact that C_2 is charged by means of C_1 in series with U_{in}, in the discharge phase. This process is less lossy than charging a capacitor merely with another capacitor, as is done in the discharge phase of a series-parallel step-down converter.

Monolithic integration of the series-parallel step-up charge pump converter topology is, similar to its step-down variant, feasible. The common limitation for any given charge-pump is the fact that η_{SW} reaches its maximum at only in particular value of k_{SW}, which is 2 for this topology. In practical applications, where k_{SW} often needs to be variable, this might prove to be a limitation. Nevertheless, this could be overcome by using a gear-box topology. An overview of other charge-pump step-up topologies and a comparison, with respect to their required area for monolithic integration, is given in [Bre08]. These and other topologies will not be discussed in this work. Charge-pump step-up converters are proven to be feasible, for the purpose of monolithic integration [Bre09a, Bre09b]. Moreover, individual converter stages can achieve high values of η_{SW}, in the order of 80% and more. This is beneficial for converters that have a moderately low value (< 2) of k_{SW} [Bre09a]. However, when higher values (> 2) of η_{SW} are required different converters need to be cascaded, having a negative impact on the overall η_{SW} [Bre09b]. Finally, a potential drawback of this step-up topology is that it requires four switches, whereas some other topologies discussed in Sect. 3.2 require only two switches.

2.3 Inductive Type DC-DC Converters

The third way to achieve DC-DC voltage conversion is by means of inductive type DC-DC converters. This type of DC-DC converters belongs, together with the charge-pump DC-DC converters discussed in Sect. 2.2, to the group of switched-mode DC-DC converters. Inductive DC-DC converters consist of one or more inductor(s) and capacitor(s) and at least two switches.[2] In the galvanically separated variant of inductive DC-DC converters the inductor is replaced by a transformer, which can be seen as two, or more, mutually magnetically coupled inductors. A selection of inductive type DC-DC converter topologies is provided in Chap. 3.

The basic theory for understanding the processes of storing energy by means of inductor and inductor-capacitor circuits is explained in the respective Sects. 2.3.1 and 2.3.2. The traditional calculation methods for the steady-state behavior of inductive type DC-DC converters are discussed in Sect. 2.3.3.

2.3.1 On Inductors

Unlike the charging of capacitors, discussed in Sect. 2.2.1, the process of charging inductors is not intrinsically lossy. In other words, the charging of an ideal induc-

---

[2] For DC-DC converters with off-chip components, one or more of the switches is often implemented as a diode. In doing so, the functionality of the converter is not altered.

Fig. 2.12 (a) The circuit for charging an inductor L in series with a resistor R by means of a voltage source U_{in}. (b) The voltage $u_L(t)$ over and the current $i_L(t)$ through L, as a function of time. (c) The energy $E_{U_{in}\to RL}(t)$ delivered by U_{in}, the energy $E_{U_{in}\to L}(t)$ stored in L and the energy $E_{U_{in}\to R}(t)$ dissipated in R, as a function of time

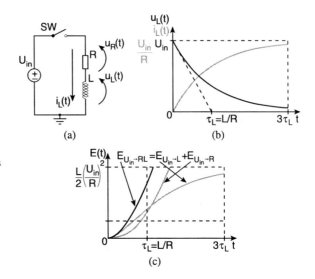

tor involves no energy loss. In order to understand the energy loss mechanism of charging non-ideal inductors, the circuit for charging an inductor L in series with a resistor R by means of a voltage source U_{in} is considered in Fig. 2.12(a). The voltage $u_L(t)$ and the current $i_L(t)$ through the inductor are plotted as a function of time t in Fig. 2.12(b). For these plots it is assumed that the initial current $i_L(0)$ through L is zero. $u_L(t)$ and $i_L(t)$ are calculated by means of (2.35) and (2.36), respectively. The time constant τ_L denotes the time where $u_L(t)$ and $i_L(t)$ reach 63% of their steady-state value. When a time equal to $3\tau_L$ is elapsed, $u_L(t)$ and $i_L(t)$ reach 95% of their steady-state value.

$$u_L(t) = (U_{in} - RI_L(0))e^{-\frac{tR}{L}} \tag{2.35}$$

$$i_L(t) = \frac{U_{in}}{R} + \left(I_L(0) - \frac{U_{in}}{R}\right)e^{-\frac{tR}{L}} \tag{2.36}$$

Figure 2.12(c) shows the energy $E_{U_{in}\to RL}(t)$ that is delivered by U_{in}, the energy $E_{U_{in}\to R}(t)$ that is dissipated in R and the energy $E_{U_{in}\to L}(t)$ that is stored in L as a function of time and for $I_L(0) = 0$. These energies are calculated by (2.37), (2.38) and (2.39), respectively, of which the results are not shown due to their complexity. It is observed that $E_{U_{in}\to R}(t)$ and $E_{U_{in}\to L}(t)$ are equal at a charging time t of about $1.15 \cdot \tau_L$. When $t < 1.15$, $E_{U_{in}\to L}(t)$ is larger than $E_{U_{in}\to R}(t)$, which is beneficial for the energy charging efficiency $\eta_{L\_charge}(t)$ of the charging of an inductor. For a charging time $t > 1.15$, $E_{U_{in}\to R}$ becomes larger than $E_{U_{in}\to L}(t)$. The value of $E_{U_{in}\to R}(t)$ can be infinitely large, as $i_L(t)$ keeps flowing in steady-state, causing continuous Joule-losses in R. The value of $E_{U_{in}\to L}(t)$, on the other hand, saturates to a finite value in steady-state for a non-ideal inductor, where $0 < R < \infty$. The steady-state value of $E_{U_{in}\to L}(t)$ is calculated by applying the limit where $t/\tau_L \to \infty$, yielding (2.40). This equals (1.6), for $I_L(0) = 0$.

2.3 Inductive Type DC-DC Converters

Fig. 2.13 The energy charging efficiency η_{L_charge} of an inductor with a series resistance charged by a voltage source U_{in}, as a function of the initial current $I_C(0)$ through the inductor, for three different charge time t

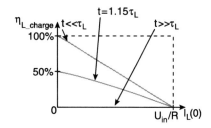

$$E_{U_{in} \to RL}(t) = \int_0^t U_{in} i_L(t)\,dt \qquad (2.37)$$

$$E_{U_{in} \to R}(t) = \int_0^t u_R(t) i_L(t)\,dt \qquad (2.38)$$

$$E_{U_{in} \to L}(t) = E_{U_{in} \to RL}(t) - E_{U_{in} \to R}(t) \qquad (2.39)$$

$$\lim_{\frac{tR}{L} \to \infty} \left(E_{U_{in} \to L}(t) \right) = \frac{L}{2}\left(\left(\frac{U_{in}}{R}\right)^2 - I_L(0)^2 \right) \qquad (2.40)$$

Obviously, the maximal energy that can additionally be stored in an inductor, given by (2.40), is not a practical value since it is associated with high losses in the (parasitic) series resistance. In order to gain a better idea of the relation between the stored and the dissipated energy, the energy charging efficiency $\eta_{L\_charge}(t)$ is calculated by means of (2.41). The result of this calculation is not shown in its mathematical form, due to its rather complex nature. Therefore, (2.41) is graphically represented in Fig. 2.13, where $\eta_{L\_charge}(t)$ is plotted as a function of $I_L(0)$, for three different cases of t: $t \ll \tau_L$, $t = 1.15 \cdot \tau_L$ and $t \gg \tau_L$.

$$\eta_{L\_charge}(t) = \frac{E_{U_{in} \to L}(t)}{E_{U_{in} \to RL}(t)} = \frac{(2.39)}{(2.37)} \qquad (2.41)$$

In Fig. 2.13 it is observed that, for equal values of $I_L(0)$, $\eta_{L\_charge}(t)$ decreases upon increasing values of t. This was already concluded for the case where $I_L(0) = 0$, in Fig. 2.12(c). Furthermore, it can be seen that $\eta_{L\_charge}(t)$ decreases as $I_L(0)$ increases, for a fixed value of t. This knowledge leads to the following conclusions for the charging of an inductor with a series resistance by means of a voltage source:

- The highest values $\eta_{L\_charge}(t)$ are obtained at $t \ll \tau_L$, meaning that the charge time t of inductors should be minimized, thereby minimizing the resistive losses. However, as the value of t becomes smaller, for a certain constant value of τ_L, the energy $E_{U_{in} \to L}(t)$ stored in L becomes smaller as well, ultimately becoming infinitesimal.
- The value of $I_L(0)$ should be minimized and ideally be zero, in order to obtain the highest values for $\eta_{L\_charge}(t)$.
- The charging of an ideal inductor, which has an infinitesimal series resistance, by means of a voltage source is lossless, regardless of the value of $I_L(0)$. The formal prove of this statement is omitted, but it can be intuitively understood by considering Fig. 2.13. If $R = 0$ then $\tau_L = \infty$ and as a consequence $t \ll \tau_L$, which

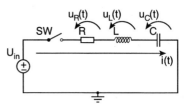

Fig. 2.14 The circuit for charging a series inductor L and a series capacitor C with a series resistor R, by means of a voltage source U_{in}

is illustrated by the upper curve. Moreover, the intersection with the X-axis U_{in}/R will be at infinity, implying that curve for $t \ll \tau_L$ will be parallel to the X-axis, intersecting the Y-axis at the value $\eta_{L\_charge}(t) = 100\%$.

It is clear that inductors have an intrinsic advantage over capacitors for the purpose of storing energy out of a voltage source. Indeed, the process of charging capacitors can only reach a high value of $\eta_{C\_charge}(t)$ if $U_{in} - U_C(0)$ is kept significantly smaller than U_{in}, regardless of the value of the parasitic series resistance. This is in contrast to the process of charging inductors, which can reach a high value of $\eta_{L\_charge}(t)$ for any given value of U_{in} and $I_C(0)$, as long as the parasitic series resistance is kept small. Therefore, inductors are theoretically well suited for the purpose of DC-DC converters, as they allow for high power conversion efficiencies over a broad range of voltage conversion levels.

2.3.2 Inductors and Capacitors: The Combination

Any given switched-mode (inductive) DC-DC converter topology has an output capacitor to smooth out the current/voltage transients, generated by the switching part of the converter, and also to act as an energy reservoir for the load when the converter is not delivering power to the output. However, as the direct charging of a capacitor through a voltage source is intrinsically lossy (see Sect. 2.2.1), an inductor needs to be added to avoid these losses. This process of charging a capacitor C with a series inductor L through a voltage source U_{in} is illustrated in Fig. 2.14, where the series resistor R represents the sum of the parasitic series resistances of the voltage source U_{in}, the inductor L and the capacitor C. This circuit is described by its differential equation, which is omitted in this discussion. Its time-domain solution can be either aperiodically damped, critically damped or periodically damped. Only the latter case, for which (2.42) is valid, is considered in this discussion as the total parasitic series resistance in practical inductive DC-DC converters is always kept small enough to fulfill this requirement.

$$R \ll 2\sqrt{\frac{L}{C}} \qquad (2.42)$$

For the periodically damped *RLC*-circuit, the current $i(t)$, the voltage $u_L(t)$ over L, the voltage $u_R(t)$ over R and the voltage $u_C(t)$ over C are given by (2.43),

2.3 Inductive Type DC-DC Converters

(2.44),[3] (2.45) and (2.46), respectively. In (2.43), $I_L(0)$ denotes the initial current, $U_C(0)$ is the initial voltage over C, ω_{LC} is the angular frequency and τ_{LC} is the time constant.

$$i(t) = \big(I_L(0)Q\cos(\omega_{LC}t) + (I_L(0)(R-2LR) + 2(U_{in} - U_C(0)))\sin(\omega_{LC}t)\big)$$
$$\cdot \frac{e^{-\frac{t}{\tau_{LC}}}}{Q} \tag{2.43}$$

with
$$\begin{cases} Q = \sqrt{\dfrac{4L}{C} - R^2} \\ \tau_{LC} = \dfrac{2L}{R} \\ \omega_{LC} = \dfrac{Q}{2L} \end{cases}$$

$$u_L(t) = L\frac{di(t)}{dt} \tag{2.44}$$
$$u_R(t) = Ri(t) \tag{2.45}$$
$$u_C(t) = U_{in} - u_L(t) - u_R(t) \tag{2.46}$$

The current and the voltages of an ideal series *RLC*-circuit as a function of time are plotted in Fig. 2.15(a). In this figure it is assumed that the resistance of R is infinitesimal, $I_L(0) = 0$ and $U_C(0) = 0$. The most important observations in this plot are that the ideal series *RLC*-circuit oscillates undamped at an angular frequency $\omega_{LC} = 2\pi f = 2\pi/T$ and that the maximal voltage U_{C_max} over C is $2U_{in}$. For the non-ideal, periodically damped *RLC*-circuit, where R is non-zero and complies with (2.42), the current and voltages are plotted in Fig. 2.15(b), where it is also assumed that $I_L(0) = 0$ and $U_C(0) = 0$. This circuit oscillates at the same angular frequency ω_{LC} as its ideal counterpart, it is however damped with a time-constant τ_{LC}. In steady-state $i_L(t)$, $u_L(t)$ and $u_R(t)$ become zero and $u_C(t)$ becomes equal to U_{in}.

The energy $E_{U_{in} \to RLC}(t)$ delivered by U_{in}, the energies $E_L(t)$ stored in L, $E_R(t)$ dissipated in R and $E_C(t)$ stored in C are given by (2.47), (2.48), (2.49) and (2.50).

$$E_{U_{in} \to RLC}(t) = \int_0^t U_{in} i(t)\,dt \tag{2.47}$$

$$E_L(t) = \int_0^t u_L(t)i(t)\,dt \tag{2.48}$$

$$E_R(t) = \int_0^t u_R(t)i(t)\,dt \tag{2.49}$$

$$E_C(t) = \int_0^t u_C(t)i(t)\,dt \tag{2.50}$$

These energies are plotted in Fig. 2.16(a) for an ideal series *RLC*-circuit, where it is assumed that the resistance of R, $I_L(0)$ and $U_C(0)$ are zero. During the first

---
[3] This equation is known as Lenz's law.

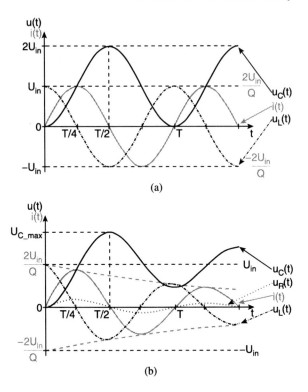

Fig. 2.15 (a) The current $i(t)$, the voltage $u_L(t)$ over L, the voltage $u_R(t)$ over R and the voltage $u_C(t)$ over C as a function of time, for in ideal ($R = 0$) and (b) a non-ideal ($R \neq 0$) series RLC-circuit

one-fourth part of the period T energy from U_{in} is stored in L and C. Afterwards, during the following one-fourth part of T energy from both U_{in} and L are stored in C. Finally, during the last half part of T this process is repeated in reversed order and the cycle recommences. It can be seen that at exactly the half of T, $E_C(t)$ is maximal and equal to $E_{U_{in} \rightarrow RLC}(t)$, implying that the energy charging efficiency $\eta_{RLC\_charge}(t)$, for charging a capacitor in an ideal series RLC-circuit by means of a voltage source, is equal to 100%. This is in contrast with the case were a capacitor is charged directly by a voltage source, where for the same case of a zero initial voltage ($U_C(0) = 0$) $\eta_{C\_charge}(t) = 50\%$, as can be seen in Fig. 2.5(c). Moreover, in the latter case C is only charged to a voltage equal to U_{in} in steady-state, whereas for the ideal RLC-circuit C is charged to $2 \cdot U_{in}$ at $T/2$ and therefore contains four times more energy. Figure 2.16(b) shows the plotted energies of a periodically damped series RLC-circuit, where it is assumed that R complies to (2.42) and both $I_L(0)$ and $U_C(0)$ are zero. Similar to the current and voltages of this system, which are shown in Fig. 2.15(b), the oscillation of these energies is damped with the time-constant τ_{LC}. To gain insight into this circuit the steady-state region, where $t \gg \tau_{LC}$, is examined first. It can be intuitively seen that in steady-state $i(t)$ is zero, thus $E_L(t)$ will also be zero. The steady-state values of $E_{U_{in} \rightarrow RLC}(t)$, $E_C(t)$ and $E_R(t)$ are positive finite values given by (2.51), (2.52) and (2.53), respectively. For the specific cases where $I_L(0)$ and $U_C(0)$ are zero these energies are equal to those of a capacitor that is charged directly by means of a voltage source.

2.3 Inductive Type DC-DC Converters

Fig. 2.16 (a) The energy $E_{U_{in} \to RLC}(t)$ delivered by U_{in} the energy $E_L(t)$ stored in L, the energy $E_R(t)$ dissipated in R and the energy $E_C(t)$ stored in C as a function of time, for an ideal ($R=0$) and (b) a non-ideal ($R \neq 0$) series RLC-circuit

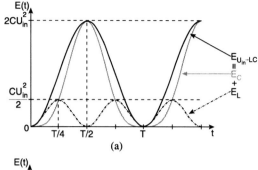

$$\lim_{\frac{t}{\tau_{LC}} \to \infty} \left(E_{U_{in} \to RLC}(t) \right) = CU_{in}(I_L(0)R(1-L) - U_C(0) + U_{in})|_{I_C(0)=U_C(0)=0}$$

$$= CU_{in}^2 \qquad (2.51)$$

$$\lim_{\frac{t}{\tau_{LC}} \to \infty} \left(E_C(t) \right) = \frac{C}{2}((I_L(0)R(1-L) - U_C(0))^2 + U_{in}^2)|_{I_C(0)=U_C(0)=0}$$

$$= \frac{CU_{in}^2}{2} \qquad (2.52)$$

$$\lim_{\frac{t}{\tau_{LC}} \to \infty} \left(E_R(t) \right) = \frac{1}{2}\big(2RCI_L(0)(L-1)(U_C(0) - U_{in}) + C(U_C(0) - U_{in})^2$$

$$+ I_L(0)^2(L + R^2C(L-1)^2)\big)\big|_{I_C(0)=U_C(0)=0}$$

$$= \frac{CU_{in}^2}{2} \qquad (2.53)$$

The periodically-damped series RLC-circuit shows no advantage for charging a capacitor, in terms of $\eta_{RLC\_charge}(t)$, in steady-state. Nevertheless, this advantage exits during specific moments of the transient region. When considering Fig. 2.16(b), it shows that the maximal $\eta_{RLC\_charge}(t)$ will occur after the first half period $T/2$ has elapsed. This $\eta_{RLC\_charge}(t)$ is calculated through (2.54), where the initial energy $E_L(0)$ stored in L is added to $E_{U_{in} \to RLC}(t)$ because $E_L(0)$ is added in a previous step that is not considered here. Only the solution for the special case where $I_L(0)$ and $U_C(0)$ are zero is shown, as the general solution of this calcula-

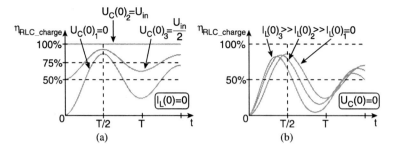

Fig. 2.17 (a) The energy charging efficiency $\eta_{RLC\_charge}(t)$ of a capacitor in a periodically-damped series *RLC*-circuit as a function of time, for different values of the initial voltages $U_C(0)$ over C and (b) for different initial currents $I_L(0)$ through L

tion is too complex. When the resistance of R is infinitesimal, $\eta_{RLC\_charge}(t)$ will be 100%.

$$\eta_{RLC\_charge}\left(\frac{T}{2}\right) = \eta_{RLC\_charge}\left(\frac{2\pi L}{Q}\right) = \frac{E_C\left(\frac{2\pi L}{Q}\right)}{E_{U_{in} \to RLC}\left(\frac{2\pi L}{Q}\right) - E_L(0)}$$

$$= \frac{E_C\left(\frac{2\pi L}{Q}\right)}{E_{U_{in} \to RLC}\left(\frac{2\pi L}{Q}\right) + \frac{LI_L(0)^2}{2}}\bigg|_{I_C(0)=U_C(0)=0}$$

$$= \frac{1 + e^{-\frac{\pi R}{Q}}}{2} \blacksquare \tag{2.54}$$

The influence of the values of $U_C(0)$ and $I_L(0)$ on $\eta_{RLC\_charge}(t)$ is graphically represented in Fig. 2.17(a) and (b). First, Fig. 2.17(a) is considered, where $\eta_{RLC\_charge}(t)$ is plotted as a function of time. Also, the value of $I_C(0)$ is assumed to be zero and three different values of $U_C(0)$ are considered: $U_C(0)_1 = 0$, $U_C(0)_2 = U_{in}/2$ and $U_C(0)_3 = U_{in}$. It is observed that at time $T/2$ the maximum values of $\eta_{RLC\_charge}(t)$ are reached and that these values increase upon increasing values of $U_C(0)$. For the case where $U_C(0) = U_{in}$, $\eta_{RLC\_charge}(t)$ reaches the value of 100%. Obviously, no energy is added to C in this case. Please note that the maximum value of $\eta_{RLC\_charge}(t)$ is also dependent on the resistance of R. This dependency is not shown for sake of simplicity. However, it can be proven that the maximum value of $\eta_{RLC\_charge}(t)$ is inversely proportional to R. In order to gain more insight into the dependency of $\eta_{RLC\_charge}(t)$ on $I_L(0)$, Fig. 2.17(b) is explained. In this figure $\eta_{RLC\_charge}(t)$ is plotted as a function of time. The value of $U_C(0)$ is assumed to be zero and three different cases for $I_L(0)$ are considered: $I_L(0)_1 = 0 \gg I_L(0)_2 \gg I_L(0)_3$. In this figure is observed that the maximum value of $\eta_{RLC\_charge}(t)$ is only reached at $T/2$ for the case where $I_L(0) = 0$. When $I_L(0) > 0$, the maximum value for $\eta_{RLC\_charge}(t)$ is reached faster than $T/2$. It can also be seen that this maximum value tends to decrease upon increasing values of $I_L(0)$, which is associated with increasing Joule-losses in R. Finally, it is noted that, similar to Fig. 2.17(a), the maximum value of $\eta_{RLC\_charge}(t)$ is inversely proportional to R.

2.3 Inductive Type DC-DC Converters

To conclude this dissertation on the charging of capacitors in a periodically-damped *RLC*-circuit by means of a voltage source, the most important observations and their consequences are listed:

- The maximum value of $\eta_{RLC\_charge}(t)$ is reached at a charge time t exactly equal to $T/2$, assuming that $I_L(0)$ is zero.
- The total parasitic series resistance R is inversely proportional to $\eta_{RLC\_charge}(t)$ and $\eta_{RLC\_charge}(t)$ reaches the value of 100% for $R = 0$, independent of the values of $U_C(0)$ and $I_L(0)$. This implies that the charging of a capacitor in an ideal *RLC*-circuit ($R = 0$) is lossless, as opposed to the charging of a capacitor by means of a voltage source.
- The maximal $E_C(t)$ is ideally four times larger than for a capacitor charged by means of a voltage source, implying that $u_C(t)$ can reach the maximum value of $2 \cdot U_{in}$.
- The smaller the difference between the values of U_{in} and $U_C(0)$, the larger the maximum value of $\eta_{RLC\_charge}(t)$, reaching 100% for the case where $U_C(0) = U_{in}$. Obviously for the latter case $E_C(t)$ is zero.
- A smaller value of $I_L(0)$ yields a larger maximum value for $\eta_{RLC\_charge}(t)$.

As both the charging of inductors by means of a voltage source and the charging of capacitors in periodically-damped series *RLC*-circuits by means of a voltage source can ideally be lossless, their combination is promising for the realization of inductive DC-DC voltage converters. The result of this is that ideal inductive DC-DC converters are able to achieve power conversion efficiencies η_{SW} of 100% for their entire voltage conversion ratio k_{SW} range, which is in contrast to ideal charge-pump DC-DC converters, which is explained in Sect. 2.2.

2.3.3 Reflections on Steady-State Calculation Methods

The behavior of ideal inductive DC-DC converters in terms of the voltage conversion ratio $k(\delta)$ as a function of the duty-cycle δ and the conduction mode (CM) is discussed in this section. For this purpose the general small-ripple approximation method is provided here [Eri04]. The method is explained by means of an ideal boost DC-DC converter, which is shown in Fig. 2.18(a), but can be used for any given inductive DC-DC converter topology. As this method is only useful for calculations of ideal converters, it will prove to be of limited value in the design of monolithic inductive DC-DC converters. An accurate model for this purpose is provided in Sect. 4.

Continuous Conduction Mode

First, the basic steady-state operation of an ideal boost converter, shown in Fig. 2.18, is explained. In continuous conduction mode (CCM) the current through the inductor $i_L(t)$ has a finite, positive value which is not zero and the operation mode consists of two phases:

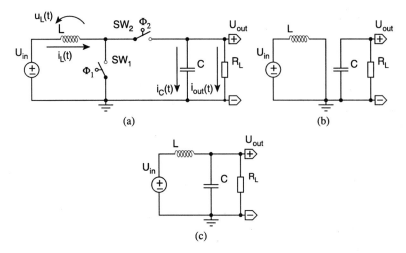

Fig. 2.18 (a) The circuit of an ideal boost DC-DC converter. (b) The equivalent circuit of the inductor charge phase and (c) the inductor discharge phase

1. *The inductor charge phase* Φ_1: The equivalent circuit for Φ_1 is shown in Fig. 2.18(b), which is achieved by closing SW_1 and opening SW_2 for a certain on-time t_{on}. During Φ_1 the inductor L is charged by the voltage source U_{in}, causing the inductor current $i_L(t)$ to increase from its minimal value I_{L_min} to its maximal value I_{L_max}, as illustrated in Fig. 2.20. Simultaneously the output capacitor C is discharged through the load R_L.
2. *The inductor discharge phase* Φ_2: The equivalent circuit for Φ_2 is shown in Fig. 2.18(c), which is achieved by opening SW_1 and closing SW_2 for a certain off-time t_{off}. During Φ_2 L is discharged into C and R_L, causing $i_L(t)$ to decrease from I_{L_max} to I_{L_min}, as can be seen in Fig. 2.20. As a result $i_L(t)$ is divided over C and R_L, thereby charging C and providing power to R_L.

Because L is discharged in series with U_{in} it can be intuitively seen that the output voltage U_{out} will always be higher than U_{in}. This will be formally proven by means of the small-ripple approximation method. For this method it is assumed that:

- All the converter components are ideal and lossless.
- The output voltage ripple $U_{out\_max} - U_{out\_min} = \Delta U_{out}$ is infinitesimal, which implies that U_{out} is equal to its mean value $\overline{U}_{out}$.
- The inductor current ripple $\Delta I_L = I_{L\_max} - I_{L\_min}$ is infinitesimal.
- All the currents and voltages are linearized, according to Fig. 2.20.
- The converter operates in steady-state.

First, the voltage $u_L(t)$ over L is considered. In Fig. 2.20 it can be seen that, for Φ_1 and Φ_2, $u_L(t)$ is given by (2.61).

$$\begin{cases} \Phi_1 : 0 \to t_{on} & \Longrightarrow \quad u_L(t) = U_{in} \\ \Phi_2 : t_{on} \to t_{off} & \Longrightarrow \quad u_L(t) = U_{in} - U_{out} \end{cases} \quad (2.61)$$

2.3 Inductive Type DC-DC Converters

⊛ CAPACITOR CHARGE BALANCE & INDUCTOR VOLT-SECOND BALANCE

For a capacitor C in a DC-DC converter, operating in steady-state, the energy E_{C_in} stored in it is equal to the energy E_{C_out} delivered by it, as stated by (2.55). Because the capacitance of C is constant, the net voltage change, during one period T, over C is zero. This is translated into (2.56), where $U_C(0)$ and $U_C(T)$ are the respective voltages over C at the beginning and end of T. For the example of Fig. 2.19(b), where the current $i_C(t)$ through C as a function of time is shown, this yields (2.56). This implies that the stored Q_{in} and the delivered charge Q_{out} in C are equal, hence the capacitor charge is in balance. Graphically, the positive area A_C^+ of Fig. 2.19(b) is equal to the negative area A_C^-.

$$E_{C\_in} - E_{C\_out} = \frac{CU_C(T)^2}{2} - \frac{CU_C(0)^2}{2} = 0 \tag{2.55}$$

$$\implies U_C(T) - U_C(0) = 0 = \frac{1}{C}\int_0^T i_C(t)\,dt \tag{2.56}$$

$$\implies \int_0^T i_C(t)\,dt = I_{C1}t_1 + I_{C2}t_2 = Q_{in} - Q_{out} = A_C^+ - A_C^- \ \blacksquare \tag{2.57}$$

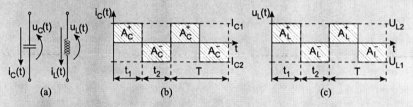

Fig. 2.19 (a) The convention of the voltage over and the current through a capacitor C and an inductor L. (b) The current $i_C(t)$ through C and (c) the voltage $u_L(t)$ over L, both in energetic equilibrium

For an inductor L in a DC-DC converter, operating in steady-state, the energy E_{L_in} stored in it is equal to the energy E_{L_out} delivered by it, as stated by (2.58). Because the inductance of L is constant, the net current change, during one period T, through L is zero. This is translated into (2.59), where $I_L(0)$ and $I_L(T)$ are the respective currents through L at the beginning and end of T. For the example of Fig. 2.19(c), where the voltage $u_L(t)$ over L as a function of time is shown, this yields (2.60). This implies that the inductor's volt-second product is in balance. Graphically, the positive area A_L^+ of Fig. 2.19(c) is equal to the negative area A_L^-.

$$E_{L\_in} - E_{L\_out} = \frac{LI_L(T)^2}{2} - \frac{LI_L(0)^2}{2} = 0 \tag{2.58}$$

$$\implies I_L(T) - I_L(0) = 0 = \frac{1}{L}\int_0^T u_L(t)\,dt \tag{2.59}$$

$$\implies \int_0^T u_L(t)\,dt = U_{L1}t_1 + U_{L2}t_2 = A_L^+ - A_L^- \ \blacksquare \tag{2.60}$$

Fig. 2.20 The linearized current $i_L(t)$ through L, the linearized voltage $u_L(t)$ over L, the linearized current $i_C(t)$ through C and the linearized output voltage $u_{out}(t)$ as a function of time, for an ideal boost converter in CCM

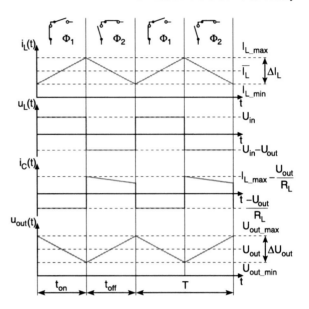

In steady-state operation the net energy change in the inductor over one switching period T is zero, thus the volt-second balance of the inductor is also zero. This yields (2.62), giving the voltage conversion ratio $k(\delta)$ as a function of the duty cycle δ. It is observed that $k(\delta)$ only depends on δ and not on the load R_L or the switching frequency f_{SW}. This follows from the fact that the converter components are assumed to be ideal. Another implication of this assumed ideality is that the power conversion efficiency η_{SW} will always be 100%, independent of the converter's operation parameters. Therefore, (2.62) will prove little value in the design process, it can however be used to understand the operation principle. A more accurate design model for a boost converter will be deduced in Chap. 4.

$$\int_0^T u_L(t)\,dt = U_{in}t_{on} + (U_{in} - U_{out})t_{off} = 0$$

$$\implies k(\delta) = \frac{U_{out}}{U_{in}} = \frac{t_{on} + t_{off}}{t_{off}} = \frac{1}{1-\delta} \blacksquare \quad (2.62)$$

The duty-cycle δ is defined by (2.63).

$$\delta = \frac{t_{on}}{t_{on}+t_{off}} = \frac{t_{on}}{T} \quad \text{with} \quad 0 \leqslant \delta \leqslant 1 \quad (2.63)$$

The relation between δ and $k(\delta)$, given by (2.62), is graphically represented in Fig. 2.21. It can be seen that U_{out} is always higher than U_{in}, as explained earlier. Also, U_{out} can be infinitely high, as δ approaches to 1. In practical implementations the maximum value of U_{out} will naturally be limited by losses.

The small-ripple approximation, used to calculate $k(\delta)$, assumes that the output ripple voltage ΔU_{out} is infinitesimal. This is obviously not the case when the values of L, C and f_{SW} are not infinitely large, even when the converter components are

Fig. 2.21 The voltage conversion ratio $k(\delta)$ as a function of the duty-cycle δ, for an ideal boost converter in CCM

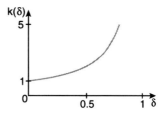

assumed ideal. Therefore, the calculation of ΔU_{out} is also performed. In Fig. 2.20 it can be seen that during Φ_1 C is discharged through R_L. If $t_{on} \ll \tau_C = CR_L$ the discharge current $i_C(t)$ of C can be approximated to have a constant value, which is formally described by (2.64).

$$\Phi_1: 0 \to t_{on} \implies i_C(t) = C\frac{du_{out}(t)}{dt} = -\frac{\overline{U_{out}}}{R_L} \qquad (2.64)$$

Thus, the net change of $u_{out}(t)$ follows from the integration of $i_C(t)$. The result of this calculation yields ΔU_{out}, which is given by (2.65).

$$\Delta U_{out} = -\frac{1}{C}\int_0^{t_{on}} i_C(t)\,dt = \frac{\overline{U_{out}}}{CR_L}t_{on} = \frac{\overline{I_{out}}}{C}t_{on} \quad \blacksquare \qquad (2.65)$$

From (2.65) it follows that ΔU_{out} will increase upon decreasing R_L, which is equal to an increasing mean output current $\overline{I_{out}}$, because C is discharged faster during Φ_1. ΔU_{out} will also increase upon increasing values of t_{on}, as this is equal to the discharge time of C. Furthermore, it is observed that ΔU_{out} is inversely proportional to the capacitance of C and independent of the inductance of L. The fact that there is no dependency on L can be intuitively understood because it is not a part of the output filter. For DC-DC converters where both L and C are a part of the output filter, this method for calculating ΔU_{out} will not be applicable. The alternative method, which is based on the charge balance of C, is explained for the DC-DC buck converter in Sect. 3.1.

Discontinuous Conduction Mode

In discontinuous conduction mode (DCM) $i_L(t)$ varies between zero and a finite positive value. In other words, $i_L(t)$ does not flow continuously. The small-ripple approximation method, for analyzing the voltage conversion ratio k_{SW} of a DC/DC converter, is somewhat more complex for DCM. Before elaborating upon this method for the ideal boost DC-DC converter example, of which the circuit is shown in Fig. 2.18(a), the three phases of this converter in DCM are explained:

1. *The inductor charge phase* Φ_1: The equivalent circuit for Φ_1 is shown in Fig. 2.18(b), which is achieved by closing SW_1 and opening SW_2 for a certain on-time t_{on}. During Φ_1 L is charged by U_{in}, causing $i_L(t)$ to increase from zero to I_{L_max}, as illustrated in Fig. 2.22. Simultaneously C is discharged through R_L.

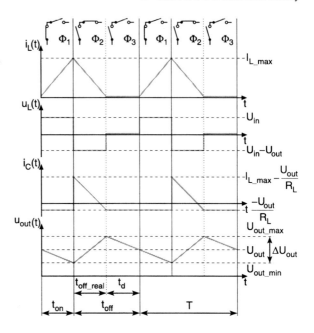

Fig. 2.22 The linearized current $i_L(t)$ through L, the linearized voltage $u_L(t)$ over L, the linearized current $i_C(t)$ through C and the linearized output voltage $u_{out}(t)$ as a function of time, for an ideal boost converter in DCM

2. *The inductor discharge phase* Φ_2: The equivalent circuit for Φ_2 is shown in Fig. 2.18(c), which is achieved by opening SW_1 and closing SW_2 for a certain real off-time t_{off_real}. This t_{off_real} is the exact duration for $i_L(t)$ to reach the value zero, which is shown in Fig. 2.22. Thus, during t_{off_real} L is discharged into C and R_L, thereby charging C and providing power to R_L.

3. *The dead-time phase* Φ_3: The equivalent circuit for Φ_3 is illustrated by the right-hand part of Fig. 2.18(b), which is achieved by both opening SW_1 and SW_2 for a certain dead-time t_d. This prevents $i_L(t)$ to become negative, discharging C into U_{in}. Thus, during Φ_3 C is discharged through R_L. Finally, it is noted that the sum of t_{off_real} and t_d equals t_{off}.

Due to the fact that, in DCM, $i_L(t)$ is zero during t_d, it cannot be approximated as an infinitesimal small-ripple. For this reason the calculation method for $k(\delta)$ in CCM, which was described earlier, will not suffice for DCM. Therefore, the calculation method for the latter mode, by means of the small-ripple approximation, method is explained. The same assumptions as for CCM are made for this calculation, except for the inductor current ripple $\Delta I_L = I_{L\_max} - I_{L\_min} = I_{L\_max}$, which cannot be considered infinitesimal.

First, the voltage $u_L(t)$, which is plotted in Fig. 2.22, is considered. For Φ_1, Φ_2 and Φ_3 this yields (2.66).

$$\begin{cases} \Phi_1 : 0 \to t_{on} & \implies u_L(t) = U_{in} \\ \Phi_2 : t_{on} \to t_{off\_real} & \implies u_L(t) = U_{in} - U_{out} \\ \Phi_3 : t_{off\_real} \to t_{off} & \implies u_L(t) = 0 \end{cases} \quad (2.66)$$

2.3 Inductive Type DC-DC Converters

In order to determine the current $i_C(t)$, an expression for $i_L(t)$ is needed. Both these currents are plotted for DCM in Fig. 2.22. By means of Lenz's law (2.44) the tangents of $i_L(t)$ during Φ_1 and Φ_2 are calculated, as stated in (2.67).

$$\begin{cases} \Phi_1 : 0 \to t_{on} & \implies \dfrac{di_L(t)}{dt} = \dfrac{U_{in}}{L} \\ \Phi_2 : t_{off\_real} \to t_{off} & \implies \dfrac{di_L(t)}{dt} = \dfrac{U_{in} - U_{out}}{L} \end{cases} \quad (2.67)$$

The linear approximation of $i_L(t)$ during Φ_2 is calculated by (2.67), yielding (2.68).

$$\begin{aligned} \Phi_2 : t_{off\_real} \to t_{off} \implies i_L(t) &= I_{L\_max} + \frac{U_{in} - U_{out}}{L} t \\ &= \frac{U_{in}}{L} t_{on} + \frac{U_{in} - U_{out}}{L} t \end{aligned} \quad (2.68)$$

By means of Fig. 2.22 and (2.68), $i_C(t)$ is calculated by (2.69), for Φ_1, Φ_2 and Φ_3.

$$\begin{cases} \Phi_1 : 0 \to t_{on} & \implies i_C(t) = -\dfrac{U_{out}}{R_L} \\ \Phi_2 : t_{on} \to t_{off\_real} & \implies i_C(t) = i_L(t) - \dfrac{U_{out}}{R_L} \\ & \qquad\qquad = \dfrac{U_{in}}{L} t_{on} + \dfrac{U_{in} - U_{out}}{L} t - \dfrac{U_{out}}{R_L} \\ \Phi_3 : t_{off\_real} \to t_{off} & \implies i_C(t) = -\dfrac{U_{out}}{R_L} \end{cases} \quad (2.69)$$

Analogue to CCM, the volt-second balance of L in DCM is zero when the converter operates in steady-state. This volt-second balance, in combination with (2.66) for $i_L(t)$, is used to calculate $k(\delta)$, yielding (2.70).

$$\int_0^T u_L(t)\,dt = U_{in}t_{on} + (U_{in} - U_{out})t_{off\_real} = 0$$
$$\implies k(\delta) = \frac{U_{out}}{U_{in}} = \frac{t_{on} + t_{off} - t_d}{t_{off} - t_d} \quad (2.70)$$

The dead-time t_d in (2.70) is an unknown output parameter and therefore an additional equation is needed to substitute it with known input parameters. For this purpose the charge balance of C is calculated, which is zero when the converter operates in its steady-state region. This is achieved by using (2.69) for $i_C(t)$, yielding (2.71) which is a quadratic equation of t_d. Solving t_d out of this quadratic equation results in two solutions, of which only the positive solution has a physical meaning.

$$\begin{aligned} \int_0^T i_C(t)\,dt &= 0 \\ &= -\frac{U_{out}}{R_L} t_{on} + \left(\left(\frac{U_{in}}{L} t_{on} - \frac{U_{out}}{R_L}\right)(t_{off} - t_d) \right. \\ &\quad \left. + \frac{U_{in} - U_{out}}{2L}(t_{off} - t_d)^2\right) - \frac{U_{out}}{R_L} t_d \end{aligned} \quad (2.71)$$

In the last step $k(\delta)$ is calculated by substituting the positive solution for t_d of (2.71) into (2.70). This results in a quadratic equation of U_{out}. The positive, physical relevant, solution of this equation yields $k(\delta)$, given by (2.72).

$$k(\delta) = \frac{U_{out}}{U_{in}} = \frac{1 + \sqrt{1 + \frac{2R_L}{L(t_{on}+t_{off})}t_{on}^2}}{2} = \frac{1 + \sqrt{1 + \frac{4\delta^2}{\frac{2L}{R_L T}}}}{2} \blacksquare \quad (2.72)$$

It is observed that $k(\delta)$ is not independent of the value of R_L, L and T in DCM. This stands in contrast to CCM where $k(\delta)$, given by (2.62), is only dependent on δ. However, both equations do not provide any information on whether the converter operates continuous or DCM and therefore it is not yet clear when to use which equation. The formal method to determine the CM for an ideal converter and a graphic representation of (2.72) is provided in the following section.

The output voltage ripple ΔU_{out} for DCM is calculated analogue to CCM. This ripple is assumed to be infinitesimal for the small-ripple approximation method, which is not the case when the values of L, C and f have a finite value. For this purpose $i_C(t)$ is considered during Φ_3 and Φ_1, as illustrated in Fig. 2.22. During Φ_3 and Φ_1 C is discharged through R_L. Thus, for the case where $t_{on} + t_d \ll \tau_C = CR_L$ the discharge current $i_C(t)$ of C can be approximated to be constant, as described by (2.73).

$$\Phi_3 \: \& \: \Phi_1 : 0 \to t_{on} + t_d \implies i_C(t) = C\frac{du_{out}(t)}{dt} = -\frac{\overline{U_{out}}}{R_L} \quad (2.73)$$

Analogue to (2.65), for CCM, ΔU_{out} is calculated by integrating $i_C(t)$ over $t_{on} + t_d$, yielding (2.73). The dead-time t_d is calculated by solving the quadratic equation (2.71).

$$\begin{aligned}
\Delta U_{out} &= -\frac{1}{C}\int_0^{t_{on}+t_d} i_C(t)\,dt \\
&= \frac{\overline{U_{out}}}{CR_L}(t_{on}+t_d) = \frac{\overline{I_{out}}}{C}(t_{on}+t_d) \\
&= \left(t_{on} + \frac{R_L T U_{in} - R_L t_{off}\overline{U_{out}} + \sqrt{R_L(R_L t_{on}^2 U_{in}^2 + 2LT(U_{in}-\overline{U_{out}})\overline{U_{out}})}}{R_L(U_{in}-\overline{U_{out}})}\right) \\
&\quad \cdot \frac{\overline{I_{out}}}{C} \blacksquare
\end{aligned} \quad (2.74)$$

As can be seen, in DCM ΔU_{out} is dependent on the same parameters as for CCM, which is described by (2.65), in addition of t_d. As a consequence ΔU_{out} is also dependent on the inductance of L, which is not the case for CCM. This dependency of L is not expected because L is not a part of the output filter, however the dependency follows from the fact that t_d is dependent on L. This dependency is quite complex and it is more interesting to understand that the inductance of L influences the duration of t_d. An increasing value of t_d, meaning that the converter is driven

2.3 Inductive Type DC-DC Converters

deeper into DCM, will increase the discharge time of C and as a result the value of ΔU_{out} will increase. An insight into the boundary between the two conduction modes is required to clarify this, which is provided in the next section. Finally, it noticed that, similar to CCM, this calculation method will not be applicable for DC-DC converters where L is also a part of the output filter. For this type of converters the charge balance method, explained in Sect. 3.1 for a DC-DC buck converter, will be required.

Continuous-Discontinuous Boundary

The calculation of $k(\delta)$ and ΔU_{out} for both CMs of an ideal boost converter, shown in Fig. 2.18(a), is already performed in the previous sections, by using the small-ripple approximation method. This same method is used in this section to determine the conduction boundary (CB) between CCM and DCM, for the ideal boost converter example. The conditions for the converter to work in either of the two modes are found by means of Fig. 2.20 and are given by (2.75), where $\overline{I_L}$ is the mean value of $i_L(t)$.

$$\begin{cases} \overline{I_L} > \dfrac{I_{L\_max} - I_{L\_min}}{2} = \dfrac{\Delta I_L}{2} \implies \text{CCM} \\ \overline{I_L} < \dfrac{I_{L\_max} - I_{L\_min}}{2} = \dfrac{\Delta I_L}{2} \implies \text{DCM} \end{cases} \quad (2.75)$$

In order to find an expression for $\overline{I_L}$, the charge balance for C in CCM is to be solved. For this purpose, $i_C(t)$ for Φ_1 and Φ_2 is determined by means of the graph is Fig. 2.20, yielding (2.76).

$$\begin{cases} \Phi_1 : 0 \to t_{on} \implies i_C(t) = -\dfrac{U_{out}}{R_L} \\ \Phi_2 : t_{on} \to t_{off} \implies i_C(t) = i_L(t) - \dfrac{U_{out}}{R_L} \end{cases} \quad (2.76)$$

When assuming that ΔI_L is infinitesimal, which is the case for the small-ripple approximation, it can be seen that $i_L(t)$ equals $\overline{I_L}$. Thus, solving the charge balance of C and substituting $\overline{I_L}$, yields (2.77).

$$\int_0^T i_C(t)\,dt = -\dfrac{U_{out}}{R_L} t_{on} + \left(\overline{I_L} - \dfrac{U_{out}}{R_L}\right) t_{off} = 0$$
$$\implies \overline{I_L} = \dfrac{U_{out}}{R_L} \cdot \dfrac{t_{on} + t_{off}}{t_{off}} = \dfrac{U_{out}}{R_L(1-\delta)} \quad (2.77)$$

Substituting the unknown output parameter U_{out} out of (2.62) and replacing it in (2.77) results in (2.78), which is only dependent on known input parameters.

$$\overline{I_L} = \dfrac{U_{in}}{R_L(1-\delta)^2} \quad (2.78)$$

Fig. 2.23 The *upper graph* shows the voltage conversion ratio $k(\delta)$ as a function of the duty-cycle δ, where the *black curve* is valid for CCM and the *gray curves* for DCM. In the *lower graph* the *black curve* shows the boundary between the two CMs and the *gray curves* illustrate three numerical examples. These graphs are valid for an ideal DC-DC boost converter

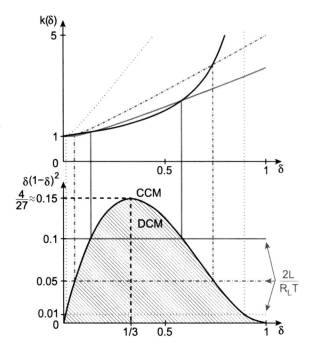

The inductor current-ripple ΔI_L is equal to the current increase of $i_L(t)$ during Φ_1. By using (2.67), this results in (2.79).

$$\Phi_1 : 0 \to t_{on} \implies \frac{di_L(t)}{dt} = \frac{U_{in}}{L} \implies \Delta I_L = \frac{U_{in}}{L} t_{on} = \frac{U_{in}}{L} T\delta \quad (2.79)$$

Finally, the condition for CCM of (2.75) can be reformulated into (2.80), by means of (2.78) and (2.79).

$$\overline{I_L} > \frac{\Delta I_L}{2} \iff \frac{U_{in}}{R_L(1-\delta)^2} > \frac{U_{in}}{2L} T\delta$$

$$\iff \frac{2L}{R_L T} > \delta(1-\delta)^2 \quad \blacksquare \quad (2.80)$$

It can be seen that the CM of an ideal DC-DC boost converter depends on the duty-cycle δ, the switching period T, the inductance of L and the load R_L. The boost converter will tend to operate in DCM for low inductances L, high values of the load resistance R_L (low loads) or longer switching periods T (lower frequencies f). Furthermore, the CM does not depend on the capacitance of C.

To get more insight into this matter, the CB is plotted in Fig. 2.23. The upper graph shows $k(\delta)$ as a function of δ, where the black curve is valid for CCM (2.62) and the gray curves for DCM (2.72). The curves for discontinuous mode are plotted for three different values of $(2L)/(R_L T)$: 0.01, 0.05 and 0.1. The CM valid for the upper graph is determined by the lower graph. In this graph the CB (2.80) is plotted by the black curve. The gray horizontal lines are the values for $(2L)/(R_L T)$, used in the upper graph. For the values of δ, at a certain value of $(2L)/(R_L T)$, which are underneath the black CB curve, the converter will operate in DCM and vice versa. From these graphs it can also be concluded that the converter will tend to operate towards CCM for values of δ approaching zero or one. Finally, it is noted that in DCM $k(\delta)$ is quasi linear dependent on δ. This is due to the fact that the $i_L(t)$ will rise in a linear fashion during Φ_1, according to (2.67), and therefore the amount of energy $E_L(t)$ stored in L will increase quadratically with δ, as determined by (1.6). The amount of energy delivered to R_L also increases quadratically with U_{out} and because this energy is equal to $E_L(t)$, U_{out} increases linear with δ.

The mathematical expressions obtained for the characterization of an ideal DC-DC boost converter will be validated with SPICE simulations in Sect. 3.2.1. A more sophisticated and accurate model, that takes all the significant losses of a monolithic DC-DC boost converter into account, is derived in Chap. 4.

2.4 INTERMEZZO: The Efficiency Enhancement Factor

Virtually any given energy converter has the purpose of performing its task with minimal energy losses.[4] For DC-DC converters this is translated into maximizing the overall power conversion efficiency η, as stated by (1.3). The advantages of attaining the highest possible η are obvious:

- Less energy consumption leads to longer battery autonomies and lower energy costs, hence greener applications.
- Less heat dissipation leads to smaller required contact surfaces and/or heatsinks, in turn resulting into smaller and more cost efficient applications.
- Less heat dissipations leads to reduced conduction losses in metals and (MOS) transistors, due to Positive Temperature Coefficient (PTC) resistive behavior.

This section describes the formal method for comparing switched DC-DC step-down converters with each other, with respect to their power conversion efficiency η_{SW}. In order to achieve this, a new figure of merit is introduced [Wen08a]: The Efficiency Enhancement Factor (*EEF*). The method takes the voltage conversion ratio k_{SW} into account, which is a crucial parameter for the comparison.

The basic idea behind the *EEF* is explained in Sect. 2.4.1 and the possible interpretations and variations of this figure of merit are described in Sect. 2.4.2.

2.4.1 The Concept

Consider two switched-mode DC-DC step-down converters DC-DC$_1$ and DC-DC$_2$, having both the same P_{out}, a different k_{SW} and η_{SW}. The value of these parameters

---

[4]By referring to energy losses the energy converter is implicitly regarded as an open system, where the energy losses are in the form of waste/dissipated heat (Joule-losses).

Table 2.1 A comparison to clarify the concept of the *EEF* figure of merit for step-down DC-DC converters. Each comparison is made between a linear series voltage converter and a switched DC-DC step-down voltage converter, having the same voltage conversion ratio $k_{lin} = k_{SW}$

DC-DC$_1$		DC-DC$_2$	
$P_{out} = 1$ W	$k_{lin} = k_{SW} = 0.8$	$P_{out} = 1$ W	$k_{lin} = k_{SW} = 0.5$
$\eta_{lin} = 80\%$	$\eta_{SW} = 85\%$	$\eta_{lin} = 50\%$	$\eta_{SW} = 55\%$
$\Longrightarrow \Delta\eta = \eta_{SW} - \eta_{lin} = 5\%$		$\Longrightarrow \Delta\eta = \eta_{SW} - \eta_{lin} = 5\%$	
$P_{in\_lin} = 1.25$ W	$P_{in\_SW} = 1.18$ W	$P_{in\_lin} = 2$ W	$P_{in\_SW} = 1.82$ W
$\Longrightarrow \Delta P_{in} = P_{in\_lin} - P_{in\_SW} = 0.07$ W		$\Longrightarrow \Delta P_{in} = P_{in\_lin} - P_{in\_SW} = 0.18$ W	
$EEF = \frac{\Delta P_{in}}{P_{in\_lin}}\big|_{k_{lin}=k_{SW}} = 5.6\%$		$EEF = \frac{\Delta P_{in}}{P_{in\_lin}}\big|_{k_{lin}=k_{SW}} = 9\%$	

is listed in Table 2.1. At first glance DC-DC$_1$ is superior to DC-DC$_2$, in terms of η_{SW}.

In order to examine this more closely, both converters are compared to their ideal series linear voltage converter equivalents, having the same voltage conversion ratios $k_{lin} = k_{SW}$ and P_{out}. The power conversion efficiency η_{lin} of these ideal linear series voltage converters is equal to k_{lin} and independent of P_{out}, which is proven by (2.2). For both examples this yields the same difference $\Delta\eta$ of power conversion efficiencies. From this point of view it is already clear that DC-DC$_1$ is not better than DC-DC$_2$, with regard to η_{SW}.

The main question still to be answered is how the improvement of the switched-mode DC-DC converters, compared to their linear series voltage converter equivalents, can be measured. In essence, it all comes down to P_{in}, which is larger than P_{out} due to the losses in a real converter. Thus, the input power P_{in_SW} of DC-DC$_1$ and DC-DC$_2$ is to be compared to the input power P_{in_lin} of their respective linear series converter equivalents, which is done by means of their difference ΔP_{in}. When comparing ΔP_{in} for both examples, it can be seen that this difference is smaller for DC-DC$_1$ than for DC-DC$_2$. This indicates that the net gain over a linear series voltage converter, in terms of P_{in}, has the highest value for DC-DC$_2$, despite its lower overall η_{SW}.

It is clear that the comparison of ΔP_{in} can only be made for equal P_{out}, which is the case in this example. In order to be able to perform this comparison independently of P_{out}, ΔP_{in} is normalized over P_{in_lin}, yielding the *EEF* for switched-mode DC-DC step-down converters. This figure of merit allows for comparing different DC-DC step-down converters in terms of their η_{SW}, independent of P_{out} and k_{SW}. A more convenient form of the *EEF*, which is equivalent to the one of Table 2.1, is given by (2.81).

$$EEF = \frac{\Delta P_{in}}{P_{in\_lin}}\bigg|_{k_{lin}=k_{SW}} = 1 - \frac{P_{in\_lin} - P_{in\_SW}}{P_{in\_lin}}\bigg|_{k_{lin}=k_{SW}} = 1 - \frac{\frac{P_{out\_SW}}{\eta_{SW}}}{\frac{P_{out\_lin}}{\eta_{lin}}}\bigg|_{k_{lin}=k_{SW}}$$

$$\Longrightarrow EEF = 1 - \frac{\eta_{lin}}{\eta_{SW}}\bigg|_{k_{lin}=k_{SW}} \blacksquare \qquad (2.81)$$

Fig. 2.24 The *upper graph* shows the power conversion efficiencies η_{SW} and η_{lin} of a switched-mode DC-DC converter and a linear series voltage converter having the same voltage conversion ratio $k_{lin} = k_{SW}$, as a function of the output power P_{out}. The *lower graph* shows the corresponding *EEF* and $\overline{EEF}$, as a function of P_{out}

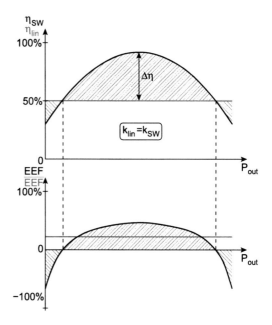

In the last step of for obtaining (2.81) it is assumed that $P_{out\_lin} = P_{out\_SW}$. The premise for this assumption is the fact that η_{lin} is not dependent on P_{out_lin}.

2.4.2 Interpretations

From the definition of the *EEF*, given by (2.81), it follows that the *EEF* will have a positive value for $\eta_{SW} > \eta_{lin}$ (wanted) and vice versa (unwanted). This is graphically illustrated in Fig. 2.24. The upper graph shows η_{SW} and η_{lin} for both a switched and a linear series DC-DC converter, for $k_{lin} = k_{SW}$, as a function of P_{out}. The lower graph shows both the *EEF* and its mean value $\overline{EEF}$ as a function of P_{out}.

Calculating the *EEF* can essentially be performed in two alternative ways, leading to different interpretations. The first method is to calculate the *EEF* for a single value of η_{SW}, usually the highest/lowest one. This yields either the best or worst case scenario.[5]

The second way to perform the calculation of the *EEF* is by calculating it for multiple values η_{SW}, each at different values of P_{out}, and applying a power activity probability distribution $\alpha(P_{out})$. This $\alpha(P_{out})$ may follow from the targeted load application or it can also be a user-defined function. An example of such a function is (linear) ramp, indicating the weight of the *EEF*, and thus η_{SW}, becomes more

---

[5] Not to be confused with the album of dEUS.

important at higher values of P_{out}. Eventually this yields the weighted Efficiency Enhancement Factor $\widetilde{EEF}$, which is given by (2.82).

$$\widetilde{EEF} = \sum_{i=1}^{n} \frac{EEF(P_{out\_i})\alpha(P_{out\_i})}{i} \qquad (2.82)$$

The $\widetilde{EEF}$ gives a more realistic view on the performance of a switched-mode DC-DC step-down converter, with regard to η_{SW}, as it provides information on the overall performance over a certain load region. Obviously, when $\alpha(P_{out}) = 1$ (2.82) yields the mean Efficiency Enhancement Factor $\overline{EEF}$, as shown in the lower graph of Fig. 2.24.

Note that the main purpose of switched-mode DC-DC step-down converters is to provide a higher value for η_{SW} than achievable with a linear series DC-DC converter, which in essence is a resistive divider. If this constraint is not fulfilled, the additional complexity and cost of a switched-mode DC-DC converter, whether it is off- or on-chip, converter cannot be justified.

2.5 Conclusions

This chapter discusses the three DC-DC conversion methods and the associated principles of energy storage in capacitors, inductors and their combination:

1. *Capacitors*: (Sect. 2.2.1) The charging of a capacitor by means of a voltage source or another capacitor is intrinsically prone to losses. The energy charging efficiency η_{C_charge} is higher when the steady-state region is more closely approximated and if the initial voltage over the capacitor being charged is as close as possible to that of the voltage source or charging capacitor.
2. *Inductors*: (Sect. 2.3.1) The charging of an inductor by means of a voltage source is ideally lossless, unless when a finite, non-zero series resistance is assumed. In the latter case η_{L_charge} is higher when the charge time of the inductor is shorter and/or if the initial current through the inductor is lower.
3. *Capacitors and inductors*: (Sect. 2.3.2) The charging of a capacitor in series with an inductor by means of a voltage source is ideally lossless, unless when a finite, non-zero series resistance is assumed. In either case the energy stored in the capacitor will be maximal after the first half period, for a periodically damped system. At this point η_{RLC_charge} is also maximal and it will be higher when the initial voltage over the capacitor is closer to that of the voltage source, or when the initial current through the inductor is lower.

By using this knowledge the two methods for switched-mode DC-DC conversion are explained, together with the linear conversion method:

1. *Linear voltage converters*: (Sect. 2.1) This converter can only decrease the output voltage and is based on dissipating the excess energy. The power conversion efficiency is proportional to the voltage conversion ratio, for a series converter, and in addition proportional to the output power, for a shunt converter.

2.5 Conclusions

2. *Charge-pump DC-DC converters*: (Sect. 2.2) This converter can both decrease or increase the output voltage. The energy conversion efficiency will ideally only reach 100% at the optimal voltage conversion ratio, which depends on the used topology, and is independent of the output power.
3. *Inductive type DC-DC converters*: (Sect. 2.3) This converter can also both decrease or increase the output voltage. The energy conversion efficiency will ideally always reach 100%, regardless of the voltage conversion ratio and output power. The mathematical methods for calculating the voltage conversion ratio and output ripple, both for CCM and DCM, for an ideal boost converter example are explained in Sect. 2.3.3.

Finally a figure of merit for comparing DC-DC step-down converters, called the Efficiency Enhancement Factor (*EEF*), is introduced in Sect. 2.4. The EEF allows comparison in terms of power conversion efficiency that is independent of the voltage conversion ratio.

Chapter 3
Inductive DC-DC Converter Topologies

The focus of this work is on the design and implementation of monolithic inductive DC-DC converters into standard CMOS IC technologies. Therefore, a more extensive discussion on inductive converter topologies is provided in this chapter. This involves obtaining insight into the basic operation principles of the different converter topologies and a comparison of a selection of these topologies. Thus allowing the determination the fundamental and intrinsic advantages and drawbacks of the different converter topologies, with the aim towards monolithic integration. These advantages and drawbacks may well differ from converter topologies that are not intended for monolithic integration, as the on-chip area requirement and available devices are much more restricted. Please note that a full coverage on inductive type DC-DC converter topologies is not intended and that merely the most promising and practical converter topologies are discussed.

The primary classes of non-galvanically separated step-down, step-up and step-up/down inductive converter topologies are discussed. These topologies are categorized according to their non-inverting voltage conversion range in the respective Sects. 3.1, 3.2 and 3.3. This categorization differs from the traditional one in the sense that only positive output voltages are considered, as the on-chip conversion towards negative voltages is omitted in this work. The converters of each of the three categories are compared in terms of their circuit topology, their basic operation and their area requirement. The comparison of the area requirement is conducted by means of SPICE-simulations.

The derived classes of DC-DC converter topologies, such as galvanically separated converters and resonant converters, are addressed in Sect. 3.4. Topology variations on the primary classes of converter topologies, incorporating multi-phase converters and Single-Inductor Multiple Output (SIMO) converters, are discussed in Sect. 3.5. Finally, the chapter is concluded in Sect. 3.6.

3.1 Step-Down Converters

Inductive DC-DC step-down converters are used to convert the input voltage U_{in} to a lower output voltage U_{out}. The application principle of DC-DC step-down converters in battery-operated systems is explained in Sect. 1.2.2. In this section five different ideal inductive DC-DC step-down converter topologies are discussed and compared with one another. This is done in view of monolithic integration, where the occupied area of the converter is a crucial parameter that is to be minimized for cost reasons. The topologies explained in this section are:

- The buck converter
- The bridge converter
- The three-level buck converter
- The buck^2 converter
- The Watkins-Johnson converter

The dominant parameters, which determine the required converter area, are the values of the passives: the inductor(s) and capacitor(s). Because ideal inductive DC-DC converters are lossless, as explained in Sect. 2.3, the power conversion efficiency η_{SW} is not a suitable parameter for a comparison. In contrast, the output voltage ripple ΔU_{out} is not infinitesimal for ideal inductive DC-DC converters, having finite values for L, C and f_{SW}. Indeed, in the previous Sect. 2.3.3 it is deduced, for a boost converter example, that the dependency of ΔU_{out} includes several input and output parameters of the converter. This is formally described by (2.65) and (2.74), for CCM and DCM respectively. Thus, different ideal step-down converter topologies can be designed for equal specifications, including ΔU_{out}, allowing them to be compared by means of the required values of the passives and indirectly by their required area.

In Sect. 2.3.3 it is mathematically proven that the capacitance of the output capacitor C of a boost converter is always a determining parameter for ΔU_{out}, the evidence that this is true for all inductive DC-DC converters is trivial. Thus, by keeping the input and output parameters of the converter constant, except for the capacitance of C, the value of C can be determined for the different topologies having the same specifications. The parameters which are kept constant and their values are listed in Table 3.1. These values are chosen such that they are in the same order of magnitude as the values used in real monolithic implementations, which can be verified in Chap. 6.

The calculations for this comparison are executed through SPICE simulations only, except for the buck converter for which the calculations are also performed by means of the small-ripple approximation, charge-balance and volt-second balance. This allows for validating these calculation methods through simulations. These simulations and calculations allow for an area-driven comparison, which is mandatory for monolithic integration. However, the required area is not the sole property that needs to be taken into account when choosing a topology. Therefore, a brief circuit-level discussion is given for each converter topology.

3.1.1 Buck Converter

The circuit of an ideal buck-converter, shown in Fig. 3.1(a), is explained more in detail than the other DC-DC step-down converter topologies, because this converter

3.1 Step-Down Converters

Table 3.1 The input and output parameters, together with their values, used to compare different DC-DC step-down converter topologies

Input/output parameter	Value
Input voltage U_{in}	2 V
Output voltage U_{out}	1 V
Output voltage ripple ΔU_{out}	50 mV
Total inductance L_{tot}	10 nH
Switching frequency f_{SW}	100 MHz
Output power 1 P_{out_1}	1 mW
Load resistance 1 R_{L_1}	1 kΩ
Output power 2 P_{out_2}	10 mW
Load resistance 2 R_{L_2}	100 Ω
Output power 3 P_{out_3}	100 mW
Load resistance 3 R_{L_3}	10 Ω
Output power 4 P_{out_4}	1 W
Load resistance 4 R_{L_4}	1 Ω

is the base of many of the implementations, discussed in Chap. 6. First the principle of operation for CCM is explained. In this CM the current $i_L(t)$ through the inductor L always has a positive, finite value and it is described in two phases:

1. *The inductor charge phase* Φ_1: The equivalent circuit for Φ_1, shown in Fig. 3.1(b), is achieved by closing SW_1 and opening SW_2 for a certain on-time t_{on}. During Φ_1 L is charged in series with C and R_L by U_{in}, causing $i_L(t)$ to increase from its minimal value I_{L_min} to its maximal value I_{L_max}, as shown in Fig. 3.2. When $i_L(t)$ becomes larger than the output current $i_{out}(t)$, C is also

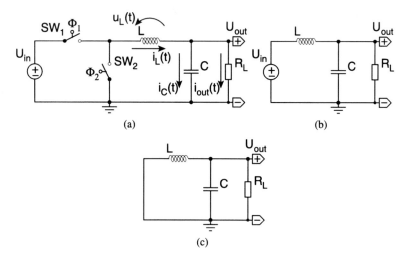

Fig. 3.1 (a) The circuit of an ideal buck DC-DC converter. (b) The equivalent circuit of the inductor charge phase and (c) the inductor discharge phase

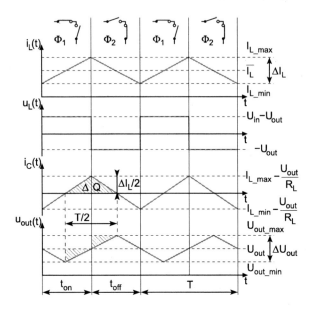

Fig. 3.2 The linearized $i_L(t)$, the linearized $u_L(t)$, the linearized $i_C(t)$ and the linearized $u_{out}(t)$ as a function of time, for an ideal buck DC-DC converter in CCM

being charged. Before this point R_L is powered through C and afterwards R_L is powered through U_{in}, because the DC-component of $i_L(t)$ flows through R_L.

2. *The inductor discharge phase* Φ_2: The equivalent circuit for Φ_2 is shown in Fig. 3.1(c), which is achieved by opening SW_1 and closing SW_2 for a certain off-time t_{off}. During Φ_2 L is discharged into C and R_L, causing $i_L(t)$ to decrease from I_{L_max} to I_{L_min}, as shown in Fig. 3.2. During the first part of Φ_2, $i_L(t)$ is larger than $i_{out}(t)$, causing C to be further charged by L and R_L to be powered by the DC-component of $i_L(t)$. After this first part of Φ_2, R_L is powered by discharging L and C.

Because L is not discharged in series with U_{in}, it can intuitively be seen that U_{out} will always be lower than U_{in}. For CCM this is confirmed by (3.1), which gives the voltage conversion ratio $k(\delta)$. It is observed that, similar to an ideal boost converter, $k(\delta)$ only depends on the duty-cycle δ for CCM. This relation is plotted by the black curve in the upper graph of Fig. 3.4.

$$k(\delta) = \delta \qquad (3.1)$$

In DCM $i_L(t)$ varies between a finite, positive value and zero. The operation of an ideal DC-DC buck converter in this CM consists of three phases:

1. *The inductor charge phase* Φ_1: The equivalent circuit for Φ_1, shown in Fig. 3.1(b), is achieved by closing SW_1 and opening SW_2 for a certain on-time t_{on}. During Φ_1 L is charged in series with C and R_L by U_{in}, causing $i_L(t)$ to increase from zero to its maximal value I_{L_max}, as shown in Fig. 3.3. From the point where $i_L(t)$ becomes larger than the output current $i_{out}(t)$ C is also being charged. Before this point R_L is powered through C and after this point R_L is powered through U_{in}, because the DC-component of $i_L(t)$ flows through R_L.

3.1 Step-Down Converters

Fig. 3.3 The linearized $i_L(t)$, the linearized $u_L(t)$, the linearized $i_C(t)$ and the linearized $u_{out}(t)$ as a function of time, for an ideal buck DC-DC converter in DCM

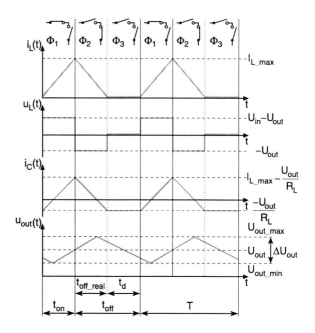

2. *The inductor discharge phase* Φ_2: The equivalent circuit for Φ_2 is shown in Fig. 3.1(c), which is achieved by opening SW_1 and closing SW_2 for a certain real off-time t_{off_real}. During Φ_2 L is discharged into C and R_L, causing $i_L(t)$ to decrease from I_{L_max} to zero, as can be seen in Fig. 3.2. During the first part of Φ_2 $i_L(t)$ is larger than $i_{out}(t)$, causing C to be further charged by L and R_L to be powered by the DC-component of $i_L(t)$. After this first part of Φ_2 R_L is powered by both discharging L and C.
3. *The dead-time phase* Φ_3: The equivalent circuit for Φ_3 consists of the series connection of C and R_L, which is achieved by both opening SW_1 and SW_2 for a certain dead-time t_d. This prevents $i_L(t)$ from becoming negative, thereby shorting C to the ground.

For of a buck converter in DCM $k(\delta)$ can be calculated analogue to a boost converter, which is explained in Sect. 2.3.3, yielding (3.2). The graphical representation of $k(\delta)$ for DCM is plotted by the gray curves of the upper graph in Fig. 3.4, for different values of $(2L)/(R_L T)$.

$$k(\delta) = \frac{2}{1 + \sqrt{1 + \frac{4\frac{2L}{R_L T}}{\delta^2}}} \quad (3.2)$$

The CB is calculated similar to a boost converter and it is defined by (3.3). When this condition is true, the buck converter operates in CCM.

$$\frac{2L}{R_L T} > 1 - \delta \quad (3.3)$$

Fig. 3.4 The *upper graph* shows $k(\delta)$ as a function of δ, where the *black curve* is valid for CCM and the *gray curves* for DCM. In the *lower graph* the *black curve* shows the boundary between the two CMs and the *gray curves* illustrate three numerical examples. These graphs are valid for an ideal DC-DC buck converter

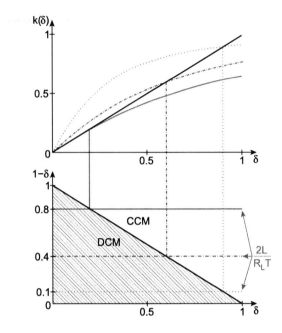

This boundary is plotted by the black curve of the lower graph in Fig. 3.4. The gray curves in this graph denote the three different values of $(2L)/(R_L T)$: 0.1, 0.4 and 0.8, for which $k(\delta)$ is plotted in the upper graph. It is observed that the buck converter will have the tendency to work in CCM for higher inductance values L, lower load resistance values R_L (higher loads) and shorter switching periods T (higher frequencies f_{SW}). When $(2L)/(R_L T)$ reaches a value that is higher than one, the converter will always operate in CCM.

The method explained in Sect. 2.3.3 for the calculation of the output voltage ripple ΔU_{out} of a boost converter is not applicable for the buck converter, because L is a part of the output filter. As a consequence, the waveform of U_{out} will be continuous, rather than piecewise linear as illustrated in Figs. 3.2 and 3.3. Therefore, an alternative method, based on the charge balance of C, is used to calculate ΔU_{out}.

This method is explained for the buck converter in CCM. The current $i_L(t)$ is divided over $i_C(t)$ and $i_{out}(t)$, as shown in Fig. 3.1(a). When the converter operates in steady-state the net change in $u_{out}(t)$ is zero, hence the DC-component of $i_C(t)$ is also zero. This implies that the DC-component of $i_L(t)$ flows through R_L. The AC-component of $i_L(t)$ entirely flows through C and is equal to the current ripple ΔI_L through L, as shown in Fig. 3.2. ΔI_L is calculated through (3.4), where the unknown mean output voltage $\overline{U_{out}}$ can be substituted by the known parameters of (3.1).

$$\Phi_1 : 0 \to t_{on} \implies \frac{di_L(t)}{dt} = \frac{U_{in} - \overline{U_{out}}}{L}$$

$$\implies \Delta I_L = \frac{U_{in} - \overline{U_{out}}}{L} t_{on} = \frac{U_{in} - \delta U_{in}}{L} t_{on} \quad (3.4)$$

3.1 Step-Down Converters

In Fig. 3.2 it can also be seen that the positive portion of $i_C(t)$ causes u_{out} to rise and vice versa. This is due to the fact that the injected positive charge change ΔQ into C causes ΔU_{out}, which follows from (3.5).

$$\Delta U_{out} = \frac{\Delta Q}{C} \tag{3.5}$$

This positive portion of $i_C(t)$ always has a duration of $T/2$ because the DC-component of $i_C(t)$ is zero and because the waveform of $i_C(t)$ is symmetrical, implying that the amplitude of $i_C(t)$ is equal to $\Delta I_L/2$. With this knowledge ΔQ can be calculated as the area underneath the positive portion of $i_C(t)$, yielding (3.6).

$$\Delta Q = \int_0^{\frac{T}{2}} i_C(t)\,dt = \frac{\frac{\Delta I_L}{2} \cdot \frac{T}{2}}{2} \tag{3.6}$$

Finally, ΔU_{out} for CCM is found by substituting (3.6) into (3.5), which yields (3.7).

$$\Delta U_{out} = \frac{(\delta - \delta^2)T^2 U_{in}}{8CL} = \frac{t_{on} t_{off} U_{in}}{8CL} \blacksquare \tag{3.7}$$

It can be seen that ΔU_{out} is inversely proportional to both the values of C and L, which is due to the fact that they are both part of the output filter. There is also a linear dependency on T, following from the fact that T influences ΔI_L, which in-turn influences ΔU_{out}. Remarkably, ΔU_{out} is not dependent on R_L, because of the fact that ΔU_{out} is only dependent on ΔI_L, which is in-turn not influenced by R_L.

The method for calculating ΔU_{out} can also be used for DCM, yielding (3.8).

$$\Delta U_{out} = \frac{U_{in}(L\overline{U_{out}} + R_L t_{on}(\overline{U_{out}} - U_{in}))^2}{2CLR_L^2 \overline{U_{out}}(U_{in} - \overline{U_{out}})} \tag{3.8}$$

Similar dependencies on parameters as in (3.7) are observed in (3.8). The parameter $\overline{U_{out}}$ can be substituted by the small-ripple approximation for U_{out}, given by (3.2), which is not performed here to minimize the complexity. The dependency on t_{off} is not explicitly visible in (3.8), nevertheless this follows from the dependency on $\overline{U_{out}}$. In addition ΔU_{out} is dependent on R_L in DCM, which is not the case for CCM. This is due to the dead-time period t_d where the converter is idle and C is being discharged through R_L, thereby lowering $u_{out}(t)$.

Table 3.2 shows the results of the SPICE simulations for the ideal buck converter, in order to comply with the specification of Table 3.1, for four different P_{out}. The two remaining degrees of freedom, namely the capacitance of C and δ, are provided together with t_{off_real} and the CM. As expected, the required capacitance, for equal L, f_{SW}, U_{in} and U_{out}, increases upon an increasing P_{out}. Also, the required δ in DCM, to maintain U_{out}, increases upon an increasing P_{out}. This follows from the fact that in DCM U_{out} is dependent on R_L. Although not explicitly shown in Table 3.2, it is noted that both the equations for $k(\delta)$, (3.1) and (3.2), and both the equations for ΔU_{out}, (3.7) and (3.8), yield exactly the same results as obtained with the SPICE-simulations of the ideal DC-DC buck converter.

To conclude the discussion on the ideal DC-DC buck converter possible benefits (✔) and drawbacks (✘), in view of monolithic integration, are provided:

Table 3.2 The SPICE-simulations results for the required capacitance C, to comply with the specifications of Table 3.1, of an ideal DC-DC buck converter, for four different output powers P_{out}. The required duty-cycle δ and the CM are also provided

P_{out}	C	δ	t_{off_real}	CM
1 mW	0.19 nF	3.15%	0.315 ns	DCM
10 mW	1.6 nF	10%	1 ns	DCM
100 mW	9.5 nF	31.5%	3.15 ns	DCM
1 W	12.5 nF	50%	5 ns	CCM

✔ Both the switches have one terminal connected to a fixed potential: SW_1 to U_{in} and SW_2 to the ground *GND* of the circuit. This is beneficial for the implementation of the switches with MOSFETs, which are switched on and off by controlling the gate voltage relative to the source voltage (see Sect. 1.3.1).

✔ Only one inductor and capacitor is used, which can be beneficial for the area requirement of the converter.

✔ The output current $i_{out}(t)$ is continuously provided by L, in CCM. Thus, no current-peaks are fed to the output which would increase ΔU_{out} for real output capacitors, having a finite parasitic series resistance (*ESR*) and parasitic series inductance (*ESL*). Although the current delivered by L in DCM for the output is not continuous, no sudden transients occur in it. This follows from the fact that the output filter consists of both and inductor and a capacitor.

✔ The converter delivers a non-inverted U_{out}, of which the *GND*-references of in- and output are physically connected with each other. This would otherwise cause problems in standard CMOS-technologies, where the substrate is connected with the *GND*-potential.

✘ The input current through U_{in} is discontinuous, having a negative impact on the overall performance of a real converter due to non-zero parasitic input resistances and inductances. Therefore, real implementations might require an additional on-chip decouple capacitor, resulting in an increasing overall area requirement.

✘ CCM is only reached for a sufficiently high P_{out}. Therefore, the advantage of the continuous output current is lost for low values of P_{out}.

3.1.2 Bridge Converter

The circuit of an ideal bridge DC-DC converter is shown in Fig. 3.5. This converter is capable of delivering a positive or a negative U_{out}, of which the absolute value is always lower than U_{in}. This is confirmed by (3.9), which shows that $k(\delta)$ is only dependent on δ for CCM. The calculation method to obtain (3.9) is explained in Sect. 2.3.3. The equation and explanation for DCM is omitted, as this provides limited added value.

$$k(\delta) = 2\delta - 1 \qquad (3.9)$$

3.1 Step-Down Converters

Fig. 3.5 The circuit of an ideal bridge DC-DC converter

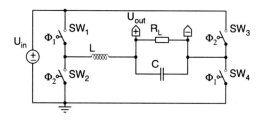

Fig. 3.6 The voltage conversion ratio $k(\delta)$ as a function of the duty-cycle δ, for a bridge converter in CCM

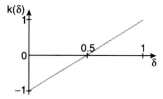

The graphical representation of (3.9) is illustrated in Fig. 3.6, where $k(\delta)$ is shown as a function of δ. It is observed that $k(\delta)$ becomes positive for values of δ larger than 50%. This discussion is limited to this region only, because on-chip negative voltages are rarely used. It can be intuitively understood that the operation of the converter in the negative region of $k(\delta)$ is dual to the positive region of $k(\delta)$.

The basic operation of the bridge converter in CCM, with a positive U_{out}, consists of the following two phases:

1. *The inductor charge phase* Φ_1: During Φ_1 SW_1 and SW_4 are closed while SW_2 and SW_3 are opened, for a time t_{on}. The positive U_{in} is applied over the series connection of L and C. This causes the, already flowing, current through L to increase, thereby charging both L and C and providing power to R_L.
2. *The inductor discharge phase* Φ_2: During Φ_2 SW_1 and SW_4 are closed while SW_2 and SW_3 are opened, for a time t_{off}. The negative $-U_{in}$ is applied over the series connection of L and C. This causes the current through L to decrease, thereby inverting the current through C for powering R_L.

In DCM a third phase occurs where the current through L becomes zero a the end of Φ_2. At this point SW_2 and SW_3 are also opened, preventing the current trough L from becoming negative and discharging C. This results in a certain t_d, during which the converter is idle.

The results of the SPICE simulations for the ideal bridge converter, in order for it to comply with the specifications of Table 3.1, are listed in Table 3.3. It is observed that in order to maintain ΔU_{out} constant, the required capacitance of C increases upon an increasing P_{out}. This results in a larger required capacitance compared to ideal buck converter. Also, the δ-range is larger for the ideal bridge converter than for the buck converter, for the same P_{out}-range.

The discussion on the ideal DC-DC bridge converter is concluded with its possible benefits and drawbacks:

Table 3.3 The SPICE-simulations results for the required capacitance C, to comply with the specifications of Table 3.1, of an ideal DC-DC bridge converter, for four different output powers P_{out}. The required duty-cycle δ and the CM are also provided

P_{out}	C	δ	t_{off_real}	CM
1 mW	0.19 nF	3.85%	0.125 ns	DCM
10 mW	1.69 nF	12.1%	0.405 ns	DCM
100 mW	11.5 nF	38.5%	1.3 ns	DCM
1 W	19 nF	75%	2.5 ns	CCM

✔ The four switches have one of their terminals connected to a fixed potential: SW_1 and SW_3 to U_{in}, SW_2 and SW_4 to GND. This simplifies the drive circuits for the MOSFETs gates, which will be used to implement the switches.

✔ In CCM L provides a continuous current to the output, avoiding current peaks through C and the output. These would cause ΔU_{out} to increase, due to the *ESR* and *ESL* of a non-ideal C.

✔ In CCM the current drawn from U_{in} is continuous. Thus, the on-chip U_{in} will be less influenced by the voltage drop over parasitic resistance and inductances at the input. In DCM the current draw from U_{in} will not be continuous, however there are also no sudden current steps. This is in contrast to a buck converter.

✘ Four switches are required, implying more area, a more complex driver and increased conduction and switching losses, compared to the buck converter.

✘ The output cannot be referred to the *GND* of the converter, which is associated with the chip's substrate. The on-chip circuitry to be supplied with the converter's output will therefore not be allowed to have a substrate reference. This can be a problem in standard CMOS IC technologies, where the bulk and source of a standard n-MOSFET are physically inextricably connected to the substrate.

✘ It shows that the bridge converter requires a larger output capacitance C than a buck converter, to obtain the same ΔU_{out}. This difference is minimal at low values of P_{out}, nevertheless it becomes significant at high values of P_{out}.

✘ CCM is only reached for a sufficiently high P_{out}. Therefore, the advantage of the continuous output current is lost for low values of P_{out}.

3.1.3 Three-Level Buck Converter

Figure 3.7 shows the circuit of an ideal three-level buck[1] DC-DC converter. Multi-level DC-DC converters were originally introduced for high-voltage conversion applications [Mey92], nevertheless monolithic integration in a CMOS technology has been attempted [Vil08].

---

[1] Multi-level implementations are also possible for other DC-DC converter topologies, such as the bridge converter. These topology variants are however not discussed in this work, as they increase the complexity by adding switches and capacitors.

3.1 Step-Down Converters

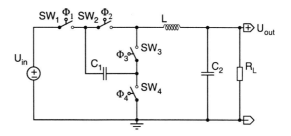

Fig. 3.7 The circuit of an ideal three-level buck DC-DC converter

The three-level buck converter is capable of converting U_{in} to a lower, non-inverted U_{out}. It can be proven that for CCM $k(\delta)$ is equal to that of a regular ideal buck converter, which is given by (3.1). The graphical representation of $k(\delta)$ as a function of time, is illustrated in Fig. 3.4 by the black curve in the upper graph.

The timing of the four switches in CCM is shown in Fig. 3.8(a) for $\delta < 0.5$ and in Fig. 3.8(b) for $\delta > 0.5$. The operation in CCM for $\delta < 0.5$ consists of the following four phases:

1. *The inductor charge, capacitor charge phase* Φ_1: During Φ_1 SW_1 and SW_3 are closed and SW_2 and SW_4 are opened, during a time t_{on}. Both L and C_1 are charged in series by U_{in}, thereby also charging C_2 and powering R_L.
2. *The first inductor discharge phase* Φ_2: During Φ_2 SW_3 and SW_4 are closed and SW_1 and SW_2 are opened, during a time $(t_{off} - t_{on})/2$. L is discharged through R_L and partially through C_2. C_1 is disconnected.
3. *The inductor charge, capacitor discharge phase* Φ_3: During Φ_3 SW_2 and SW_4 are closed and SW_1 and SW_3 are opened, during a time t_{on}. C_1 is discharged, thereby charging L, C_2 and R_L.
4. *The second inductor discharge phase* Φ_4: This phase is identical to Φ_2.

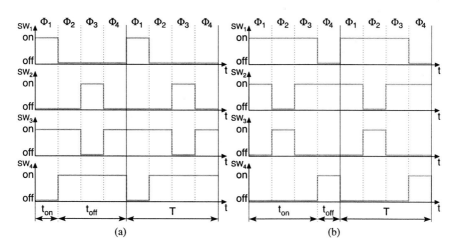

Fig. 3.8 (a) The timing of the four switches of an ideal three-level buck DC-DC converter in CCM, for $\delta < 0.5$ and (b) for $\delta > 0.5$

Table 3.4 The SPICE-simulations results for the required capacitances C_1 and C_2, to comply with the specifications of Table 3.1, of an ideal three-level buck DC-DC converter, for four different output powers P_{out}. The required duty-cycle δ and the CM are also provided

P_{out}	C_1	C_2	$C_{tot} = C_1 + C_2$	δ	t_{off_real}	CM
1 mW	0.3 nF	0.06 nF	0.36 nF	27%	0 ns	DCM
10 mW	0.5 nF	0.16 nF	0.66 nF	50%	5 ns	BCM
100 mW	5 nF	0 nF	5 nF	50%	5 ns	CCM
1 W	50 nF	0 nF	50 nF	50%	5 ns	CCM

The operation in CCM for $\delta > 0.5$ consists of the following four phases:

1. *The first inductor charge phase* Φ_1: During Φ_1 SW_1 and SW_2 are closed and SW_3 and SW_4 are opened, during a time $(t_{on} - t_{off})/2$. L is charged by U_{in}, thereby charging C_2 and R_L. C_1 is disconnected.
2. *The inductor discharge, capacitor charge phase* Φ_2: During Φ_2 SW_1 and SW_3 are closed and SW_2 and SW_4 are opened, during a time t_{off}. The series connection of U_{in}, C_1 and L causes L to be discharged and C_1 to be charged. The current from this series circuit powers R_L and also partially C_2.
3. *The second inductor charge phase* Φ_3: This phase is identical to Φ_1.
4. *The inductor discharge, capacitor discharge phase* Φ_4: During Φ_4 SW_2 and SW_4 are closed and SW_1 and SW_3 are opened, during a time t_{off}. The series connection of C_1 and L causes both of them to be discharged into R_L and partially into C_2.

It can be proven that the mean voltage over the flying capacitor C_1 is always $U_{in}/2$, for steady-state operation. It is also observed that the input node of the output filter can have three different voltage levels: GND, $U_{in}/2$ or U_{in}. Hence resulting in a three-level converter.[2] Please note that DCM is not considered in this discussion.

The results of the SPICE simulations for the ideal three-level buck converter, in order for it to comply with the specifications of Table 3.1, are listed in Table 3.4. The value of C_1 was chosen such that the voltage swing over it equals 100 mV. This value is chosen such that the voltage drop over each of the four switches is limited to $U_{out} + 10\%$. Assuming that U_{out} equals the nominal CMOS technology supply voltage, this allows for the switches to be implemented as single MOSFET transistors. When allowing a larger ripple voltage over C_1 the switches would have to be implemented as stacked transistors or thick-oxide devices, to cope with the high voltage. This assumption follows from realized DC-DC step-down designs, which will be elaborated upon in Chap. 6. The value of C_1 can also not be chosen to be smaller than 0.3 nF, because the resonance frequency of C_1 and L will approach the switching frequency f_{SW}. This would cause the current through L to become negative before the ending of the switching period, as explained in Sect. 2.3.2, which is undesired.

---

[2]A regular buck-converter is a two-level converter, because the voltage on the input node of the output filter can be either GND or U_{in}.

3.1 Step-Down Converters

Fig. 3.9 The circuit of an ideal buck[2] DC-DC converter

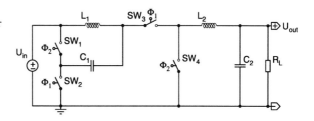

It is observed that the converter requires no output capacitor C_2 for $P_{out} = 1$ W and $P_{out} = 100$ mW, which yields respective ΔU_{out} of 34 mV and 50 mV. However, for lower P_{out} a small value of C_2 is required. For the two largest P_{out} the converter operates in CCM and at $P_{out} = 10$ mW it operates at the boundary between the two CMs, which is also referred to as boundary condition mode (BCM). This is in contrast with a buck and bridge converter, which only tend to operate in CCM for the highest P_{out}. Therefore, the δ-range for this converter is smaller, for the same P_{out}-range.

To conclude this discussion the benefits and drawbacks of this converter are provided:

✔ In CCM L provides a continuous current to the output, relaxing the specifications of C_2. When the minimal P_{out} is large enough, C_2 can even be omitted.
✔ U_{out} is not inverted and referred to GND.
✔ CCM is maintained for a broad range of P_{out}, which is beneficial for ΔU_{out}.
✔ The total required capacitance C_{tot} is lower compared to a buck or bridge converter, at low values for P_{out}.
✘ SW_2 and SW_3 do not have a terminal which is connected to a fixed potential, implying a more complex driver is required.
✘ The mean voltage on C_1 tends to drift above U_{in} or GND, when the timing of the switches is not exactly as prescribed by Fig. 3.8. Thus a feedback mechanism to keep it in the safe-operating limits of the technology is required.
✘ The current drawn from U_{in} is discontinuous, likely causing the need for additional on-chip input decoupling.
✘ Four switches are required, requiring more area than the two switches of a buck converter, for similar conduction losses. These will also cause increased switching losses and a more complex driver.

3.1.4 Buck[2] Converter

The circuit of an ideal buck[2] DC-DC converter is shown in Fig. 3.9 [Mak91]. This converter converts U_{in} to a lower, non-inverted U_{out}. The relation between $k(\delta)$ and δ, for CCM, is given by (3.10), which shows the quadratic dependency. The explanation of the DCM is omitted in this dissertation.

$$k(\delta) = \delta^2 \qquad (3.10)$$

Fig. 3.10 The voltage conversion ratio $k(\delta)$ as a function of the duty-cycle δ, for an ideal buck^2 converter in CCM

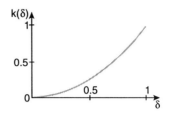

The graphical representation of (3.10) is illustrated in Fig. 3.10. The buck^2 converter has an improved performance for low values of (δ), which is equivalent to large differences between the values of U_{in} and U_{out}, compared to a standard buck converter. This is due to the presence of two filters, each consisting of an inductor and capacitor, as opposed to one filter in a standard buck converter.

The operation of the buck^2 converter in CCM consists of the following two phases:

1. *The inductors charge phase* Φ_1: During Φ_1 SW_1 and SW_3 are closed and SW_2 and SW_4 are opened, during a time t_{on}. Both L_1 and L_2 are charged in series with U_{in}. L_2 also receives energy from C_1, thereby discharging C_1. At the output, C_2 is charged and R_L is powered from the current through L_2.
2. *The inductors discharge phase* Φ_2: During Φ_2 SW_2 and SW_4 are closed and SW_1 and SW_3 are opened, during a time t_{off}. L_1 is discharged through C_1, thereby charging C_1. L_2 is discharged through the output, thereby powering R_L through both C_2 and L_2.

In DCM SW_2 and SW_4 are opened when the respective currents through L_2 and L_4 become zero, which does not necessarily occurs simultaneously. This causes the whole converter, or either the part L_1–C_1 or the part L_2–C_2, to be idle for the remaining part of t_{off}.

Table 3.5 lists the results of the SPICE simulations for the ideal buck^2 converter, in order for it to comply with the specifications of Table 3.1. All the parameters are chosen such that the total required capacitance is minimized, with some restrictions. C_1 is chosen to be the smallest possible value, as it has less effect on ΔU_{out} than C_2. However, the capacitance of C_1 is inversely proportional to the voltage swing over SW_3 and SW_4. Thus, its minimum required value is limited such that $U_{SW_3} \leqslant 2 \cdot U_{in}$ and $U_{SW_4} \leqslant 2 \cdot U_{in}$. It can be proven that for the desired specifications of U_{in} and

Table 3.5 The SPICE-simulations results for the required capacitances C_1 and C_2, to comply with the specifications of Table 3.1, of an ideal buck^2 DC-DC converter, for four different output powers P_{out}. The required duty-cycle δ and the CM are also provided

P_{out}	C_1	C_2	$C_{tot} = C_1 + C_2$	δ	$t_{off\_real\_SW2}$	$t_{off\_real\_SW4}$	CM
1 mW	0.0035 nF	0.2 nF	0.2035 nF	3.5%	0.06 ns	0.24 ns	DCM
10 mW	0.035 nF	1.7 nF	1.735 nF	11%	0.22 ns	0.7 ns	DCM
100 mW	0.3 nF	10 nF	10.3 nF	33%	0.55 ns	2.4 ns	DCM
1 W	3 nF	10 nF	13 nF	74%	2.6 ns	2.6 ns	CCM

3.1 Step-Down Converters

U_{out}, these values cannot be made significantly smaller. The inductances for L_1 and L_2, which respectively are 3 nH and 7 nH, are chosen such that the required specifications of Table 3.1 are met with the minimal total required capacitance.

From Table 3.5 it can be concluded that the required C_{tot} decreases with decreasing P_{out}. However, the difference between C_{tot} for $P_{out} = 1$ W and $P_{out} = 100$ mW is rather small. This is due to the fact that the converter switches from CCM to DCM. It is also observed that the required capacitance for C_1 is much smaller than for C_2.

Finally, the benefits and drawbacks of the buck^2 converter are:

✔ In CCM L_2 provides a continuous current to the output, relaxing the specifications of C_2.
✔ U_{out} is not inverted and referred to *GND*.
✘ The total required capacitance is quasi the same as for the buck converter.
✘ CCM is only reached at high P_{out}.
✘ Increased complexity, compared to the buck converter, due to four required switches and an additional inductor and capacitor.
✘ The voltage over SW_3 can easily reach $2 \cdot U_{in}$. When assuming that U_{out} is the nominal technology supply voltage, SW_3 would have to be implemented by means of four stacked standard transistors. This will cause increased conduction and switching losses and a will require a more complex driver.
✘ SW_3 does not have a terminal that is connected to a fixed voltage, causing the need for a more complex driver.
✘ The current drawn from U_{in} is discontinuous, likely causing the need for additional on-chip input decoupling.

3.1.5 Watkins-Johnson Converter

The circuit of an ideal Watkins-Johnson DC-DC converter, using one inductor and four switches, is shown in Fig. 3.11(a). The equivalent circuit, using two coupled inductors and two switches, is shown in Fig. 3.11(b). It can be proven that both circuits yield the same functionality, which is decreasing U_{in} or increasing and inverting it. This behavior, $k(\delta)$ as a function of δ, is formally described for CCM by (3.11), which is valid for both the versions of the converter.

$$k(\delta) = \frac{2\delta - 1}{\delta} \qquad (3.11)$$

Figure 3.12 shows the plot of (3.11). It is observed that for $0 < \delta < 0.5$ the converter yields an inverting step-up function and that for $0.5 < \delta < 1$ the converter yields a non-inverting step-down function. Therefore, the Watkins-Johnson converter can be regarded as a step-up/down converter. However, the conversion towards negative voltages on-chip has virtually no applications. For this purpose the inverting step-up functionality will be neglected in this work, thereby only considering it as a step-down converter. Moreover, only the topology from Fig. 3.11(a) will

Fig. 3.11 (a) The circuit of an ideal Watkins-Johnson DC-DC converter, using an inductor and (b) using two coupled inductors

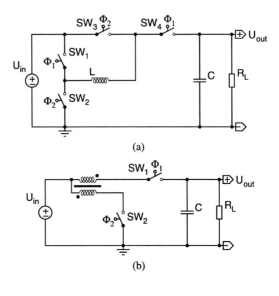

Fig. 3.12 The voltage conversion ratio $k(\delta)$ as a function of the duty-cycle δ, for a Watkins-Johnson converter in CCM

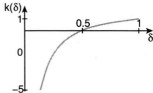

be discussed in this work, as the coupled inductors from the alternative topology would pose a difficulty for the overall comparison of DC-DC step-down converters.

The basic operation of the Watkins-Johnson converter in CCM, with a positive U_{out}, consists of the following two phases:

1. *The inductor charge phase* Φ_1: During Φ_1 SW_1 and SW_4 are closed and SW_2 and SW_3 are opened, during a time t_{on}. L is charged in series with C and R_L through U_{in}, thereby also charging C and providing power to R_L.
2. *The inductor discharge phase* Φ_2: During Φ_2 SW_2 and SW_3 are closed and SW_1 and SW_4 are opened, during a time t_{off}. L is discharged through U_{in} and simultaneously C is discharged through R_L.

In DCM SW_2 and SW_3 are opened at the end of Φ_2, preventing the current through L from becoming negative. This would cause C to be discharged through L and U_{in} during Φ_1. During the time when all the four switches are open the converter is idle.

The results of the SPICE simulations for the ideal Watkins-Johnson converter, in order for it to comply with the specifications of Table 3.1, are listed in Table 3.6. It is observed that at $P_{out} = 1$ W the required capacitance of C is about six times higher compared to the buck converter, in CCM. For low values of P_{out} the required capacitance of C has the same order of magnitude than for a buck converter.

3.1 Step-Down Converters

Table 3.6 The SPICE-simulations results for the required capacitance C, to comply with the specifications of Table 3.1, of an ideal DC-DC Watkins-Johnson converter, for four different output powers P_{out}. The required duty-cycle δ and the CM are also provided

P_{out}	C	δ	t_{off_real}	CM
1 mW	0.19 nF	4.4%	0.22 ns	DCM
10 mW	1.7 nF	14%	0.7 ns	DCM
100 mW	12 nF	44.4%	2.22 ns	DCM
1 W	66 nF	66.6%	3.34 ns	CCM

To conclude the discussion on the ideal Watkins-Johnson DC-DC converter the possible benefits and drawbacks of this converter are provided:

✔ The four switches have one terminal which is connected to a fixed potential: SW_1 and SW_3 to U_{in}, SW_2 to GND and SW_4 to U_{out}. This simplifies the drivers of these switches.

✔ The output is not inverted and referred to the GND potential of the converter.

✘ In neither one of the CMs a continuous current is delivered to C and R_L. Thus, the parasitic ESL and ESR will cause ΔU_{out} to increase in real converters, putting more stringent specifications on C.

✘ In neither one of the CMs a continuous current is drawn from U_{in}. Moreover, the current through U_{in} is reversed in polarity, between Φ_1 and Φ_2. During Φ_2 current is fed into U_{in}, which is likely not to be allowed in the majority of applications. Therefore an on-chip decouple capacitor with a large capacitance will be required to cope with this current.

✘ The converter requires four switches. Compared to a buck converter, this implies that the driver logic will be more complex, the conduction and switch losses will increase and the required area will increase.

✘ At high P_{in} the converter requires a larger output capacitance C, compared to a buck converter.

3.1.6 Step-Down Converter Summary

The most significant properties and parameters of the five discussed step-down DC-DC converters topologies are summarized in Table 3.7. It is observed that the buck converter requires the smallest number of components: one inductor, one capacitor and two switches. Because it only uses two switches, it has the potential to achieve the highest power conversion efficiency, since these switches cause conduction and switching losses. This is especially the case for monolithic DC-DC converters, due to their high switching frequencies. These losses will be explained more into detail in Chap. 4, where an accurate model is proposed which takes all the significant losses into account. It can also be seen that none of the converter topologies provides a continuous current to the output filter in DCM, implying that more stringent

Table 3.7 The comparison of key properties and parameters of five types of step-down DC-DC converters, with respect to monolithic integration (✔ = yes, ✘ = no)

	Buck	Bridge	3-Level buck	Buck^2	Watkins-Johnson
# Switches	2	4	4	4	2–4
# Capacitors	1	1	2	2	1
# Inductors	1	1	1	2	1
Floating switches	✘	✘	✔	✔	✘
Inverted output	✘	✘	✘	✘	✘
Floating output	✘	✔	✘	✘	✘
Complex timing	✘	✘	✔	✔	✘
Continuous I_{out}	CCM	CCM	CCM	CCM	✘
Continuous I_{in}	✘	✘	✘	✘	✘
C_{tot} @ $P_{out} = 1$ mW	0.19 nF	0.19 nF	0.36 nF	0.2035 nF	0.19 nF
C_{tot} @ $P_{out} = 10$ mW	1.6 nF	1.69 nF	0.66 nF	1.735 nF	1.7 nF
C_{tot} @ $P_{out} = 100$ mW	9.5 nF	11.5 nF	5 nF	10.3 nF	12 nF
C_{tot} @ $P_{out} = 1$ W	12.5 nF	19 nF	50 nF	13 nF	66 nF

specifications will have to be set to the output filter. The comparison does not account for on-chip input decoupling, which is potentially necessary as none of the converter topologies draws a continuous supply current.

Figure 3.13 shows the required C_{tot} as a function of P_{out} for each of the five step-down DC-DC converter topologies, such that these converters meet the specifications of Table 3.1. In this comparison the buck converter also proves to be the best choice for monolithic integration for $P_{out} > 1$ W, where it requires the lowest C_{tot}. The buck^2 is the second best choice at high values of P_{out}, as its required amount of C_{tot} does not differ much from the buck converter. However, the buck^2 needs two additional switches. For $P_{out} < 100$ mW the required C_{tot} does not differ significantly amongst the converter topologies, except for the three-level buck converter. The latter converter requires about 50% less total capacitance than its counterparts, for low values of P_{out}. However, it requires four switches and also a quite complex timing scheme.

The emphasis of this work is on maximizing P_{out}, minimizing the total required area and maximizing η_{SW}. Therefore, the buck converter, and its variants, will be the converter of choice for the realizations in this work. The possible variations of this converter will be explained in Sects. 3.5.1 and 3.5.2.

3.2 Step-Up Converters

Inductive DC-DC step-up converters are used to convert the input voltage U_{in} to a higher output voltage U_{out}. The application principle of DC-DC step-up converters in battery-operated systems is explained in Sect. 1.2.2. In this section three different

3.2 Step-Up Converters

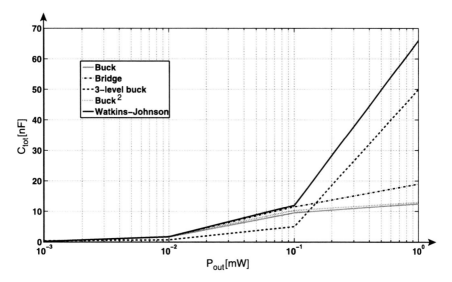

Fig. 3.13 The total required capacitance C_{tot} of five step-down DC-DC converter topologies as a function of the output power P_{out}. These values are obtained by means of SPICE-simulations, such that the five converters meet with the specifications of Table 3.1

ideal inductive DC-DC step-up converter topologies, are discussed and compared with one another. This is done in view of monolithic integration, where the occupied area of the converter is a crucial parameter that is to be minimized for cost-reasons. The topologies explained in this section are:

- The boost converter
- The current-fed bridge converter
- The inverse Watkins-Johnson converter

The comparison of DC-DC step-up converters will be performed analogue to the comparison of DC-DC step-down converters, by means of ΔU_{out}, as explained in Sect. 3.1. For this purpose, the different converter topologies are designed to meet with the specifications of Table 3.8, yielding the total required capacitance. The values of Table 3.8 are chosen in such a way that they are in the same order of magnitude as the values used in real implementations, which are discussed in Chap. 6.

The calculations for this comparison are executed through SPICE simulations only, except for the boost converter for which the calculations are also performed by means of the small-ripple approximation, charge balance and volt-second balance, which are discussed in Sect. 2.3.3. This allows for validating these calculation methods through simulations. These simulations and calculations will allow for an area-driven comparison, which is mandatory for monolithic integration. However, the required area is not the sole property that needs to be taken into account when choosing a topology. Therefore, a brief circuit-level discussion is also given for each converter topology.

Table 3.8 The input and output parameters, together with their values, used to compare different DC-DC step-up converter topologies

Input/output parameter	Value
Input voltage U_{in}	1 V
Output voltage U_{out}	2 V
Output voltage ripple ΔU_{out}	100 mV
Total inductance L_{tot}	10 nH
Switching frequency f_{SW}	100 MHz
Output power 1 P_{out_1}	1 mW
Load resistance 1 R_{L_1}	4 kΩ
Output power 2 P_{out_2}	10 mW
Load resistance 2 R_{L_2}	400 Ω
Output power 3 P_{out_3}	100 mW
Load resistance 3 R_{L_3}	40 Ω
Output power 4 P_{out_4}	1 W
Load resistance 4 R_{L_4}	4 Ω

3.2.1 Boost Converter

The circuit of an ideal boost DC-DC converter is shown in Fig. 2.18. The basic operation of this converter is already discussed as an example for calculation methods, in Sect. 2.3.3. These calculation methods where used to deduce $k(\delta)$ for continuous (2.62) and DCM (2.72) and also ΔU_{out} for continuous (2.65) and DCM (2.74).

The results of the SPICE simulations for the ideal boost converter, in order for it to comply with the specifications of Table 3.8, are listed in Table 3.9. It can be seen that the required capacitance of C in CCM is proportionally much larger than in DCM. This is due to the fact that the output filter solely consists of C and that the current delivered to the output is discontinuous. This discontinuous current causes large current transients at the transitions between Φ_1 and Φ_2. It is also observed that the converter only works in CCM at high values of P_{out}.

The discussion on the ideal DC-DC boost converter is concluded with its possible benefits and drawbacks for monolithic integration:

✔ Both switches have one terminal that is connected to a fixed potential: SW_1 to GND and SW_2 to U_{out}. This simplifies the drivers of the switches.

Table 3.9 The SPICE simulations results for the required capacitance C, to comply with the specifications of Table 3.8, of an ideal DC-DC boost converter, for four different output powers P_{out}. The required duty-cycle δ and the CM are also provided

P_{out}	C	δ	t_{off_real}	CM
1 mW	0.048 nF	3.15%	0.315 ns	DCM
10 mW	0.45 nF	10%	1 ns	DCM
100 mW	3.4 nF	31.5%	3.15 ns	DCM
1 W	25 nF	50%	5 ns	CCM

3.2 Step-Up Converters

Fig. 3.14 The circuit of an ideal current-fed bridge DC-DC converter

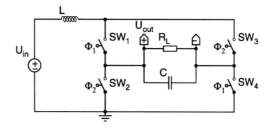

Fig. 3.15 The voltage conversion ratio $k(\delta)$ as a function of the duty-cycle δ, for a current-fed bridge converter in CCM

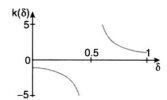

✔ This converter uses only one inductor and capacitor, which can be beneficial for its area requirement.
✔ In CCM the converter draws a continuous current from U_{in}, this relaxes the specifications for a potential on-chip decouple capacitor.
✔ The converter delivers a non-inverted U_{out}, referred to *GND* of the circuit.
✘ The current delivered to the output filter is always discontinuous, putting more stringent specifications to the output capacitor C.

3.2.2 Current-Fed Bridge Converter

The circuit of an ideal current-fed bridge DC-DC converter is shown in Fig. 3.14 [Sev79]. This converter is capable of converting U_{in} into a certain U_{out} of which the absolute value is equal or higher than U_{in}. This is formally confirmed for CCM by (3.12), which gives $k(\delta)$.

$$k(\delta) = \frac{1}{2\delta - 1} \tag{3.12}$$

The graphical representation of (3.12) is illustrated in Fig. 3.15, where $k(\delta)$ is plotted as a function of δ. The value of $k(\delta)$ is negative for $0 < \delta < 0.5$ and positive for $0.5 < \delta < 1$, in CCM. This discussion will only consider the latter case, because the generation of on-chip negative voltages is rarely used in practice.

The basic operation of the current-fed bridge converter in CCM and for positive values of $k(\delta)$, consists of the following two phases:

1. *The inductor discharge phase* Φ_1: During Φ_1 SW_1 and SW_4 are closed and SW_2 and SW_3 are opened, during a time t_{on}. This causes L to be discharged, thereby charging C and providing power to R_L.

86　　　　　　　　　　　　　　　　　　　　3　Inductive DC-DC Converter Topologies

Table 3.10 The SPICE simulations results for the required capacitance C, to comply with the specifications of Table 3.8, of an ideal DC-DC current-fed bridge converter, for four different output powers P_{out}. The required duty-cycle δ and the CM are also provided

P_{out}	C	δ	t_{off_real}	CM
1 mW	0.075 nF	3.8%	0.13 ns	DCM
10 mW	0.7 nF	12%	0.42 ns	DCM
100 mW	6.2 nF	39%	1.4 ns	DCM
1 W	37.5 nF	75%	2.5 ns	CCM

2. *The inductor charge phase* Φ_2: During Φ_2 SW_2 and SW_3 are closed and SW_1 and SW_4 are opened, during a time t_{off}. This causes L to be charged by U_{in}, thereby discharging C into R_L.

In DCM SW_1 and SW_4 are opened when the current through L becomes zero. This prevents the current through L from reversing and discharging C through L and U_{in}.

The results of the SPICE simulations for the ideal current-fed bridge converter, in order for it to comply with the specifications of Table 3.8, are listed in Table 3.10. In contrast to what might be expected from Fig. 3.15, δ does not increase upon decreasing values of R_L in DCM, for the same U_{out}. Furthermore, it is observed that the converter only operates in CCM at high values of P_{out}.

The discussion on the ideal DC-DC current-fed bridge converter is concluded with its possible benefits and drawbacks:

✔ In CCM the converter draws a continuous current from U_{in}, relaxing the specifications of a potential on-chip input decoupling capacitor.

✘ The current delivered by L reverses between Φ_1 and Φ_2, making it discontinuous. This puts more stringent specifications to C.

✘ The converter requires four switches, as opposed to the boost converter which only needs two switches. This will cause an increased area requirement and also increased conduction and switching losses.

✘ SW_1 and SW_3 do not possess a terminal that is connected with a fixed potential. Therefore, the complexity of the drivers for these switches will increase.

✘ The output cannot be referred to *GND*, which is associated with the chip's substrate. This is a drawback for the on-chip load circuits, which are referred to the substrate.

✘ It shows that the current-fed bridge converter requires a higher capacitance for C compared to a boost converter, thus a larger area will be required.

3.2.3 Inverse Watkins-Johnson Converter

The circuit of an ideal inverse Watkins-Johnson converter, implemented with an inductor and four switches, is shown in Fig. 3.16(a). The equivalent circuit of this

3.2 Step-Up Converters

Fig. 3.16 (a) The circuit of an ideal inverse Watkins-Johnson DC-DC converter, using an inductor and (b) using two coupled inductors

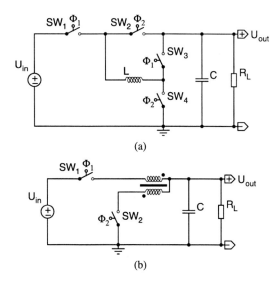

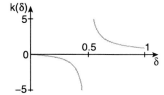

Fig. 3.17 The voltage conversion ratio $k(\delta)$ as a function of the duty-cycle δ, for an inverse Watkins-Johnson converter in CCM

converter, implemented with two coupled inductors and two switches, is shown in Fig. 3.16(b). In order to perform a fair and transparent comparison, only the implementation of Fig. 3.16(a) will be discussed in this work. The inverse Watkins-Johnson converter is capable of converting U_{in} in either a positive U_{out}, larger than U_{in}, or in a negative U_{out}, of which the absolute value can either be lower or higher than U_{in}. As a consequence the inverse Watkins-Johnson converter can be regarded as an inverting step-up/down converter or as a non-inverting step-up converter. In this work the latter case will be discussed. The formal description of $k(\delta)$ for CCM is given by (3.13).

$$k(\delta) = \frac{\delta}{2\delta - 1} \qquad (3.13)$$

The graphical representation of (3.13) is illustrated by Fig. 3.17, where $k(\delta)$ is plotted as a function of δ. It shows that $0 < U_{out} < -\infty$ for $0 < \delta < 0.5$ and that $\infty < U_{out} < 1$ for $0.5 < \delta < 1$.

The basic operation of the inverse Watkins-Johnson converter in CCM, with a positive U_{out}, consists of the following two phases:

1. *The inductor discharge phase* Φ_1: During Φ_1 SW_1 and SW_3 are closed and SW_2 and SW_4 are opened, during a time t_{on}. This causes L to be discharged, thereby charging C and providing power to R_L.

Table 3.11 The SPICE simulations results for the required capacitance C, to comply with the specifications of Table 3.8, of an ideal DC-DC inverse Watkins-Johnson converter, for four different output powers P_{out}. The required duty-cycle δ and the CM are also provided

P_{out}	C	δ	t_{off_real}	CM
1 mW	0.01 nF	4.6%	0.24 ns	DCM
10 mW	0.9 nF	14.2%	0.71 ns	DCM
100 mW	8.2 nF	46%	2.3 ns	DCM
1 W	66 nF	66.6%	3.34 ns	CCM

2. *The inductor charge phase* Φ_2: During Φ_2 SW_2 and SW_4 are closed and SW_1 and SW_3 are opened, during a time t_{off}. This causes L to be charged by C, simultaneously C is also discharged into R_L.

In DCM SW_2 and SW_4 are opened when the current through L is zero. This prevents the current through L from becoming negative, which would cause energy from C to be fed back into U_{in}. The converter is idle during this period.

The results of the SPICE simulations, for the converter to comply with the specifications of Table 3.8, are listed in Table 3.11. The required capacitance of C is more than two times larger compared to a boost converter and also significantly larger than for a current-fed bridge converter.

The discussion on the ideal DC-DC inverse Watkins-Johnson converter is concluded with its possible benefits and drawbacks:

✔ Each of the four switches has one of its terminals connected to a fixed potential: SW_1 to U_{in}, SW_2 and SW_3 to U_{out} and SW_4 to *GND*, simplifying the drivers.
✔ U_{out} is non-inverting and referred to the *GND* of the converter.
✘ The $i_L(t)$ reverses polarity between Φ_1 and Φ_2, resulting in a discontinuous output current and setting more stringent limits to the specifications of C.
✘ The current drawn from U_{in} is not continuous. Therefore, an on-chip decouple capacitor is likely to be required.
✘ Unlike a boost converter, this converter requires four switches, resulting in an increased area requirement and also increased conduction and switching losses.
✘ This converter requires roughly two times the amount of output capacitance compared to a boost converter.

3.2.4 Step-Up Converter Summary

The most significant properties and parameters of the three discussed DC-DC step-up converter topologies are summarized in Table 3.12. It is observed that all the topologies require only one capacitor and inductor. The boost converter requires the smallest number of components by using only two switches, while the other two converters require four switches. Thus, the boost converter has the potential of achieving the highest power conversion efficiency, since each switch will cause

3.2 Step-Up Converters

Table 3.12 The comparison of key properties and parameters of three types of DC-DC step-up converters, with respect to monolithic integration (✔ = yes, ✘ = no)

	Boost	Current-fed bridge	Inverse Watkins-Johnson
# Switches	2	4	4
# Capacitors	1	1	2
# Inductors	1	1	1
Floating switches	✘	✔	✘
Inverted output	✘	✘	✘
Floating output	✘	✔	✘
Complex timing	✘	✘	✘
Continuous I_{out}	✘	✘	✘
Continuous I_{in}	CCM	CCM	✘
C_{tot} @ $P_{out} = 1$ mW	0.048 nF	0.075 nF	0.01 nF
C_{tot} @ $P_{out} = 10$ mW	0.45 nF	0.7 nF	0.9 nF
C_{tot} @ $P_{out} = 100$ mW	3.4 nF	6.2 nF	8.2 nF
C_{tot} @ $P_{out} = 1$ W	25 nF	37.5 nF	66 nF

significant conduction and switching losses. Neither one of the converter topologies provide a continuous current to the output filter, thus putting more stringent specifications to the output capacitor. However, the boost converter, for instance, requires less capacitance than a buck converter in DCM (see Table 3.2), despite the fact that the buck converter delivers no sudden current transients to the output capacitor. This is due to the $U_{out} = 1$ V specification for step-down converters, compared to 2 V for step-up converters. This implies that the output current for the step-up converters is less, for the same P_{out}, relaxing the specifications on the output capacitance. The reason for choosing different specifications for step-down and step-up converters is that they should be closely related to the real implementations of these respective converters, as discussed in Chap. 6. These different specifications also imply that the step-up converters will demand a higher input current, as U_{in} is chosen 1 V, compared to 2 V for the step-down converters. Therefore, when this input current is discontinuous an on-chip decouple capacitor may be required, which is however not taken into account for this comparison. The specifications of this decouple capacitor for the boost converter and the current-fed bridge converter can be relaxed, compared to step-down converters, because their input current never shows steep transients.

Figure 3.18 shows the required C_{tot} is a function of P_{out} for each of the three DC-DC step-up converter topologies, such that these converters meet with the specifications of Table 3.8. It follows that the boost converter requires the smallest capacitance, followed by the current-fed bridge converter and the inverse Watkins-Johnson converter. The amount of required capacitance differs the most at $P_{out} > 10$ mW. For $P_{out} = 1$ mW the inverse Watkins-Johnson converter requires the lowest capacitance, of which the area will be insignificant compared to the two additional switches.

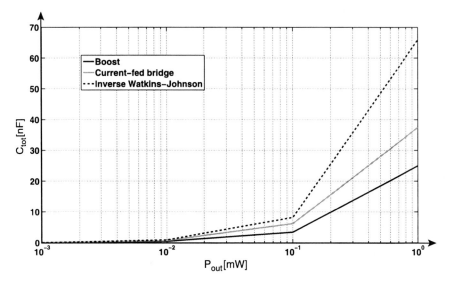

Fig. 3.18 The total required capacitance C_{tot} of three DC-DC step-up converter topologies as a function of the output power P_{out}. These values are obtained by means of SPICE-simulations, such that the three converters meet with the specifications of Table 3.8

As the emphasis of this work is to maximize P_{out}, minimize the required area and maximize η_{SW}, the boost converter is chosen for implementations in this work.

3.3 Step-Up/Down Converters

Inductive DC-DC step-up/down converters are used to convert the input voltage U_{in} to a higher or lower output voltage U_{out}. The application domain for DC-DC step-down converters is mostly situated in battery-operated systems, where the battery voltage is about equal to the required supply voltage of the application. In this section five different ideal inductive DC-DC step-up/down converter topologies are discussed and compared with one another. This is done in view of monolithic integration, where the occupied area of the converter is a crucial parameter that is to be minimized for cost-reasons. The topologies explained in this section are:

- The buck-boost converter
- The non-inverting buck-boost converter
- The Ćuk converter
- The SEPIC converter
- The Zeta converter

The comparison of DC-DC step-up/down converters will be performed analogue to the comparison of DC-DC step-down and step-up converters, by means of ΔU_{out}, as explained in Sect. 3.1. For this purpose, the different converter topologies are

3.3 Step-Up/Down Converters

Table 3.13 The input and output parameters, together with their values, used to compare different DC-DC step-up/down converter topologies

Input/output parameter	Value
Input voltage U_{in}	1 V
Output voltage U_{out}	1 V
Output voltage ripple ΔU_{out}	50 mV
Total inductance L_{tot}	10 nH
Switching frequency f_{SW}	100 MHz
Output power 1 P_{out_1}	1 mW
Load resistance 1 R_{L_1}	1 kΩ
Output power 2 P_{out_2}	10 mW
Load resistance 2 R_{L_2}	100 Ω
Output power 3 P_{out_3}	100 mW
Load resistance 3 R_{L_3}	10 Ω
Output power 4 P_{out_4}	1 W
Load resistance 4 R_{L_4}	1 Ω

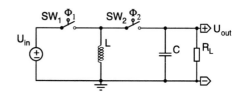

Fig. 3.19 The circuit of an ideal buck-boost DC-DC converter

designed to meet with the specifications of Table 3.13, yielding the total required capacitance C_{tot}. The calculations for this comparison are executed through SPICE simulations, allowing for an area-driven comparison which is mandatory for monolithic integration.

3.3.1 Buck-Boost Converter

The circuit of an ideal buck-boost DC-DC converter is shown in Fig. 3.19. This converter is capable of converting U_{in} into a negative U_{out}, of which the absolute value can either be lower or higher than U_{in}. The mathematical relation between $k(\delta)$ and δ for CCM is given by (3.14).

$$k(\delta) = -\frac{\delta}{1-\delta} \qquad (3.14)$$

The graphical representation of $k(\delta)$ as a function of δ is illustrated in Fig. 3.20. It shows that for $0 < \delta < 0.5$, $0 < U_{out} < -1$ and for $0.5 < \delta < 1$, $-1 < U_{out} < -\infty$. Because the on-chip conversion to negative voltages is beyond the scope of this work, the buck-boost converter will not be taken into account for this comparison.

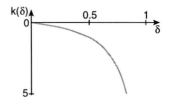

Fig. 3.20 The voltage conversion ratio $k(\delta)$ as a function of the duty-cycle δ, for an ideal buck-boost converter in CCM

3.3.2 Non-inverting Buck-Boost Converter

The circuit of an ideal non-inverting buck-boost DC-DC converter is shown in Fig. 3.21. This converter is capable of converting U_{in} in to a non-inverted U_{out}, which can be either lower or higher than U_{in}. The mathematical relation between $k(\delta)$ and δ for CCM is given by (3.15).

$$k(\delta) = \frac{\delta}{1-\delta} \quad (3.15)$$

The graphical representation of $k(\delta)$ as a function of δ is illustrated in Fig. 3.22. It shows that for $0 < \delta < 0.5$, $0 < U_{out} < 1$ and for $0.5 < \delta < 1$, $1 < U_{out} < \infty$.

The basic operation of the non-inverting buck-boost converter in CCM, with $U_{in} = U_{out}$, consists of the following two phases:

1. *The inductor charge phase* Φ_1: During Φ_1 SW_1 and SW_3 are closed and SW_2 and SW_4 are opened, during a time t_{on}. This causes L to be charged by U_{in} and simultaneously C is discharged through R_L.
2. *The inductor discharge phase* Φ_2: During Φ_2 SW_2 and SW_4 are closed and SW_1 and SW_3 are opened, during a time t_{off}. This causes L to be discharged through C and R_L, thereby providing power to R_L and charging C.

In DCM SW_2 and SW_4 are opened when the current through L becomes zero, during Φ_2. This prevents the current through L from becoming negative, which would cause C to be discharged.

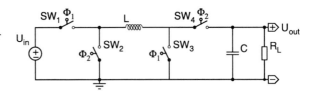

Fig. 3.21 The circuit of an ideal non-inverting buck-boost DC-DC converter

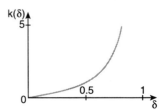

Fig. 3.22 The voltage conversion ratio $k(\delta)$ as a function of the duty-cycle δ, for an ideal non-inverting buck-boost converter in CCM

3.3 Step-Up/Down Converters

Table 3.14 The SPICE simulations results for the required capacitance C, to comply with the specifications of Table 3.13, of an ideal DC-DC non-inverting buck-boost converter, for four different output powers P_{out}. The required duty-cycle δ and the CM are also provided

P_{out}	C	δ	t_{off_real}	CM
1 mW	0.19 nF	4.47%	0.447 ns	DCM
10 mW	1.7 nF	14.1%	1.41 ns	DCM
100 mW	12 nF	44.7%	4.47 ns	DCM
1 W	98 nF	50%	5 ns	CCM

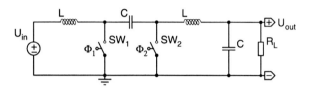

Fig. 3.23 The circuit of an ideal Ćuk DC-DC converter

The results of the SPICE simulations for the ideal non-inverting buck-boost converter, in order for it to comply with the specifications of Table 3.13, are listed in Table 3.14. It is observed that this converter requires a high output capacitance C, especially in CCM, compared to a buck or boost converter. This is due to the fact that the current delivered to the output by L is discontinuous and has a steep transient. This is similar to the boost converter, however as $U_{out} = 1$ V the output current in this case is larger compared to the boost converter, for the same P_{out}.

To conclude the discussion on the ideal DC-DC non-inverting buck-boost converter, a summary of the possible benefits and drawbacks is provided:

✔ The four switches have one terminal connected to a fixed potential: SW_1 to U_{in}, SW_2 and SW_3 to GND and SW_4 to U_{out}, which simplifies the driver circuits.
✔ The output of the converter is non-inverting and referred to GND.
✘ The current delivered by L to the output filter of the converter is always discontinuous, putting more stringent specifications to C.
✘ The current drawn from U_{in} is always discontinuous. Therefore, an on-chip decouple capacitor might be required.

3.3.3 Ćuk Converter

The circuit of an ideal Ćuk DC-DC converter is shown in Fig. 3.23. This converter is used for converting U_{in} into a negative U_{out} of which the absolute value is lower or higher than U_{in}. The mathematical relation between $k(\delta)$ and δ for CCM is identical to that of the buck-boost converter, which is given by (3.14). The graphical representation of (3.14) is illustrated in Fig. 3.20, where $k(\delta)$ is plotted as a function of δ.

The Ćuk converter will not be used for this comparison, as the on-chip conversion to negative voltages is not considered in this work.

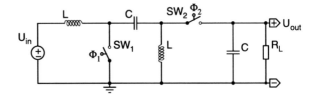

Fig. 3.24 The circuit of an ideal SEPIC DC-DC converter

3.3.4 SEPIC Converter

The circuit of an ideal DC-DC Single-Ended Primary-Inductance Converter (SEPIC) is shown in Fig. 3.24 [Mas77]. The SEPIC converter is capable of converting U_{in} into a non-inverted U_{out}, which has a lower or higher value than U_{in}. For CCM the relation between $k(\delta)$ and δ is identical to that of a non-inverting buck-boost converter, given by (3.15). The graphical representation of (3.15) is illustrated in Fig. 3.22, where $k(\delta)$ is plotted as a function of δ.

The basic operation of the SEPIC converter in CCM, with $U_{in} = U_{out}$, consists of the following two phases:

1. *The inductors charge phase* Φ_1: During Φ_1 SW_1 is closed and SW_2 is opened, during a time t_{on}. This causes L_1 and L_2 to be charged by U_{in} and C_1 respectively, thereby discharging C_1. Simultaneously, C_2 is discharged through R_L.
2. *The inductors discharge phase* Φ_2: During Φ_2 SW_2 is closed and SW_1 is opened, during a time t_{off}. This causes L_1 to discharge, thereby charging C_1 and C_2 and providing power to R_L. L_2 is also discharged, causing C_2 to be charged and R_L to be powered.

In DCM the currents through L_1 and L_2 do not stop flowing and are allowed to change polarity. This is not the case for all the converter topologies discussed until now, where the current through the inductor(s) stops flowing in DCM. In this mode SW_2 is opened at the moment when the currents through L_1 and L_2 are equal and flow towards the output. This prevents C_2 to be discharged through the inductors. When both switches are opened C_2 continues delivering power to R_L. Simultaneously, L_1, L_2, C_1 and U_{in} are series connected, forming a series LC-circuit as described in Sect. 2.3.2.

The results of the SPICE simulations for the ideal SEPIC converter, in order for it to comply with the specifications of Table 3.13, are listed in Table 3.15. All the parameters are chosen such that the total required capacitance is minimized, with the following restriction: the respective voltages over SW_1 and SW_2 are not allowed to rise above $3 \cdot U_{in}$. This restriction is made for the assumption that the converter will be implemented in a technology having U_{in} as nominal supply voltage. Therefore, SW_1 and SW_2 might be realized each by using three standard stacked-transistors, in total requiring six transistors. When comparing with the non-inverting buck-boost converter it can be seen that, when U_{in} is assumed to rise to some extend above the nominal technology voltage, also here six transistors are needed: two for SW_1 and SW_2 and one for SW_3 and SW_4. This restriction sets a lower limit for the capacitance

3.3 Step-Up/Down Converters

Table 3.15 The SPICE simulations results for the required capacitances C_1 and C_2, to comply with the specifications of Table 3.13, of an ideal SEPIC DC-DC converter, for four different output powers P_{out}. The required duty-cycle δ and the CM are also provided

P_{out}	C_1	C_2	$C_{tot} = C_1 + C_2$	δ	t_{off_real}	CM
1 mW	0.006 nF	0.19 nF	0.195 nF	2.1%	0.18 ns	DCM
10 mW	0.033 nF	1.8 nF	1.833 nF	7.5%	0.6 ns	DCM
100 mW	0.37 nF	16.5 nF	16.87 nF	25.5%	1.95 ns	DCM
1 W	2.5 nF	105 nF	107.5 nF	52.5%	4.75 ns	CCM

of C_1. The capacitance of C_2, on the other hand, is dominant for the value of ΔU_{out}. The inductances of L_1 and L_2 are respectively 3 nH and 7 nH. These values are chosen such that the total required capacitance C_{tot} is minimized.

From Table 3.15 it is concluded that the required C_{tot} is higher compared to a non-inverting buck-boost converter. This becomes more significant for high values of P_{out}.

To conclude the discussion on the ideal SEPIC DC-DC converter, a summary of the possible benefits and drawbacks is provided:

✔ The two switches each have one terminal connected to a fixed potential: SW_1 to GND and SW_2 to U_{out}. This simplifies the driver circuits for these switches.
✔ In CCM the input current is continuous, relaxing the specifications of the input decouple capacitor. In DCM the input current is also continuous, but it has a varying polarity. Thus, energy will be exchanged with U_{in}. However, as U_{in} may not be able to receive energy, the on-chip decouple capacitor should be dimensioned large enough to cope with this.
✔ The output is non-inverted and referred to the GND of the converter.
✘ The current delivered to the output filter is not continuous in neither CM, putting more stringent specifications to the output capacitor C_2.
✘ The total required capacitance C_{tot} is higher compared to a non-inverting buck-boost converter.
✘ In DCM two current-sensing is required for L_1 and L_2 in order to determine when the currents through L_1 and L_2 are equal.

3.3.5 Zeta Converter

The circuit of an ideal zeta[3] DC-DC converter is shown in Fig. 3.25. The zeta converter is capable of converting U_{in} into a non-inverted U_{out}, having either a lower or higher value than U_{in}. For CCM the relation between $k(\delta)$ and δ is given by (3.15), as it is equal to that of a non-inverting buck-boost converter. The graphical representation of (3.15) is shown in Fig. 3.22, where $k(\delta)$ is plotted as a function of δ.

---

[3]This converter is also referred to as inverse-SEPIC converter.

Fig. 3.25 The circuit of an ideal zeta DC-DC converter

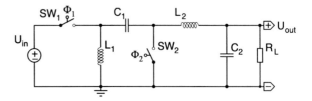

The basic operation of the zeta converter in CCM, with $U_{in} = U_{out}$, consists of the following two phases:

1. *The inductors charge phase* Φ_1: During Φ_1 SW_1 is closed and SW_2 is opened, during a time t_{on}. This causes L_1 and L_2 to be charged by U_{in} and C_1 respectively, thereby discharging C_1. Simultaneously, C_2 is charged and R_L is powered by U_{in} and C_1.
2. *The inductors discharge phase* Φ_2: During Φ_2 SW_2 is closed and SW_1 is opened, during a time t_{off}. This causes L_1 to discharge, thereby charging C_1. Simultaneously, L_2 and C_2 are also discharged, powering R_L.

In DCM the currents through L_1 and L_2 do not stop flowing and are allowed to change polarity, analogue to the SEPIC converter. In this CM SW_2 is opened at the moment when the currents through L_1 and L_2 are equal and flow towards the output. This prevents C_2 to be discharged through L_2. When both switches are opened C_2 continues to deliver power to R_L. At the same time L_1, L_2, C_1 and C_2 are connected in series, forming an ideal series LC-circuit as described in Sect. 2.3.2. In this LC-circuit a continuous exchange of energy takes place until the next switch cycle commences.

The results of the SPICE simulations for the ideal SEPIC converter, in order for it to comply with the specifications of Table 3.13, are listed in Table 3.16. All the parameters are chosen such that the total required capacitance is minimized, with the following restriction: the respective voltages over SW_1 and SW_2 are not allowed to rise above $3 \cdot U_{in}$. This restriction is made for the assumption that the converter will be implemented in a technology having U_{in} as nominal supply voltage. Therefore, SW_1 and SW_2 might be realized each by using three standard stacked-transistors, in total requiring six transistors. When comparing with the non-inverting buck-boost converter it can be seen that, when U_{in} is assumed to rise to some extend above the nominal technology voltage, also here six transistors are needed: two for SW_1 and SW_2 and one for SW_3 and SW_4. This restriction sets a lower limit for the capacitance of C_1. The capacitance of C_2 on the other hand is dominant for the value of ΔU_{out}. The inductances of L_1 and L_2 are respectively 3 nH and 7 nH. These values are chosen such that the total required capacitance C_{tot} is minimized.

From Table 3.16 it can be concluded that the required C_{tot} is about five times lower compared to a non-inverting buck-boost converter and a SEPIC converter, for $P_{out} = 1$ W. The reason for this fact is that a constant current is being delivered to the output, as the converter operates in CCM. For lower P_{out} the required C_{tot} is in the same order of magnitude compared to a non-inverting buck-boost converter

3.3 Step-Up/Down Converters

Table 3.16 The SPICE simulations results for the required capacitances C_1 and C_2, to comply with the specifications of Table 3.13, of an ideal zeta DC-DC converter, for four different output powers P_{out}. The required duty-cycle δ and the CM are also provided

P_{out}	C_1	C_2	$C_{tot} = C_1 + C_2$	δ	t_{off_real}	CM
1 mW	0.006 nF	0.26 nF	0.266 nF	2.15%	0.18 ns	DCM
10 mW	0.033 nF	2.8 nF	2.833 nF	7.7%	0.6 ns	DCM
100 mW	0.37 nF	14 nF	14.37 nF	26%	1.95 ns	DCM
1 W	2.5 nF	19.5 nF	22 nF	52.5%	4.75 ns	CCM

and a SEPIC converter. This is due to the fact that despite the current delivered to the output is still continuous, it can change polarity because the converter works in DCM at lower values of P_{out}.

To conclude the discussion on the ideal DC-DC zeta converter, a summary of the possible benefits and drawbacks is provided:

✔ Each of the two switches each has one terminal connected to a fixed potential: SW_1 to U_{in} and SW_2 to *GND*. This simplifies the driver circuits for these switches.
✔ The current delivered to the output is continuous in both CMs, relaxing the specifications for the output capacitor C_2.
✔ The output is non-inverted and referred to the *GND* of the circuit.
✔ The total required capacitance C_{tot} is roughly five times less compared to a non-inverting buck-boost converter and SEPIC converter, for $P_{out} = 1$ W. This yields a huge area reduction for high values of P_{out}.
✘ The input current is discontinuous in both CMs, putting more stringent specifications to the on-chip input decoupling.
✘ In DCM current-sensing is required for L_1 and L_2, to determine when these currents are equal.

3.3.6 Step-Up/Down Converter Summary

Table 3.17 summarizes the most significant properties and parameters of the five discussed DC-DC step-up/down converter topologies. No simulation data of C_{tot} is provided for the buck-boost converter and Ćuk converter, because of the fact that they produce an inverted U_{out}. They are added to Table 3.17 for sake of completeness and are no longer considered in this comparison. The first trade-off to be made is the number of components: a non-inverting buck-boost converter requires four switches, one capacitor and one inductor, whereas the SEPIC converter and zeta converter require two switches, two capacitors and two inductors. In terms of the conduction and switching losses caused by the switches, this puts the potentially highest power conversion efficiency advantage at the SEPIC converter and the zeta converter. However, the SPICE-simulations are performed such that the maximal allowable voltage over the switches of the SEPIC converter and zeta converter is

Table 3.17 The comparison of key properties and parameters of five types of step-down DC-DC converters, with respect to monolithic integration (✔ = yes, ✘ = no)

	Buck-boost	Non-inv. buck-boost	Ćuk	SEPIC	Zeta
# Switches	2	4	2	2	2
# Capacitors	1	1	2	2	2
# Inductors	1	1	2	2	2
Floating switches	✘	✘	✘	✘	✘
Inverted output	✔	✘	✔	✘	✘
Floating output	✘	✘	✘	✘	✘
Complex timing	✘	✘	✔	✔	✔
Continuous I_{out}	✘	✘	✔	✘	✔
Continuous I_{in}	✘	✘	✔	✔	✘
C_{tot} @ $P_{out} = 1$ mW	–	0.19 nF	–	0.195 nF	0.266 nF
C_{tot} @ $P_{out} = 10$ mW	–	1.7 nF	–	1.833 nF	2.833 nF
C_{tot} @ $P_{out} = 100$ mW	–	12 nF	–	16.87 nF	14.37 nF
C_{tot} @ $P_{out} = 1$ W	–	98 nF	–	107.5 nF	22 nF

$3 \cdot U_{in}$. It is also assumed that the nominal technology supply voltage equals U_{in}, implying that a total of six transistors is required to implement these switches. This is the same amount needed for the non-inverting buck-boost converter. Therefore, it can be concluded that, as a first-order approximation, the conduction and switching losses will be similar for the three converter topologies. A drawback for the SEPIC converter and zeta converter is that they require current-sensing for both inductors in DCM, implying more complex timing and control circuitry. Finally, it is observed that the non-inverting buck-boost converter does not provide a continuous current at either the input or the output. In contrast, the SEPIC converter has a continuous input current and the zeta converter a continuous output current, implying that either the input decoupling capacitor or the output filter capacitor of these converters can have relaxed specifications.

The required C_{tot} as a function of P_{out}, for the non-inverting buck-boost converter, the SEPIC converter and the zeta converter in order to comply with the specifications of Table 3.13, is plotted in Fig. 3.26. It is readily observed that the zeta converter has the advantage for $P_{out} > 100$ mW, reaching up to a factor five at $P_{out} = 1$ W. At lower values of P_{out} the non-inverting buck-boost converter is advantageous. However, it does not provide a continuous input current, as does the SEPIC converter. When taking the associated input decouple capacitor into account this advantage may be lost.

The practical implementations of this work, discussed in Chap. 6, do not comprise step-up/down converters. At the time of writing (Q4 2010), this type of monolithic DC-DC converter has not been demonstrated in the literature.

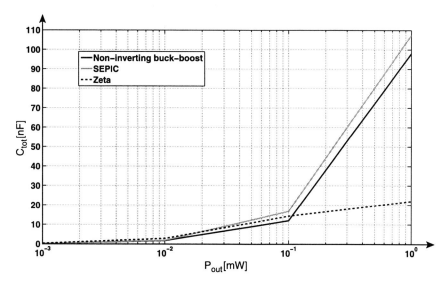

Fig. 3.26 The total required capacitance C_{tot} of three DC-DC step-up/down converter topologies as a function of the output power P_{out}. These values are obtained by means of SPICE-simulations, such that the three converters meet with the specifications of Table 3.13

3.4 Other Types of Inductive DC-DC Converters

From the primary classes of non-galvanically separated DC-DC converters, which are discussed in the respective Sects. 3.1, 3.2 and 3.3, two secondary classes of DC-DC converters can be derived. These are namely the galvanically separated DC-DC converters and the resonant DC-DC converters, of which a selection of commonly used circuit topologies is discussed in Sects. 3.4.1 and 3.4.2, respectively.

None of these DC-DC converter types have been practically realized in this work. Therefore, they will be subject to a brief discussion, giving the designer a sense of their potential for monolithic integration. A more general and conservative discussion on these converters can be found in [Eri04].

3.4.1 Galvanic Separated Converters

Galvanically separated DC-DC converters acquire a transformer to achieve galvanic[4] separation between the in- and output of the converter. This galvanic separation is required in most mains-operated[5] applications, as explained in Sect. 1.2.1, for the sake of safety.

---

[4] Named after the discoverer of galvanic electricity Luigi Galvani (1737–1798).

[5] For this purpose these converters are preceded by a (full-bridge) rectifier and smoothing capacitor.

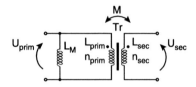

Fig. 3.27 The model of an ideal transformer Tr, together with its magnetizing inductance L_M

The calculation methods explained in Sect. 2.3.3 remain valid for galvanically separated DC-DC converters, providing the ideal model for a transformer is used. This model is shown in Fig. 3.27 and it consists of an ideal[6] transformer Tr, with n_{prim} and n_{sec} the respective number of primary and secondary winding turns, and a magnetizing inductance L_M. For the steady-state calculations L_M should comply with the volt-second balance and Tr merely converts the voltage over the primary winding U_{prim} into the secondary winding voltage U_{sec}. When assuming Tr to be ideal, this is done with the factor determined by the winding turn ratio n_{Tr}, as given by (3.16) and where L_{prim} and L_{sec} denote the respective inductances for the primary and the secondary winding.

$$U_{sec} = U_{prim} n_{Tr}|_{k_M=1} = U_{prim} \frac{n_{sec}}{n_{prim}}\bigg|_{k_M=1} = U_{prim} \sqrt{\frac{L_{sec}}{L_{prim}}}\bigg|_{k_M=1} \quad (3.16)$$

Monolithic transformers are feasible and can achieve fairly high magnetic coupling factors k_M in the order of 0.8–0.95 [Bio06]. k_M is calculated by (3.17), where M is the mutual inductance. The power conversion efficiency η_{Tr} of an ideal transformer, having no conduction or core losses, is proportional to k_M.

$$k_M = \frac{M}{\sqrt{L_{prim} L_{sec}}} \quad \text{with} \quad 0 \leqslant k_M \leqslant 1 \quad (3.17)$$

In the following sections a brief discussion is provided on some widely used galvanic separated DC-DC converter topologies, which are all derived from the primary class of DC-DC converters.

Derived Step-Down Converters

The circuit of an ideal forward DC-DC converter is shown in Fig. 3.30. It is derived from the buck converter, which is discussed in Sect. 3.1.1. The relation between $k(\delta)$ and δ for CCM is equal to that of the buck converter, given by (3.1), multiplied by ratio of the number of turns of the third and the first winding n_3/n_1, as stated in (3.22). The number of turns of the first and second winding are commonly made equal $n_1 = n_2$, limiting the maximal value of δ to 50%.

---

[6] A transformer is ideal when: $k_M = 1$ and $L_{prim}, L_{sec} \to \infty$.

3.4 Other Types of Inductive DC-DC Converters

④ ON COUPLED INDUCTORS

Figures 3.28(a) and (b) show the circuits for calculating the energy transfer of coupled inductors Tr, in the Laplace-domain. $i_{prim}(t)$ and $i_{sec}(t)$ are given by (3.18) and (3.19). The energy conversion efficiency $\eta_{Tr}(t)$ of Tr is given by (3.20). The steady-state energy E_{R_2} delivered to R_2 is given by (3.21).

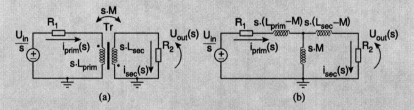

Fig. 3.28 (a) The circuit for calculating the energy transfer of Tr and (b) the equivalent T-circuit, both in the Laplace-domain

$$i_{prim}(s) = \frac{R_2 U_{in} + L_{sec} U_{in} s}{R_1 R_2 s + (L_{sec} R_1 + L_{prim} R_2)s^2 + (L_{prim} L_{sec} - M^2)s^3}$$
$$\implies i_{prim}(t) = \mathcal{L}^{-1} i_{prim}(s) \tag{3.18}$$

$$i_{sec}(s) = \frac{M U_{in}}{R_1 R_2 + (L_{sec} R_1 + L_{prim} R_2)s + (L_{prim} L_{sec} - M^2)s^2}$$
$$\implies i_{sec}(t) = \mathcal{L}^{-1} i_{sec}(s) \tag{3.19}$$

$$\eta_{Tr}(t) = \frac{E_{R_2}(t)}{E_{U_{in}}(t)} = \frac{\int_0^t U_{out}(t) i_{sec}(t)\,dt}{\int_0^t U_{in}(t) i_{prim}(t)\,dt} \tag{3.20}$$

$$E_{R_2} = \lim_{t \to \infty}\left(E_{R_2}(t)\right) = \frac{U_{in}^2}{R_1} \frac{M^2}{2(L_{sec} R_1 + L_{prim} R_2)} \tag{3.21}$$

$$k(\delta) = \frac{n_3}{n_1}\delta \quad \text{with} \quad \delta \leqslant \frac{1}{1 + \frac{n_2}{n_1}} \tag{3.22}$$

This converter is not particularly well suited for monolithic integration because of the fact that the transformer Tr requires three windings. This limits the number of metals-layers to be used in parallel for one winding, or it requires for the use of metals which are more close to the substrate. In this way the parasitic winding resistance and/or the parasitic substrate capacitance is increased, implying higher losses. In addition an inductor L is required at the output, increasing the area requirement of the converter causing additional losses.

Another galvanic separated variant of the buck converter is the full-bridge buck converter, of which the ideal circuit is shown in Fig. 3.31. The relation between $k(\delta)$ and δ in CCM is equal to that of the buck converter multiplied by n_{Tr}, as given by (3.23).

$$k(\delta) = n_{Tr}\delta \tag{3.23}$$

⊛ON COUPLED INDUCTORS (CONTINUED)

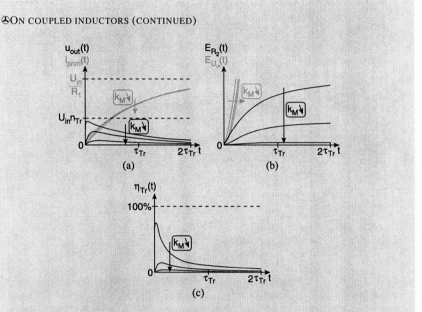

Fig. 3.29 (a) $i_{prim}(t)$ and $u_{out}(t)$, (b) $E_{U_{in}}(t)$ and $E_{R_2}(t)$ and (c) $\eta_{Tr}(t)$ as a function of time t, different values of coupling factor k_M

Figure 3.29(a) shows that $i_{prim}(t)$ roughly equal to (2.36), with $\tau_{Tr} = L_{prim}/R_1$. The maximal $u_{out}(t)$ occurs sooner and becomes larger for increasing k_M and is zero in steady-state. Figure 3.29(b) shows that $E_{U_{in}}(t)$ increases towards ∞, as i_{prim} keeps flowing. In steady-state $E_{R_2}(t)$ is defined by (3.21), which is proportional to k_M. Figure 3.29(c) shows that the maximum of $\eta_{Tr}(t)$ occurs sooner and becomes larger for higher values of k_M. Thus the primary winding of Tr should be disconnected from U_{in} at a certain fixed point in time, for obtaining the maximal $\eta_{Tr}(t)$.

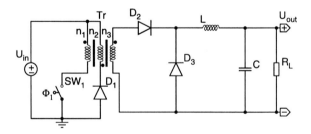

Fig. 3.30 The circuit of an ideal forward DC-DC converter

This converter exhibits a number of variants such as the half-bridge implementation, where SW_3 and SW_4 are replaced by capacitors, eliminating the need for D_3 and D_4. In this case a factor 0.5 is to be added in (3.23), as the voltage swing over the primary winding is only half of that of a full-bridge converter. The output recti-

3.4 Other Types of Inductive DC-DC Converters

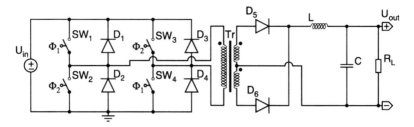

Fig. 3.31 The circuit of an ideal full-bridge buck DC-DC converter

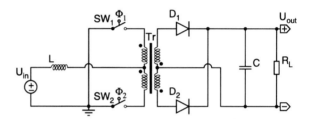

Fig. 3.32 The circuit of an ideal push-pull boost DC-DC converter

fier may also be implemented as a full-bridge rectifier, eliminating the need for the center-tapped secondary winding.

Due to the simplified transformer, compared to the forward converter, this converter has a better prospect towards monolithic integration. This is confirmed by [Del09], where a full-bridge variant with active full-bridge output rectification is implemented as a micro-converter. This micro-converter is only partly integrated on-chip and uses many non-standard techniques for passive integration such as: trench capacitors on silicon and micro-inductors/transformers with micro-machined magnetic cores.

Derived Step-Up Converters

The circuit of an ideal push-pull boost DC-DC converter is shown in Fig. 3.32. It is derived from the boost converter, which is discussed in Sect. 3.2.1. The relation between $k(\delta)$ and δ for CCM is equal to that of the boost converter, given by (2.62), multiplied by n_{Tr}, as stated by (3.24).

$$k(\delta) = n_{Tr} \frac{1}{1-\delta} \qquad (3.24)$$

Similar to the full-bridge converter this converter has potential towards monolithic integration. However, no practical monolithic implementations have yet been reported in the literature.

Fig. 3.33 The circuit of an ideal flyback DC-DC converter

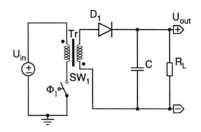

Derived Step-Up/Down Converters

The galvanic separated variant of the buck-boost converter, discussed in Sect. 3.3.1, is the flyback DC-DC converter of which the ideal circuit is shown in Fig. 3.33. The relation between $k(\delta)$ and δ for CCM is equal to that of a buck-boost converter multiplied by n_{Tr}. The inverting property of the buck-boost converter can obviously be chosen by altering the direction of the windings relative to one another.

$$k(\delta) = n_{Tr} \frac{\delta}{1-\delta} \qquad (3.25)$$

This converter has the simplest topology of the galvanic separated inductive DC-DC converters, only requiring one primary switch and one secondary rectifier (diode). The transformer is also very basic, as it does not require center-tapped windings. Also, the additional in/output inductor, acquired in the previously discussed galvanic separated converters, can be omitted in this converter. For these reasons the flyback converter proves to be suited for monolithic integration, which is confirmed by the practical realization of [Sav03]. However, this implementation suffers a relatively low power density of 50 mW/mm^2 and a low power conversion efficiency of 16.2%. The bottleneck of this type of converter is the parasitic series resistance of the air-core transformer's windings, which accounts for 40% of the total power losses of the converter.

3.4.2 Resonant DC-DC Converters

Figure 3.34 shows the circuit of an ideal series resonant DC-DC converter. This converter consists of a switch-network, formed by SW_1, SW_2, SW_3 and SW_4, to generate a square wave AC voltage. This AC voltage is fed into a series resonant LC-network, formed by L and C_1, whereafter it is fed into a full-bridge rectifier, formed by D_1, D_2, D_3 and D_4. The DC output voltage of the rectifier is filtered by output capacitor C_1 and is also the output voltage U_{out} of the converter. The voltage conversion ratio $k(f_{SW})$ of this converter is calculated by means of the transfer function $\|H(f_{SW})\|$ between the in- and output, which is given by (3.26) [Eri04].

3.4 Other Types of Inductive DC-DC Converters

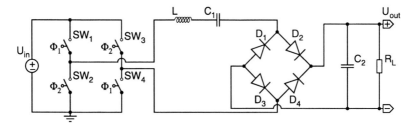

Fig. 3.34 The circuit of an ideal series resonant DC-DC converter

Fig. 3.35 The voltage conversion ratio $k(f_{SW})$ as a function of the switching frequency f_{SW}, for a series resonance DC-DC converter

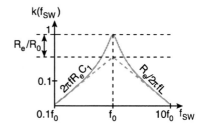

$$k(f_{SW}) = \|H(f_{SW})\| = \frac{1}{1 + \frac{R_0}{R_e}\left(\frac{f_0}{f_{SW}} - \frac{f_{SW}}{f_0}\right)^2}$$

with
$$\begin{cases} f_0 = \frac{1}{2\pi\sqrt{LC_1}} \\ R_0 = \sqrt{\frac{L}{C_1}} \\ R_e = R_L \frac{8}{\pi^2} \end{cases} \tag{3.26}$$

In (3.26) f_0 is the resonance frequency, R_e is equivalent load resistance seen from the AC input side of the full-bridge rectifier and R_0 is the output resistance at f_0 of the LC-network. The graphical representation of (3.26) is shown in Fig. 3.35, where $k(f_{SW})$ is plotted as a function of f_{SW}. It is observed that $k(f_{SW})$ varies with f_{SW} and that it can have values between zero and one. The maximum $k(f_{SW}) = 1$ occurs at $f_{SW} = f_0$ and this value decreases with either increasing or decreasing values of f_{SW}. Thus, as opposed to the inductive DC-DC converters discussed earlier in this chapter, U_{out} of resonant DC-DC converters is controlled by means of f_{SW} rather than δ. Please note that (3.26) is only valid when the odd harmonics of f_{SW}, produced by the switch-network, are sufficiently attenuated by the LC-network.

The monolithic variants of resonant DC-DC converters are not yet proven to be feasible in the literature. Nevertheless, from a practical point of view this can be possible, regarding some remarks. First, the switching network contains four switches, probably introducing significant losses. Also, the full-bridge rectifier will introduce significant losses, especially at low output voltages, which could be overcome by replacing the diodes by active MOSFETs. In the end, it is questionable if the sig-

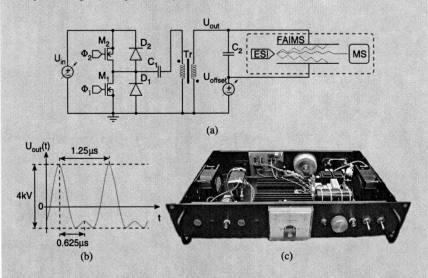

○ HIGH-FIELD ASYMMETRIC WAVEFORM ION MOBILITY SPECTROMETRY

The FAIMS technique is used for the separation of chemical components from a bulk sample. It is based on the unique non-linear mobility of ions at high electric fields [Gue10]. A mixture of ions, generated by means of Electro-Spray Ionization (ESI), is conducted between two parallel plates on which a high asymmetric AC voltage is applied, as illustrated in Fig. 3.36(b). A DC-offset voltage enables selectivity for a certain chemical compounds, thereby deflecting the other compounds towards the plates. The selected compound is then analyzed through Mass Spectrometry (MS).

Fig. 3.36 (a) The circuit of a halve-bridge galvanic separated series resonance DC-AC high-voltage converter for the FAIMS setup. (b) U_{out} of the DC-AC converter as a function of t. (c) A photograph of the realization of the DC-AC converter

The circuit of the resonant DC-AC converter which generates the FAIMS's high-voltage is shown in Fig. 3.36(a) and a photo of the implementation is shown in Fig. 3.36(c). The converter achieves a peak-to-peak output voltage U_{out_ptp} of 4 kV at a fundamental frequency of 800 kHz.

nificant amount of additional required components can compete with a simple buck converter, having the same functionality. For this reason monolithic integration of this converter is not regarded practically useful in this work.

Finally, it is noted that step-up/down functionality can also be achieved by means of a parallel resonant LC-network, or the combination of parallel and series resonant networks. Also, galvanic separated variants exits, using the primary winding of a transformer in the LC or LCC-network. These converters are widely used for driving Compact Fluorescent Lamps (CFL). The are also used for multiple-output mains power supplies, in combination with transformers with multiple secondary windings.

Fig. 3.37 The concept of multi-phase DC-DC converters

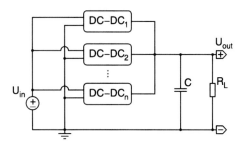

3.5 Topology Variations

The primary classes of inductive DC-DC converters, discussed in Sects. 3.1, 3.2 and 3.3, can be adopted, yielding two new subclasses. These subclasses are the multi-phase and the Single-Inductor Multiple-Output (SIMO) DC-DC converters, which are discussed in the respective Sects. 3.5.1 and 3.5.2. A new variant of a SIMO converter, intended for monolithic DC-DC converters, is discussed in Sect. 3.5.3.

3.5.1 Multi-phase DC-DC Converters

The concept of multi-phase DC-DC converters is illustrated in Fig. 3.37. Each of the DC-DC-stages contains a switching-stage and energy-storing elements, except for the output capacitor C which is shared by the DC-DC stages. The multi-phase setup has several possible advantages compared to a single-phase setup, which may be combined with one another:

1. *Output power P_{out}*: For single-phase converter the maximal P_{out} is limited, thus by adding n phases the total maximal P_{out} can be increased with this factor n. The power conversion efficiency will ideally not be affected.
2. *Power conversion efficiency η_{SW}*: η_{SW} is a function of P_{out} and will tend to drop above a certain P_{out} level, for a non-ideal converter. Adding DC-DC stages can enable the individual stages to be operated more closely to their optimal power conversion efficiency point η_{SW_max}, yielding an increased overall η_{SW}. This will of course result in an area A versus η_{SW} trade-off. This methodology is illustrated in Fig. 3.38 for a two-phase versus single-phase DC-DC converter, where the two-phase converter will achieve a higher η_{SW_max} than the single-phase converter, for the same P_{out}.
3. *Area A*: The DC-DC stages are generally operated out of phase of one another, distributing the in- an output current in the time-domain. For a certain specification of the in- and output voltage ripple this will result in relaxed specifications of the in- and output capacitor, compared to a single-phase converter with the same P_{out}. In other words the required A for these components will be reduced. However, the additional inductor(s) and switching-stage(s) will also require more A. Whether or not this will result in a decreased total A requirement will be investigated for the boost and the buck converter in the following sections.

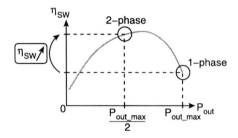

Fig. 3.38 The example of how a two-phase DC-DC converter can achieve a higher power conversion efficiency η_{SW} than a single-phase DC-DC converter, at the same output power P_{out}

4. *Output voltage ripple* ΔU_{out}: The previous trade-off can also be acquired to obtain a lower ΔU_{out} in a multi-phase converter, compared to its single-phase equivalent with the same output capacitor.

It is clear that these possible advantages are mutually non-exclusive, nevertheless they are all contradictory to each other. This will result in important design trade-offs, which will be deducted in Chap. 4.

Multi-phase DC-DC converters can also be implemented using coupled inductors, their feasibility for the purpose of monolithic integration has been demonstrated in [Wib08], which uses a capacitive coupled two-phase buck DC-DC converter topology. This topology has the advantage of eliminating ΔU_{out} for the entire voltage conversion ratio range, assuming ideal converter components. Therefore, it can achieve high power densities because the required capacitance for the output capacitor is minimized. Nevertheless, the achieved power density of this implementation is smaller than achieved with a standard four-phase buck converter [Wen09b]. Other possible advantages of coupled inductors in multi-phase topologies is the minimization of the reverse recovery losses in the rectifiers and the reduced phase current unbalance [Lee00]. However, the phase current unbalance in monolithic converters proves not to be a significant problem in monolithic DC-DC converters, because it is largely due to a physical unbalance between the different phases [Eir08]. This problem is more stringent for converters with external components and it can be largely avoided in monolithic converters by respecting the rules of symmetry in the chip lay-out. A more detailed discussion on inductive DC-DC converters with coupled inductors is omitted in this work, as it is virtually impossible to achieve significant magnetic coupling between more than two inductors on-chip, without compromising the quality-factor (Q) of the inductors.

Output Voltage Ripple in Multi-phase Converters

One of the possible benefits of multi-phase DC-DC converters is their ability to achieve a lower ΔU_{out} for the same P_{out}, providing that the individual DC-DC-stages are operated out of phase of each other. The following discussion will explain a mathematical method for the general approximation of ΔU_{out} of n-phase DC-DC converters in CCM. The method assumes that the phase-offset between the consecutive DC-DC-stages is equal to $2 \cdot \pi / n$ rad and that the mathematical relation between ΔU_{out} and δ is known for a single-phase converter.

3.5 Topology Variations

Fig. 3.39 (a) The timing signals of a two-phase converter and (b) the equivalent representation with sine waves, assuming that the converter is operating in CCM

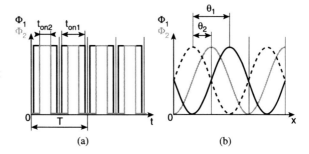

Consider the timing signals of a two-phase DC-DC converter, as illustrated in Fig. 3.39(a), where both converters are operated with a phase-offset of π rad. The gray timing signal represents a value of δ much smaller than 50% and for the black timing signal the value of δ is close to 50%. When assuming that the output filter of the DC-DC converter only passes the fundamental frequency f_{SW} of the switch-network, $u_{out}(t)$ of a single-phase converter is a pure sine wave. It follows that the total $u_{out}(t)$ of the two-phase converter will be the sum of two sine-waves. Therefore, it can intuitively be understood that ΔU_{out} will be zero when the value of δ of the individual DC-DC-stages is equal to 50% and that ΔU_{out} will have a finite value for all other values of δ. Hence, the physical effect of the value of δ on ΔU_{out} can be modeled by the phase difference θ between the two sine waves, as illustrated in Fig. 3.39(b). This leads to the projection of δ onto θ, which is stated by (3.27).

$$\delta : 0\% \rightarrow 100\%$$
$$\Downarrow$$
$$\theta : 0 \text{ rad} \rightarrow 2 \cdot \pi \text{ rad} \tag{3.27}$$

The approximation of $u_{out}(t)$ is calculated by the sum of two sine waves, having a phase difference as given by (3.27), resulting in $u_{out}(x)$ which is given by (3.28).

$$u_{out}(x) = \sin(x) + \sin(x + \theta) \tag{3.28}$$

The value of ΔU_{out} is approximated by the amplitude of (3.28): $\hat{u}_{out}$. This is calculated by equating the derivative of $u_{out}(x)$ to zero, yielding (3.29).

$$\hat{u}_{out}(x) = \frac{du_{out}(x)}{dx} = \cos(x) + \cos(x + \theta) = 0$$
$$\implies x = \arccos\left(\frac{\sin(\theta)}{2\sqrt{\cos^2\left(\frac{\theta}{2}\right)}}\right) \tag{3.29}$$

The locus of $\hat{u}_{out}$ as a function of θ is calculated by substituting x from (3.29) and replacing it in (3.28), after simplification this yields (3.30).

$$\hat{u}_{out}(\theta) = \sqrt{2}\sqrt{1 + \cos(\theta)} \tag{3.30}$$

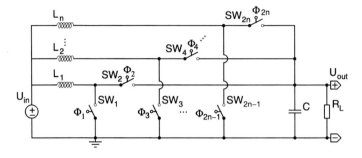

Fig. 3.40 The circuit of an ideal n-phase boost DC-DC converter

Finally, the combination of (3.30) and (3.27) yield the factor Υ, given by (3.31). Υ is to be multiplied with the function $\Delta U_{out}(\delta)$ of any given single-phase converter, resulting in the approximation of $\Delta U_{out}(\delta)$ for its n-phase equivalent, were n is a power of 2. The factor K is dependent on the form-factor of the current trough the output capacitor of the converter. It is equal to 4 for a square waveform (e.g. boost converter) and it is equal to 2 for a triangle waveform (e.g. buck converter).

$$\Upsilon = \prod_{i=1}^{n'} \frac{1}{K}\sqrt{2}\sqrt{1+\cos(\pi 2^i \delta)} \quad \text{with} \quad n = 2^{n'} \blacksquare \qquad (3.31)$$

This approximation for ΔU_{out} will be validated for a boost and a buck converter in the following sections. Please note that this approximation is only valid for CCM. For DCM the method for calculating ΔU_{out} also depends on the converter topology and a general analytical method does not exist. Moreover, it will strongly depend on the type of control strategy, as will be discussed in Chap. 5.

Multi-phase Boost Converter

Figure 3.40 shows the circuit of an ideal n-phase boost DC-DC converter, requiring n inductors, $2 \cdot n$ switches and one output capacitor. The following discussion will show the calculation method for ΔU_{out} of a two-phase boost converter example, in CCM.

The calculation of ΔU_{out} for a boost converter requires the knowledge of $i_C(t)$, as explained in Sect. 2.3.3. For this purpose, $i_C(t)$ is graphically represented in Fig. 3.41(a) and (b) for the respective cases where $\delta < 50\%$ and $\delta > 50\%$. First, the case for $\delta < 50\%$ is discussed. It is observed in Fig. 3.41(a) that for this case the currents from L_1 and L_2 show an overlap during the discharge phase of the inductors. During t_{on} of each DC-DC-stage $i_C(t)$ is given by (3.32).

$$\delta < 50\%: \quad 0 \to t_{on} \implies i_C(t) = \frac{I_{L\_max} + I_{L\_min}}{2} - \frac{\overline{U_{out}}}{R_L} \qquad (3.32)$$

3.5 Topology Variations

Fig. 3.41 (a) The current $i_C(t)$ through the output capacitor C of a 2-phase boost converter for $\delta < 50\%$ and (b) for $\delta > 50\%$, both valid for CCM. $i_C(t)$ is divided into the respective parts from the first (*black curve*) and second converter (*gray curve*)

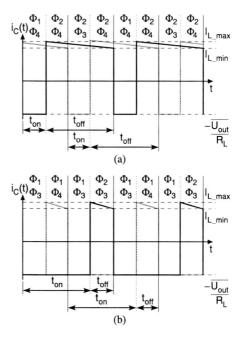

In (3.32) only the parameters I_{L_max} and I_{L_min} are unknown. They can be solved from the system of equations formed by the expressions for $\overline{I_{out}}$ and ΔI_L, as stated by (3.33).

$$\begin{cases} \overline{I_{out}} = 2\dfrac{I_{L\_max} + I_{L\_min}}{2}(1-\delta) = \dfrac{\overline{U_{out}}}{R_L} \\ \Delta I_L = I_{L\_max} - I_{L\_min} = \dfrac{U_{in}}{L}t_{on} \end{cases} \quad (3.33)$$

By means of (3.32) and (3.33) ΔU_{out} can be found, using the same method as (2.65). Finally, substituting $\overline{U_{out}}$ by (2.62), yields (3.34).

$$\delta < 50\%: \quad \Delta U_{out} = -\frac{1}{C}\int_0^{t_{on}} i_C(t)\,dt = \frac{t_{on}}{C}\left(\frac{\overline{U_{out}}}{R_L} - \frac{I_{L\_max} + I_{L\_min}}{2}\right)$$

$$= \frac{TU_{in}(\delta - 2\delta^2)}{2CR_L(\delta-1)^2} \blacksquare \quad (3.34)$$

For the second case where $\delta > 50\%$, as illustrated in Fig. 3.41(b), the currents from L_1 and L_2 do not overlap during their respective discharge phase. Thus, $i_C(t)$ is calculated through (3.35).

$$\delta > 50\%: \quad 0 \to \frac{t_{on} - t_{off}}{2} \implies i_C(t) = \frac{\overline{U_{out}}}{R_L} \quad (3.35)$$

Again by using (3.35) in (2.65) and replacing $\overline{U_{out}}$ by (2.62), yields (3.36).

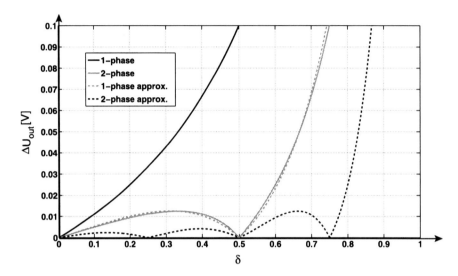

Fig. 3.42 The output voltage ripple ΔU_{out} as a function of the duty-cycle δ for an ideal 1-phase, 2-phase and 4-phase boost DC-DC converter. For the 1-phase and 2-phase boost converter both the exact and approximated functions are plotted. For the 2-phase and 4-phase boost converter the approximated functions are plotted

$$\delta > 50\%: \quad \Delta U_{out} = -\frac{1}{C}\int_0^{\frac{t_{on}-t_{off}}{2}} i_C(t)\,dt = \frac{\overline{U_{out}}}{CR_L}\left(\frac{t_{on}-t_{off}}{2}\right)$$
$$= \frac{TU_{in}(1+2\delta)}{2CR_L(1-\delta)} \blacksquare \tag{3.36}$$

To conclude the discussion on the multi-phase boost converter, ΔU_{out} is plotted as a function of δ in CCM, which is shown in Fig. 3.42, for the values of Table 3.8 and for $C = 25$ nF. In this figure (2.65) is plotted for a single-phase boost converter (black curve) and the positive parts of (3.34) and (3.36) yield the curve for a two-phase boost converter (gray curve). The curve for a two-phase boost converter can also be approximated by the product of (2.65) with (3.31) for $n = 2$, which is shown by the gray dotted curve. It can be seen that both the exact curve and the approximated curve form a good match. By using this approximation a four-phase converter can also easily be plotted, which is shown by the black dotted curve.

It is clear that the use of multiple phases dramatically reduces ΔU_{out} and that ΔU_{out} will even become zero at certain values for δ, increasing with the number of phases. The fact that ΔU_{out} is not dependent on the inductances of the inductors implies that multi-phase boost converters can achieve a significant area reduction, because the value of C can be smaller for the same ΔU_{out}, at constant P_{out}. This is of course only the case for CCM and will therefore prove to be of limited use for the purpose of monolithic integration, since these converters will tend to perform more efficient at DCM (see Chap. 4). In DCM the positive effect of multi-phase converters on ΔU_{out} will strongly depend on the control strategy, which is discussed

3.5 Topology Variations

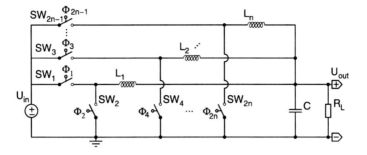

Fig. 3.43 The circuit of an ideal n-phase buck DC-DC converter

in Chap. 5. Notice that multi-phase monolithic DC-DC step-up converters have not yet been proven in the literature.

Multi-phase Buck Converter

Figure 3.43 shows the circuit of an ideal n-phase DC-DC buck converter, using n inductors, $2 \cdot n$ switches and one output capacitor. The following discussion will show the calculation method for ΔU_{out} of a two-phase buck converter example, in CCM.

The calculation of ΔU_{out} for a buck converter requires the knowledge of the total current ripple ΔI_{L_tot} through the inductors L_1 and L_2, as explained in Sect. 3.1.1. Therefore, $i_{L\_1}(t)$ and $i_{L\_2}(t)$ are plotted in Fig. 3.44(a) and (b) for the respective cases where $\delta < 50\%$ and $\delta > 50\%$. First, the case for $\delta < 50\%$ is considered. It can be seen in Fig. 3.44(a) that ΔI_{L_tot} is given by (3.37).

$$\delta > 50\%: \quad \Delta I_{L\_tot} = \Delta I_{L\_1} + \Delta I_{L\_2} = \frac{t_{off} - t_{on}}{2} 2 \frac{U_{out}}{L} \qquad (3.37)$$

By using (3.37) with the same method used to calculate (3.7) ΔU_{out} for a two-phase buck converter is obtained, as given by (3.38).

$$\delta < 50\%: \quad \Delta U_{out} = \frac{\Delta Q}{C} = \frac{\Delta I_{L\_tot} T}{8C} = \frac{(\delta - 2\delta^2)T^2 U_{in}}{8CL} \blacksquare \qquad (3.38)$$

For the second case where $\delta > 50\%$, as shown in Fig. 3.44(b), ΔI_{L_tot} is calculated through (3.39).

$$\delta > 50\%: \quad \Delta I_{L\_tot} = \Delta I_{L\_1} + \Delta I_{L\_2} = \frac{t_{on} - t_{off}}{2} 2 \frac{U_{in} - U_{out}}{L} \qquad (3.39)$$

Similar to (3.38), (3.39) yields ΔU_{out} for $\delta > 50\%$, which is given by (3.40).

$$\delta > 50\%: \quad \Delta U_{out} = \frac{\Delta Q}{C} = \frac{\Delta I_{L\_tot} T}{8C} = \frac{(\delta - \delta^2)T^2 U_{in}}{8CL} \blacksquare \qquad (3.40)$$

The discussion on the two-phase buck converter is concluded with the graphical representation of ΔU_{out} as a function of δ for CCM, which is illustrated in Fig. 3.45.

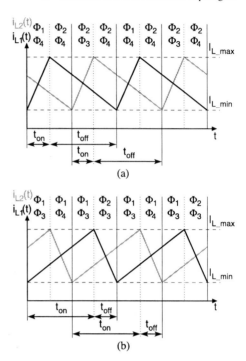

Fig. 3.44 (a) The respective currents $i_{L1}(t)$ and $i_{L2}(t)$ through inductors L_1 and L_2 of a 2-phase buck converter for $\delta < 50\%$ and (b) for $\delta > 50\%$, both valid for CCM

The parameters used for this plot are the same as in Table 3.1 and for $C = 12.5$ nF. The black curve is valid for a single-phase buck converter, as given by (3.7). The gray curve is the plot for a two-phase buck converter, of which the two sections are given by the positive values of (3.38) and (3.40). The approximation of ΔU_{out} for the two-phase buck converter, calculated by multiplying (3.7) with (3.31) for $n = 2$, is shown by the gray dotted curve. It can be seen that this approximation shows a good correspondence with the exact curve. The approximation for a four-phase buck converter is calculated similar to the two-phase converter, but for $n = 4$, and it is shown by the black dotted curve.

Clearly, the use of multi-phase buck converters can also dramatically reduce ΔU_{out}, especially for certain values of δ. However, opposed to the multi-phase boost converter, ΔU_{out} for the multi-phase buck converter is inverse proportional to the inductances of the inductors. Thus, for maintaining the same area requirement as for a single-phase converter the inductance per phase should be approximately be the n-times smaller than for the single-phase equivalent. This is not represented in Fig. 3.45, where the value of the individual inductances is always 10 nH. When applying the fixed total inductance of 10 nH, the curves of Fig. 3.45 will all reach the same maximal amplitude compared to the single-phase converter. This implies that there will still be an advantage for ΔU_{out} for certain values of δ, but for other certain values of δ the advantage might become very small and even non-existing. It is clear that these situations should be avoided. Note that, analogue to the boost converter, monolithic (multi-phase) buck converters will tend to be operated in DCM, due to the associated increased efficiency. In DCM the ΔU_{out} reduction will strongly

3.5 Topology Variations

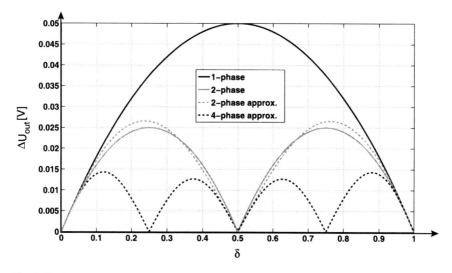

Fig. 3.45 The output voltage ripple ΔU_{out} as a function of the duty-cycle δ for an ideal 1-phase, 2-phase and 4-phase buck DC-DC converter. For the 1-phase and 2-phase buck converter both the exact functions are plotted. For the 2-phase and 4-phase buck converter the approximated functions are plotted

depend on the used control strategy. A more in depth discussion on this topic is therefore provided in Chap. 5.

Monolithic multi-phase DC-DC buck converters are proven feasible in literature, where [Abe07] discusses a two-phase implementation and [Wen09b] discusses a four-phase implementation. It shows these multi-phase converters are capable of achieving a high value P_{out} of 800 mW and that they are able to achieve a high power density of 213 mW/mm^2 [Wen09b]. At the time of writing these are the highest values reported in the literature for monolithic DC-DC converters in a standard CMOS technology. This converter, along with another multi-phase buck converter implementations, is discussed in Chap. 6.

3.5.2 Single-Inductor Multiple-Output DC-DC Converters

The concept of Single-Inductor Multiple-Output (SIMO) DC-DC converters is shown in Fig. 3.46. It comprises a single DC-DC-stage and multiple outputs, each separated from both the DC-DC-stage and from one another by means of switches. Each of the individual outputs also has a dedicated output capacitor.

The two main methods for distributing the delivered energy of the DC-DC-stage over the multiple outputs are [Kwo09]:

1. *Dedicated charge/discharge cycle*: The inductor of the DC-DC stage is consecutively charged by U_{in} and discharged through one of the outputs.

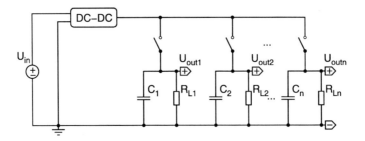

Fig. 3.46 The concept of Single-Inductor Multiple-Output (SIMO) DC-DC converters

2. *Shared charge/discharge cycles*: The inductor of the DC-DC-stage is first charged and then consecutively discharged through the different outputs.

Many variants of these two methods are possible, such as: the CM, the order in which to charge the outputs, Pulse Width Modulation (PWM) or Pulse Frequency Modulation (PFM) switching-schemes... Control methods suited for monolithic integration are discussed in Chap. 5. It is also possible to combine the multi-phase concept, explained in Sect. 3.5.1, with the SIMO concept to increase P_{out}, decrease ΔU_{out}, or combinations of both. An implementation example of such a combination is given in Chap. 6. Finally, it is noted that galvanic separated DC-DC converters, which are discussed in Sect. 3.4.1, are intrinsically suited for the SIMO concept, just by adding additional secondary windings [Ma03]. Clearly this method allows for only one output to be controlled in terms of P_{out}, since the separate secondary windings still share the same (magnetic) core.

In the following sections a SIMO boost and SIMO buck converter are briefly discussed.

SIMO Boost Converter

Figure 3.47 shows the circuit of an ideal SIMO boost DC-DC converter with n outputs. The converter requires one inductor, $n+1$ switches and n output capacitors. Similar to a standard boost converter, the SIMO boost converter draws a continuous input current in CCM and the currents delivered to the outputs are always discontinuous. Also, all the switches and outputs are non-floating.

It can be concluded that this converter is suited for the purpose of monolithic integration. It has the advantage of being able to deliver multiple output voltages with only one inductor. This can certainly be beneficial for the area requirement, compared to separate single-output converters. The main drawback will obviously be the fact that the limited amount of energy, which can be stored in the inductor is to be divided over the outputs. This type of converter has not yet been reported in the literature as being practically realized in its monolithic form. Nevertheless, a practical realization performed in this work is discussed in Chap. 6.

3.5 Topology Variations

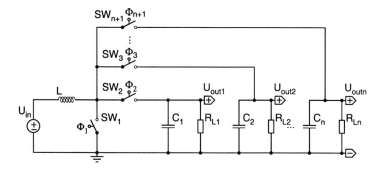

Fig. 3.47 The circuit of an ideal SIMO boost DC-DC converter with n outputs

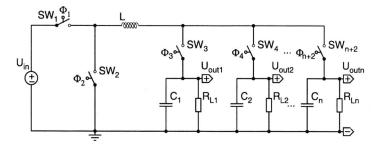

Fig. 3.48 The circuit of an ideal SIMO buck DC-DC converter with n outputs

SIMO Buck Converter

Figure 3.48 shows the circuit of an ideal SIMO buck DC-DC converter with n outputs. This converter incorporates one inductor, $n + 2$ switches and n output capacitors. Similar to a standard buck converter the SIMO buck converter draws a discontinuous input current in CCM, but opposed to a standard buck converter the currents delivered to the outputs are also always discontinuous. This implies that the individual output capacitors will require an increase capacitance, compared to a single-output buck converter with the same P_{out} and ΔU_{out}. Also, during the respective discharge cycles of the inductor through the outputs, two switches are in the current path: SW_1 and one of the output-select switches SW_3–SW_n. This will cause increased conduction losses due to these switches.

Apart from the drawbacks of this converter, compared to its single output equivalent, it is still considered suited for the purpose of monolithic integration. The area advantage over n single output converters will be lower, compared to the SIMO boost converter, because of the discontinuous output current. Also, the power conversion efficiency will tend to be somewhat lower compared to its single-output equivalent, due to the increased conduction losses. To the author's knowledge this type of converter has not yet been reported in its monolithic form in the literature.

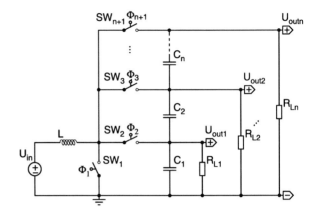

Fig. 3.49 The circuit of an ideal DC-DC boost Series Multiple Output Converter (SMOC) with n outputs

3.5.3 On-Chip Topologies

The SIMO converters, discussed in Sect. 3.5.2, all have an intrinsic disadvantage for the purpose of monolithic integration. Namely, when one or more of their output voltages is higher than the nominal supply voltage of the used IC-technology, the output capacitors will have to be implemented as the series connection of two or more capacitors. The reason for this is that the output capacitors are commonly implemented as Metal-Insulator-Metal (MIM) capacitors or as Metal-Oxide-Semiconductor (MOS) capacitors, which have a maximum operating voltage which in most is cases limited to the nominal technology supply voltage.

The following two sections provide the solution for this area-consuming problem: the Series Multiple Output Converter (SMOC). This topology variation will be discussed for a boost and a buck converter in the following sections. Please note that a similar topology variation has been proposed in [Nam09]. However, [Nam09] uses asynchronous freewheeling diodes rather than synchronous switches and it is intended for a different area of applications, incorporating external components and high voltages (few hundred volts).

SMOC Boost Converter

Figure 3.49 shows the circuit of an ideal boost SMOC with n outputs. This converter requires the same number components as a regular SIMO boost converter. In a standard CMOS implementations the required number of output capacitors is reduced with the SMOC topology, compared to the SIMO topology. This reduction can be understood by means of a simple example, comparing a two-output SIMO and SMOC converter in CCM, which is provided in Table 3.18. For this example it is assumed that all the in- and output parameters for the two converters are equal. It is also assumed that $P_{out\_1} = P_{out\_2}$, implying that the output capacitor C_{out_2} of output number two can have half the capacitance of the output capacitor C_{out_1} of output number one, for the same value of ΔU_{out}. This can be understood through

3.5 Topology Variations

Table 3.18 The comparison of the required total output capacitance for a two-output boost SIMO and SMOC converter in CCM

	SIMO	SMOC
U_{out_1}	U_{dd}	U_{dd}
U_{out_2}	$2 \cdot U_{dd}$	$2 \cdot U_{dd}$
C_{out_1}	C	C
C_{out_2}	$C/2$	$C/2$
#C_{out_1}	C	C
#C_{out_2}	$2 \cdot C$	C
#C_{out_tot}	$3 \cdot C$	$2 \cdot C$

(2.65) for CCM, as $\overline{I_{out\_1}} = 2 \cdot \overline{I_{out\_2}}$. Furthermore, it is assumed that the maximum operating voltage of a capacitor is equal to U_{dd}. The difference between the two converters is found in the total required capacitance #C_{out_2} to implement C_{out_2}. For the SIMO converter, having a dedicated output capacitor for each output, two capacitors each having a capacitance of twice the required capacitance for C_{out_2} need to be connected in series. The SMOC converter on the other hand, requires only one capacitor having twice the required capacitance. Thus, it can be concluded that the SMOC converter in this example achieves an area reduction of 1/3 for the output capacitors.

The conclusion for the novel boost SMOC topology is that it is better suited for monolithic integration compared to a SIMO boost converter topology, because it can achieve the same specifications with less area. Obviously, this statement is only true for the condition that one or more of the output voltages is higher than the nominal maximum operating voltage of the capacitors. Moreover, this area reduction will become more significant when the output voltages become higher, requiring more capacitors to be placed in series. The example from Table 3.18 is only valid for CCM, but the proof that this statement is also valid for DCM is trivial. However, for that case the area reduction is more dependent on in- and output parameters and it is therefore omitted. Note that a monolithic variant of this converter has not yet been achieved.

SMOC Buck Converter

Figure 3.50 shows the circuit of an ideal buck SMOC with n outputs. This converter requires the same number components as a regular SIMO buck converter. Similar to the boost SMOC the buck SMOC is able to achieve a reduction of the required total output capacitance in a standard CMOS implementations, compared to the SIMO topology. This reduction is again discussed by means of a simple example, comparing a two-output SIMO and SMOC converter in CCM, which is provided in Table 3.19. For this example it as assumed that all the in- and output parameters for the two converters are equal. It is also assumed that $U_{in} = 3 \cdot U_{dd}$, $U_{out\_1} = U_{dd}$ and $U_{out\_2} = 2 \cdot U_{dd}$, implying that the two output capacitors C_{out_1} and C_{out_2} have the same capacitance for the same value of ΔU_{out}. This can be understood through (3.7)

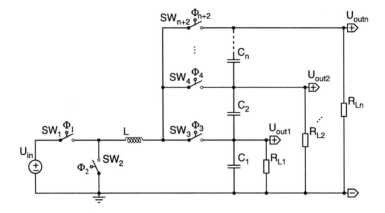

Fig. 3.50 The circuit of an ideal DC-DC buck Series Multiple Output Converter (SMOC) with n outputs

Table 3.19 The comparison of the required total output capacitance for a two-output buck SIMO and SMOC converter in CCM

	SIMO	SMOC
U_{out_1}	U_{dd}	U_{dd}
U_{out_2}	$2 \cdot U_{dd}$	$2 \cdot U_{dd}$
C_{out_1}	C	$2 \cdot C$
C_{out_2}	C	C
$\#C_{out\_1}$	C	$2 \cdot C$
$\#C_{out\_2}$	$4 \cdot C$	$2 \cdot C$
$\#C_{out\_tot}$	$5 \cdot C$	$4 \cdot C$

for CCM, which yields the same value for both $\delta = 1/3$ and $\delta = 2/3$. Furthermore, it is assumed that the maximum operating voltage of a capacitor is equal to U_{dd}. The difference between the two converters is found in the total required capacitance $\#C_{out\_2}$ to implement C_{out_2}. For the SIMO converter, having a dedicated output capacitor for each output, two capacitors each having a capacitance of twice the required capacitance for C_{out_2} need to be connected in series. The SMOC converter on the other hand, requires only one capacitor having twice the required capacitance. Thus, it can be concluded that the SMOC converter in this example achieves an effective area reduction of 1/5 for the output capacitors.

It is concluded that the new buck SMOC topology is better suited for monolithic integration compared to a SIMO buck topology, because it can achieve the same specifications with less area. Obviously, this statement is only true for the condition that one or more of the output voltages is higher than the nominal maximum operating voltage of the capacitors. Although the actual area reduction of the buck SMOC in CCM is less compared to the boost SMOC, there is also an inherent advantage of the buck SMOC. It turns out that C_{out_1} is twice the required value when using the SMOC topology, therefore ΔU_{out_1} will be only half of the value achieved with the SIMO topology. Analogue to the boost SMOC, the area reduction of the buck SMOC will become more significant when the output voltages become higher, requiring more capacitors to be placed in series. The example from Table 3.19 is only valid for CCM, but the proof that this statement is also valid for DCM is trivial. However, for that case the area reduction is more dependent on in- and output parameters and is therefore omitted. A monolithic realization of this converter, combined with multi-phase, is discussed in Chap. 6.

3.6 Conclusions

In this chapter the most used inductive DC-DC converter topologies are discussed, with respect to monolithic integration.

First, the primary classes of step-down, step-up and step-up/down converters are discussed in the respective Sects. 3.1, 3.2 and 3.3. Within each primary class a comparison is performed on various topologies, with respect to the output voltage ripple. This leads to the conclusion that for a maximal output power and a minimal reacquired area the following topologies are best suited:

- *Step-down*: Buck converter.
- *Step-up*: Boost converter.
- *Step-up/down*: Zeta converter.

Secondly, two other types of inductive DC-DC converters are discussed in Sect. 3.4:

- *Galvanic separated converters*: These converters require an on-chip transformer, which has two intrinsic drawbacks. The first drawback is that the energy conversion efficiency is proportional to the mutual coupling factor between the windings, which is never maximal. This causes the theoretical power conversion efficiency to be lower than 100%. The second drawback is the fact that the series resistance of the windings will be higher, compared to an inductor occupying the same area. This will cause increased losses. The flyback converter is considered best fitted for monolithic integration, as it requires merely one switch and one rectifier. Thus, suffering minimal losses in these components.
- *Resonant converters*: These converters are not considered practical for monolithic integration as their switch-network and rectifiers will cause high losses.

Thirdly, topology variations on the primary class of DC-DC converters are discussed in Sect. 3.5.

- *Multi-phase topologies*: These converters can be used to: increase the output power, to increase the power conversion efficiency, to decrease the output voltage ripple and/or increase the power density. All of these potential benefits are all contradictory to one another, but not mutually exclusive. Therefore, multi-phase converters are considered promising for monolithic integration.

- *SIMO converters*: These converters have multiple outputs, using only one inductor. For a certain limited output power they will require less area than a combination of single-output converters.
- *SMOC topologies*: These converters have the same functionality as SIMO converters. They require less area by placing the output capacitors in series, limiting the voltage across each capacitor to its nominal maximum supply voltage.

Chapter 4
A Mathematical Model: Boost and Buck Converter

Monolithic DC-DC converters are characterized by low values for the capacitors (nF) and inductors (nH), which are at least three orders of magnitude lower than their off-chip equivalents. This is because these values are proportional to their occupied chip-area, which is to be minimized. As a result, the switching frequency will also be three orders of magnitude higher compared to converters with external components, in the order of hundred MHz. This high switching frequency will introduce significant switching-losses due to several parasitic effects. Another important difference is the low Q-factor of the on-chip capacitors and inductors, resulting in increased conduction losses and output voltage ripple. Therefore, it is self evident that the basic equations, discussed in Sect. 2.3.3, for describing ideal DC-DC converters, will not suffice for the design of their monolithic variants.

The design of monolithic converters might be performed by means of accurate SPICE simulations, however this option is quite time consuming as it requires computational intensive transient simulations. Moreover, this design process will be executed in a multi-dimensional design space, making it considerably complex for the designer to find the optimal design. For these reasons it is understood that the deduction of a complex, but accurate, mathematical model will eventually yield a more efficient design flow, in addition to an improved understanding of the important trade-offs.

The mathematical steady-state design model is based on the differential equations of the DC-DC converter, which takes all the resistive losses into account. These equations and the method to calculate the output parameters are deduced for both a boost and a buck converter in Sect. 4.1. The resistive losses, together with dynamic losses, are modeled in Sect. 4.2. The converters in this work are all designed for maximal output power, causing the chip temperature to increase above the ambient temperature. The effects of this increased temperature are modeled in Sect. 4.3. Merging the differential equations with the additional losses yields the final model flow. This final model will be used to deduce generally valid, qualitative trade-offs for monolithic DC-DC converters. The latter two aspects are discussed in Sect. 4.4. The chapter is concluded in Sect. 4.5.

4.1 Second-Order Model: Boost and Buck Converter

The importance of an accurate steady-state model for the design of monolithic inductive DC-DC converters cannot be underestimated. The multiple input design variables, such as: the inductance, the output capacitance, the switch sizes, the switching frequency, the input voltage, the on-time..., combined with the multiple output design variables, such as: the output voltage, the maximal output power, the power conversion efficiency..., result in a tedious design process with many trade-offs. The low values of the on-chip inductance(s) and capacitance(s), combined with high switching frequencies, fast transients and numerous parasitic effects, imply that monolithic DC-DC converters often operate at the physical limits of the IC-technology in which they are implemented. From this point of view it is clear that there is a strong need for a fast and accurate model that can tackle the many difficult design trade-offs and also ensure this accuracy at the limits.

Steady-state models for inductive DC-DC converters have been introduced decades ago [Kos68] and they are mostly based on some form of small-ripple approximation [Mid76], which is also explained in Sect. 2.3.3. These methods can give indications for the design of quasi-ideal DC-DC converters that use external passives and operate at moderately low switching frequencies (up to a few hundred kHz). Extensions on these models can be made to introduce additional losses, caused by the non-ideal converter components [Eri04, Ebe09]. However, for exploring the boundaries of the design-space of monolithic inductive DC-DC converters these models simply lack accuracy. This problem is overcome by the mathematical model described throughout this chapter. This model is explained by means of a boost and a buck converter example, because these converters are the base for the practical realizations in this work. Note that the general idea behind the model can be adapted for any given inductive DC-DC converter topology.

The presented model is based on the basic differential equations for the boost and the buck converter, taking all the resistive losses into account, which is discussed in the respective Sects. 4.1.1 and 4.1.3. The time-domain solutions of these equations are used to calculate the output voltage together with all the other output parameters, for given input parameters. The analytical methods for performing this important step are explained in Sects. 4.1.2 and 4.1.4 for a boost and a buck converter, respectively.

4.1.1 Differential Equations: Boost Converter

Figure 4.1 shows the circuit of a boost converter that includes all the significant resistive losses, which are caused by:

- R_{in}: The parasitic series resistance of the input voltage source U_{in}.
- R_{Ls}: The parasitic series resistance of the inductor L.
- R_{SW1}: The parasitic series resistance of the low-side switch SW_1.
- R_{SW2}: The parasitic series resistance of the high-side switch SW_2.

4.1 Second-Order Model: Boost and Buck Converter

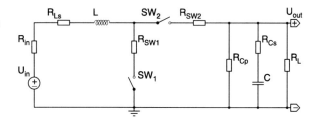

Fig. 4.1 The circuit of a boost DC-DC converter with all its resistive losses

- R_{Cp}: The parasitic parallel resistance of the output capacitor C.
- R_{Cs}: The parasitic series resistance of the output capacitor C.

Charging the Inductor

The equivalent circuit of the charging phase of L for the boost converter is shown in Fig. 4.2(a). For the detailed discussion on this circuit the reader is referred to Sect. 2.3.1. It is clear that R_a is the sum of the parasitic resistances, as stated by (4.1).

$$R_a = R_{in} + R_{Ls} + R_{SW1} \qquad (4.1)$$

The first-order differential equation for this circuit, together with the initial condition for $i_L(t)$, is given by (4.2).

$$L\frac{di_L(t)}{dt} + R_a i_L(t) - U_{in} = 0 \quad \text{with} \quad i_L(0) = I_{L\_min} \qquad (4.2)$$

In (4.2) I_{L_min} is the value of $i_L(t)$ at the end of the preceding discharge phase, in steady-state. I_{L_min} is zero for DCM and has a positive, finite value for CCM, according to (4.3).

$$\begin{cases} \text{DCM: } I_{L\_min} = 0 \\ \text{CCM: } 0 < I_{L\_min} < \infty \end{cases} \qquad (4.3)$$

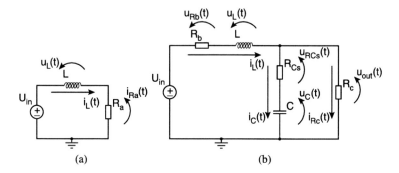

Fig. 4.2 (a) The equivalent circuit of the charge phase and (b) discharge phase of the inductor L for a boost DC-DC converter with all its resistive losses

The time-domain solution for (4.2) is in-turn given by (4.4).

$$i_L(t) = \frac{U_{in}}{R_a} + \left(i_L(0) - \frac{U_{in}}{R_a}\right)e^{-t\frac{R_a}{L}} \qquad (4.4)$$

Discharging the Inductor

The equivalent circuit of the discharge phase of L for the boost converter is shown in Fig. 4.2(b). The equivalent resistors R_b and R_c comply with (4.5) and (4.6), respectively.

$$R_b = R_{in} + R_{Ls} + R_{SW2} \qquad (4.5)$$

$$R_c = \frac{R_{Cp}R_L}{R_{Cp} + R_L} \qquad (4.6)$$

It can be formally proven that the second-order differential equation for this equivalent circuit, in addition with the two initial conditions, is given by (4.7).

$$\left(L + \frac{LR_{Cs}}{R_c}\right)\frac{d^2 i_L(t)}{dt^2} + \left(R_b + \frac{R_{Cs}}{R_c}(R_b + R_c) + \frac{L}{R_c C}\right)\frac{di_L(t)}{dt}$$
$$+ \frac{R_b + R_c}{R_c C} i_L(t) - \frac{U_{in}}{R_c C} = 0$$

with
$$\begin{cases} i_L(0) = I_{L\_max} \\ \frac{di_L(0)}{dt} = I_{L\_max}\left(R_c - \frac{(R_b + R_c)(R_c + R_{Cs})}{R_c}\right) \\ \qquad \cdot \frac{R_c}{L(R_c + R_{Cs})} + \frac{U_{in}}{L} - U_{C\_min}\frac{R_c}{L(R_c + R_{Cs})} \end{cases} \qquad (4.7)$$

In (4.7) I_{L_max} denotes the value of $i_L(t)$ at the end of the preceding charge phase, in steady-state. The value of I_{L_max} is always positive and finite. The parameters U_{C_min} denotes the value of $u_C(t)$ at the end of the preceding charge phase, in steady-state. Thus, as a consequence U_{C_min} is the lowest value of $u_C(t)$ in steady-state. The time-domain solution for (4.7) is quite complex and does not provide any additional insights, therefore it is not explicitly provided.

4.1.2 Calculating the Output Voltage: Boost Converter

The time-domain solutions of the respective first- and second-order differential equations of the equivalent charge and discharge phase for the boost converter, given by (4.3) and (4.7) in Sect. 4.1.1, are the base of the presented mathematical steady-state model for the boost converter. This base steady-state model acquires the following circuit and input parameters:

4.1 Second-Order Model: Boost and Buck Converter

- L: The inductance of the inductor.
- C: The capacitance of the output capacitor.
- R_L: The resistance of the load resistor.
- All the parasitic resistances, as denoted in Fig. 4.1: R_{in}, R_{Ls}, R_{SW1}, R_{SW2}, R_{Cp} and R_{Cs}.
- U_{in}: The input voltage of the converter.
- t_{on}: The charge time of the inductor, the on-time.
- t_{off}: The time between two consecutive charge cycles, the off-time.

The combination of these circuit and input parameters with the mathematical steady-state model for the boost DC-DC converter will yield the following primary output parameters:

- The CM of the boost converter.
- t_{off_real}: The time during the discharging of the inductor until $i_L(t)$ becomes zero in DCM, the real off-time.
- $\overline{U_{out}}$: The mean output voltage.
- U_{out_RMS}: The RMS output voltage.
- ΔU_{out}: The output voltage ripple.
- I_{L_min}: The minimum current through the inductor.
- I_{L_max}: The maximum current through the inductor.

Off all these parameters $\overline{U_{out}}$ is the key parameter, of which most of the other output parameters will be derived. The respective calculation strategy of $\overline{U_{out}}$ for both CCM and DCM is explained in the following sections, in addition with the iterative method for the determining the CM.

Discontinuous Conduction Mode

In DCM I_{L_max} is given by (4.4) for $i_L(0) = I_{L\_min} = 0$ and $t = t_{on}$, as stated by (4.8).

$$I_{L\_max} = \frac{U_{in}}{R_a}\left(1 - e^{-t_{on}\frac{R_a}{L}}\right)$$
$$\implies g_1\{U_{in}; t_{on}\} \qquad (4.8)$$

Equation (4.8) implies that the value of I_{L_max} only depends on the input parameters U_{in} and t_{on}. Therefore, both $i_{SW1}(t)$ and I_{L_max} are known.

The calculation of $i_{SW2}(t)$ is less straightforward as it also depends on output parameters, which are not yet known. Consider the graph of Fig. 4.3, where $i_{SW2}(t)$ is shown by the gray curve for steady-state DCM and its linear approximation is shown by the black curve, as a function of time. Obviously, $i_{SW2}(t)$ is equal to $i_L(t)$ during the discharge phase and it is described by the time-domain solution of (4.7). It can also be seen in Fig. 4.3 that $i_{SW2}(t)$ becomes zero after a certain time t_{off_real}. This t_{off_real} can be calculated by means of equating (4.7) to zero, for $t = t_{off\_real}$. As a result $i_{SW2}(t)$ is a function of input and output parameters, given by (4.9).

Fig. 4.3 The current $i_{SW2}(t)$ through SW_2 as a function of time t for a boost converter in steady-state DCM is shown by the *gray curve*, the *black curve* shows its linear approximation

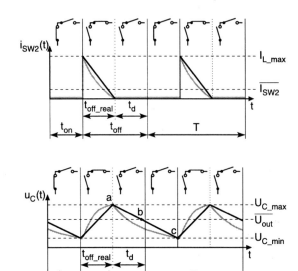

Fig. 4.4 The voltage $u_C(t)$ over C as a function of time t for a boost converter in steady-state DCM is shown by the *gray curve*, the *black curve* shows its piecewise linear approximation

$$I_{L\_min} = 0 = g_2\{U_{in}; U_{C\_min}; t_{on}; t_{off\_real}\} \quad (4.9)$$

During the discharge phase $i_L(t)$ is divided into $i_C(t)$ and $i_{R_c}(t)$, as shown in Fig. 4.2(b). In steady-state operation no DC-current will flow through C because the charge balance of C is zero in steady-state, which is explained in Sect. 2.3.3. This implies that, during the discharge phase, the DC-component of $i_L(t)$ can only flow through R_c and the AC-component of $i_L(t)$ will flow through C. The DC-component of $i_L(t)$ during the discharge phase is equal to the mathematical mean value of $i_{SW2}(t)$, denoted as $\overline{I_{SW2}}$. By linearizing $i_{SW2}(t)$, as illustrated by the black curve in Fig. 4.3, the mean output voltage $\overline{U_{out}}$ in steady state is approximated by (4.10).

$$\overline{U_{out}} = \overline{I_{R_c}} R_c = \overline{I_{SW2}} R_c = R_c \frac{1}{T} \int_0^T i_{SW2}(t)\, dt \simeq \frac{I_{L\_max}}{2} \frac{t_{off\_real}}{t_{on} + t_{off}} R_c$$

$$\implies \overline{U_{out}} = g_3\{I_{L\_max}; t_{on}; t_{off}; t_{off\_real}\} \quad (4.10)$$

It shows that $\overline{U_{out}}$ can be calculated trough (4.10) if t_{off_real} is known, which can in-turn be calculated through (4.9), assuming U_{C_min} is known. For this purpose, the voltage $u_C(t)$ over C as a function of time, for a boost converter in steady-state DCM, is shown in Fig. 4.4 by the gray curve. The piecewise linear approximation of $u_C(t)$ is represented by the black curve in Fig. 4.4. It is observed that, between the marked point a and c, C is discharged through R_c, resulting in an exponential RC discharge curve. The intersection between $\overline{U_{out}}$ and the piecewise linearized $u_C(t)$ curve is marked by point b, which is located equidistantly to point a and c. Consequentially, the time between point a and b is equal to the time between point b and c, as expressed in (4.11).

$$t_{a \to b} = t_{b \to c} = \frac{t_{a \to c}}{2} = \frac{t_{on} + t_d}{2} \quad (4.11)$$

4.1 Second-Order Model: Boost and Buck Converter

From (4.11) and Fig. 4.4 it can be concluded that when the exponential RC discharge curve is tracked, starting from the value of $\overline{U_{out}}$ in (point b) and during the time stated by (4.11), this will approximately result in the value of U_{C_min} (point c). The mathematical translation of this observation yields (4.12).

$$U_{C\_min} \simeq \overline{U_{out}} e^{-\frac{t_{on}+t_d}{2CR_c}} = \overline{U_{out}} e^{-\frac{t_{on}+t_{off}-t_{off\_real}}{2CR_c}}$$

$$\implies U_{C\_min} = g_4\{\overline{U_{out}}; t_{on}; t_{off}; t_{off\_real}\} \quad (4.12)$$

Notice that the method for obtaining (4.12) assumes that the mean value of $u_C(t)$, denoted as $\overline{U_C}$, is equal to $\overline{U_{out}}$. In reality however, there is a voltage drop $u_{RCs}(t)$ over the parasitic series resistance R_{Cs} of C. Nevertheless, it can be intuitively seen that in steady-state $u_{RCs}(t)$ will have an equal negative and positive share, due to the charge balance requirement of C, effectively canceling out the effect of R_{Cs} on $\overline{U_C}$.

From the combination of (4.8), (4.9), (4.10) and (4.12) the function stated by (4.13) is deduced.

$$g_5\{U_{in}; t_{on}; t_{off}; t_{off\_real}\} \quad (4.13)$$

In this function the only unknown output parameter is t_{off_real}. Because of the fact that it is not possible to obtain an explicit form of (4.13), the iterative method of Newton-Raphson is acquired to determine the numerical value of the roots of this function. In other words, the values of t are sought where $i_L(t)$ becomes zero. The iterative method of Newton-Raphson requires an initial starting-value which needs to be sufficiently close to the actual root. Also (4.13) will have multiple roots, as it behaves similar as a periodically-damped RLC-network described in Sect. 2.3.2. Obviously, $i_L(t)$ is not allowed to become negative during the discharge phase, implying that the smallest positive root will represent the value of t_{off_real}. Numerically, this calculation is achieved by performing the Newton-Raphson iteration ten times, starting from $t_{off}/10$ to t_{off}. When t_{off_real} is known, U_{C_min} can be calculated by obtaining its explicit form from (4.9). This explicit is form is not provided due to its high degree of complexity.

The next step is calculating U_{C_max}, by means of U_{C_min}. This is achieved by tracking the black curve of Fig. 4.4 from point c to point a, in the knowledge that this track is determined by an exponential RC curve, yielding (4.14).

$$U_{C\_max} = U_{C\_min} e^{\frac{t_{on}+t_d}{CR_c}} = U_{C\_min} e^{\frac{t_{on}+t_{off}-t_{off\_real}}{CR_c}} \quad (4.14)$$

Finally, with the knowledge of U_{C_min} and U_{C_max}, U_{out_RMS} can be calculated through (4.15), which is valid for DCM.

$$U_{out\_RMS} = \sqrt{\frac{1}{T}\int_0^T u_C^2(t)\,dt}$$

$$\simeq \frac{U_{C\_min}+U_{C\_max}}{2}\sqrt{1+\frac{1}{12}\left(2\frac{U_{C\_max}-U_{C\_min}}{U_{C\_max}+U_{C\_min}}\right)^2} \quad \blacksquare \quad (4.15)$$

Despite the fact that R_{Cs} has no influence on the assumption that $\overline{U_C} = \overline{U_{out}}$, R_{Cs} will still have an influence on ΔU_{out}. Indeed, a larger value of R_{Cs} will cause an increased voltage drop $u_{Rcs}(t)$ over R_{Cs}, which will in-turn lead to an increased value of ΔU_{out}. The value for ΔU_{out}, which takes the additional voltage drop due to R_{Cs} into account, is calculated by means of (4.16).

$$\Delta U_{out} \simeq U_{out\_max} - U_{out\_min}$$
$$= (U_{C\_max} + (I_{L\_max} - \overline{I_{out}})R_{Cs}) - (U_{C\_min} - \overline{I_{out}}R_{Cs}) \blacksquare \quad (4.16)$$

Continuous Conduction Mode

Except for the fact that I_{L_min} is not zero and that $t_{off\_real} = t_{off}$, the calculation method of U_{out_RMS} for CCM is similar to that of DCM. Therefore, the following explanation will be more briefly.

The time-domain solution of the differential equation (4.2) of the charge circuit gives an expression for I_{L_max}, for $t = t_{on}$. This is expression is stated by (4.17).

$$I_{L\_max} = \frac{U_{in}}{R_a} + \left(I_{L\_min} - \frac{U_{in}}{R_a}\right)e^{-t_{on}\frac{R_a}{L}}$$
$$\implies I_{L\_max} = g_6\{U_{in}; I_{L\_min}; t_{on}\} \quad (4.17)$$

The time-domain solution of the differential equation (4.7) for the discharge circuit and for $t = t_{off}$, yields a function which is dependent on the in and output parameters described by (4.18). The solution of (4.7) is not provided due to its rather complex nature.

$$I_{L\_min} = g_7\{U_{in}; U_{C\_min}; I_{L\_max}; t_{off}\} \quad (4.18)$$

Analogue to the method used for DCM, $\overline{U_{out}}$ is approximated trough the linearized $i_{SW2}(t)$, yielding (4.19).

$$\overline{U_{out}} = \overline{I_{Rc}}R_c = \overline{I_{SW2}}R_c = R_c\frac{1}{T}\int_0^T i_{SW2}(t)\,dt \simeq \frac{I_{L\_min} + I_{L\_max}}{2}\frac{t_{off}}{t_{on}+t_{off}}R_c$$
$$\implies \overline{U_{out}} = g_8\{I_{L\_min}; I_{L\_max}; t_{on}; t_{off}\} \quad (4.19)$$

During t_{on} C is discharged through R_L. When piecewise linearizing $u_C(t)$, similar as illustrated in Fig. 4.4, U_{C_min} can be obtained through (4.20).

$$U_{C\_min} \simeq \overline{U_{out}}e^{-\frac{t_{on}}{2CR_c}}$$
$$\implies U_{C\_min} = g_9\{\overline{U_{out}}; t_{on}\} \quad (4.20)$$

In the last step (4.17), (4.18), (4.19) and (4.20) are combined to obtain the explicit form for U_{C_min}. This explicit form for U_{C_min} is merely dependent on known input parameters, allowing it to be calculated directly. Note that the symbolic result of this calculation is very complex and does not contribute to any additional insights. It is therefore omitted.

The calculation of U_{C_max} is performed similar to (4.14) and is given by (4.21).

$$U_{C\_max} = U_{C\_min}e^{\frac{t_{on}}{CR_c}} \quad (4.21)$$

Finally, U_{out_RMS} and ΔU_{out} are obtained for CCM through the same methods used for the DCM, respectively given by (4.15) and (4.16).

4.1 Second-Order Model: Boost and Buck Converter

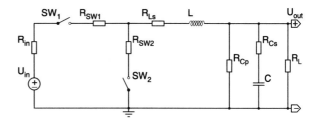

Fig. 4.5 The circuit of a buck DC-DC converter with all its resistive losses

4.1.3 Differential Equations: Buck Converter

The circuit of a buck converter, which includes all the significant resistive losses, is shown in Fig. 4.5. These resistive losses are caused by parasitic resistances, which have a similar meaning as explained in Sect. 4.1.1. This circuit will form the base of the mathematical model of the buck converter. Analogue to the base model for the boost converter, the model for the buck converter requires the fundamental differential equations for the respective charge and discharge circuits. These are explained in the following sections.

Charging the Inductor

Figure 4.6(a) shows the equivalent circuit of the charge phase of a buck converter with the resistive losses added. The equivalent resistors R_a and R_b are defined by (4.22) and (4.23), respectively.

$$R_a = R_{in} + R_{SW1} + R_{Ls} \tag{4.22}$$

$$R_b = \frac{R_{Cp} R_L}{R_{Cp} + R_L} \tag{4.23}$$

The second-order differential equation of the equivalent charge circuit of the buck converter, in addition with the two initial conditions, is given by (4.24).

$$\left(L + \frac{LR_{Cs}}{R_b}\right)\frac{d^2 i_L(t)}{dt^2} + \left(R_a + \frac{R_{Cs}}{R_b}(R_a + R_b) + \frac{L}{R_b C}\right)\frac{d i_L(t)}{dt}$$
$$+ \frac{R_a + R_b}{R_b C} i_L(t) - \frac{U_{in}}{R_b C} = 0$$

$$\text{with} \begin{cases} i_L(0) = I_{L\_min} \\ \dfrac{d i_L(0)}{dt} = I_{L\_min}\left(R_b - \dfrac{(R_a + R_b)(R_b + R_{Cs})}{R_b}\right) \\ \qquad \cdot \dfrac{R_b}{L(R_b + R_{Cs})} + \dfrac{U_{in}}{L} - \overline{U_{out}}\dfrac{R_b}{L(R_b + R_{Cs})} \end{cases} \tag{4.24}$$

In (4.24) I_{L_min} denotes the end value of $i_L(t)$ of the previous discharge phase, in steady-state. The value of I_{L_min} depends on the CM, as stated by (4.3). In the second initial condition, the initial voltage over C is approximated with $\overline{U_{out}}$. It can

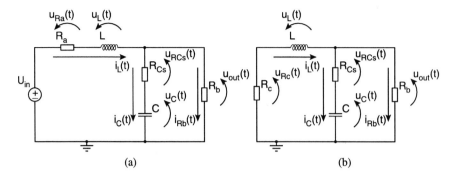

Fig. 4.6 (a) The equivalent circuit of the charge phase and (b) discharge phase of the inductor L for a buck DC-DC converter with all its resistive losses

be verified in the respective Figs. 3.2 and 3.3 that $u_C(0)$ is approximately equal to $\overline{U_{out}}$ at the beginning of the charge phase, for both CMs. The time-domain solution of (4.24) is omitted due to its high degree of complexity.

Discharging the Inductor

The equivalent discharge circuit of the buck converter with resistive losses is shown in Fig. 4.6(b). The equivalent resistor R_b complies with (4.23), while R_c corresponds to (4.25).

$$R_c = R_{SW2} + R_{Ls} \tag{4.25}$$

The second-order differential equation which describes the buck converter during the discharge phase is equal to (4.24), under the following conditions: R_a is to be replaced by R_c, I_{L_min} is to be replaced by I_{L_max} and U_{in} is equal to zero. I_{L_max} is the end value of the $i_L(t)$ at the end of the previous charge phase. Similar as for the charge phase the initial value of $u_C(0)$ is approximated with $\overline{U_{out}}$.

4.1.4 Calculating the Output Voltage: Buck Converter

The base of the mathematical steady-state model of the buck converter, which takes all the significant resistive losses into account, is based on the time-domain solutions of (4.24) and its adapted form for the discharge phase. This model requires the same circuit and input parameters as for the boost converter, explained in Sect. 4.1.2 and it obtains the same output parameters. Of all the output parameters, $\overline{U_{out}}$ is again the key parameter of which most of the other output parameters are derived. The following sections will provide the calculation method for both CMs of the buck converter. The method is similar to that of the boost converter, but not equal to it.

4.1 Second-Order Model: Boost and Buck Converter

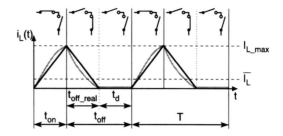

Fig. 4.7 The current $i_L(t)$ through L as a function of time t for a buck converter in steady-state DCM is shown by the *gray curve*, the *black curve* shows its linear approximation

Discontinuous Conduction Mode

In DCM I_{L_max} is determined by the time-domain solution of (4.24), which is valid for the charge phase of L, for $i_L(t) = I_{L\_min} = 0$ and $t = t_{on}$. Thus, I_{L_max} is a function of the input parameters stated by (4.26).

$$I_{L\_max} = g_{10}\{U_{in}; \overline{U_{out}}; t_{on}\} \quad (4.26)$$

The calculation of $i_L(t)$ during the discharge phase of L is performed through the variant of (4.24), which is valid for the discharge phase and is clarified in Sect. 4.1.3. I_{L_min} becomes zero for $t = t_{off\_real}$, which is stated by (4.27).

$$I_{L\_min} = 0 = g_{11}\{\overline{U_{out}}; I_{L\_max}; t_{off\_real}\} \quad (4.27)$$

Analogue to the boost converter, $i_L(t)$ is divided into $i_C(t)$ and $i_{Rb}(t)$. Since the steady-state DC-current component of $i_C(t)$ is zero, due to the charge balance of C, this DC-component flows entirely through R_b. Figure 4.7 illustrates $i_L(t)$ both for the real case (gray curve) and the piecewise linearized case (black curve). The latter calculation is used for the approximation of $\overline{U_{out}}$, which is calculated through (4.28).

$$\overline{U_{out}} = \overline{I_{Rb}}R_b = \overline{I_L}R_b = R_b \frac{1}{T}\int_0^T i_L(t)\,dt \simeq R_c \frac{I_{L\_max}}{2}\left(\frac{t_{on} + t_{off\_real}}{t_{on} + t_{off}}\right)$$
$$\Longrightarrow \overline{U_{out}} = g_{12}\{I_{L\_max}; t_{on}; t_{off}; t_{off\_real}\} \quad (4.28)$$

The combination of (4.26), (4.27) and (4.28) yields a function in which t_{toff_real} is the only unknown input parameter, as denoted by (4.29).

$$g_{13}\{U_{in}; t_{on}; t_{off}; t_{off\_real}\} \quad (4.29)$$

It is not possible to obtain an explicit form of (4.29) into t_{off_real}. Therefore, the numerical iterative method of Newton-Raphson is used for finding the first positive root of (4.29), of which the value is equal to t_{off_real}. The starting values for this iteration are found similar as for the boost converter, explained in Sect. 4.1.2. Once t_{off_real} is known I_{L_max} and $\overline{U_{out}}$ can be calculated through (4.26) and (4.28), respectively.

The calculation method of ΔU_{out}, used for the boost converter in Sect. 4.1.2, is not applicable for the buck converter. Instead, the method described in Sect. 3.1.1 is used. This method is based on the calculation of the injected positive charge change

Fig. 4.8 The voltage $u_C(t)$ over C as a function of time t for a buck converter in steady-state DCM is shown by the *gray curve*, the *black curve* shows its piecewise linear approximation

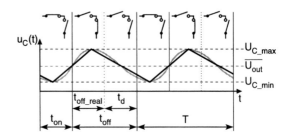

ΔQ into C, illustrated in Fig. 3.2. For this purpose $i_C(t)$ is calculated through (4.30).

$$i_C(t) = i_L(t) - i_{Rb}(t) \simeq i_L(t) - \frac{\overline{U_{out}}}{R_b} \qquad (4.30)$$

In (4.30) $i_L(t)$ is calculated by means of the time-domain solution of (4.24) for the charge phase, yielding $i_{C\_charge}(t)$. For the discharge phase $i_L(t)$ is calculated by means of the time-domain solution of the adapted form for the discharge phase of (4.24), yielding $i_{C\_discharge}(t)$. ΔQ is calculated similar to (3.6), yielding (4.31).

$$\Delta Q = \int_{t_{zero1}}^{t_{zero2}} i_C(t)\,dt = \int_{t_{zero1}}^{t_{on}} i_{C\_charge}(t)\,dt + \int_0^{t_{zero2}} i_{C\_discharge}(t)\,dt \qquad (4.31)$$

In (4.31) $i_C(t)$ is divided into the respective charge and discharge parts: $i_{C\_charge}(t)$ and $i_{C\_discharge}(t)$. The respective times at which $i_{C\_charge}(t)$ and $i_{C\_discharge}(t)$ intersect the X-axis and become zero are denoted by t_{zero1} and t_{zero2}. These times are equal to the respective first positive root of $i_{C\_charge}(t)$ and $i_{C\_discharge}(t)$, which are calculated numerically by means of the iterative method of Newton-Raphson.

The approximation of the maximal output voltage U_{out_max}, taking the voltage ripple due to the finite values of both of C and R_{Cs} into account, is given by (4.32).

$$U_{out\_max} \simeq \overline{U_{out}} + \frac{\Delta Q}{2C} + R_{Cs}\frac{I_{L\_max} - \overline{I_{out}}}{2} = U_{C\_max} + R_{Cs}\frac{I_{L\_max} - \overline{I_{out}}}{2} \qquad (4.32)$$

The according approximation of the minimal output voltage U_{out_max} is calculated similar as (4.32) and is given by (4.33).

$$U_{out\_min} \simeq \overline{U_{out}} - \frac{\Delta Q}{2C} - R_{Cs}\frac{I_{L\_max} - \overline{I_{out}}}{2} = U_{C\_min} - R_{Cs}\frac{I_{L\_max} - \overline{I_{out}}}{2} \qquad (4.33)$$

By means of (4.32) and (4.33) the expression for the approximation of ΔU_{out} can be calculated, which is given by (4.34).

$$\Delta U_{out} \simeq U_{out\_max} - U_{out\_min} = \frac{\Delta Q}{C} + R_{Cs}(I_{L\_max} - \overline{I_{out}}) \blacksquare \qquad (4.34)$$

Finally, U_{out_RMS} can be approximated through the piecewise linearized $u_C(t)$, which is illustrated in Fig. 4.8 by the black curve. This is done by using the values of U_{C_max} and U_{C_min}, respectively given by (4.32) and (4.33), into (4.15).

Continuous Conduction Mode

The calculation method for U_{out_RMS} for a buck converter in CCM is analogue to DCM, except for the fact that I_{L_min} has a positive, finite value rather than being zero. Furthermore, there is no dead-time t_d, implying that $t_{off\_real} = t_{off}$. In the following discussion the calculation method is briefly clarified.

During the charge phase in steady-state I_{L_max} is determined by the time-domain solution of (4.24), for $t = t_{on}$. As a result, I_{L_max} is dependent on the in- and output parameters described by (4.35).

$$I_{L\_max} = g_{14}\{U_{in}; \overline{U_{out}}; I_{L\_min}; t_{on}\} \qquad (4.35)$$

For the discharge phase $i_L(t)$ is calculated through the adapted form of (4.24). From this adapted form I_{L_max} is determined for $t = t_{off}$ in steady-state, yielding (4.36).

$$I_{L\_min} = g_{15}\{\overline{U_{out}}; I_{L\_max}; t_{off}\} \qquad (4.36)$$

In steady-state operation the charge balance of C is zero and the DC-component of $i_L(t)$ will only flow through R_b. The approximation for $\overline{U_{out}}$ is obtained through the piecewise linearized $i_L(t)$, stated by (4.37).

$$\overline{U_{out}} = \overline{I_{Rb}}R_b = \overline{I_L}R_b = R_b \frac{1}{T}\int_0^T i_L(t)\,dt \simeq R_b \frac{I_{L\_min} + I_{L\_max}}{2}$$
$$\implies \overline{U_{out}} = g_{16}\{I_{L\_max}; I_{L\_min}\} \qquad (4.37)$$

A closed-form expression of $\overline{U_{out}}$, which is only dependent on known input parameters, is obtained through the combination of (4.35), (4.36) and (4.37). This expression is not provided due to its high degree of complexity.

Finally, ΔU_{out} and U_{out_RMS} are calculated by means of (4.34) and (4.15), respectively.

4.2 Non-ideal Converter Components Models

Section 4.1 describes a mathematical model for a boost and buck converter which takes all the significant resistive losses into account. This section will elaborate upon the mathematical models of these resistive losses and also other types of losses, commonly referred to as switching losses or dynamic losses. For this purpose, electrical models are introduced for each of the converter components, describing their significant non-ideal aspects. As a result, the parasitic resistances, capacitances and inductances are identified. These will introduce power losses and other effects, such as an increased output voltage ripple, during the DC-DC converter's operation. These power losses are mathematically described in order to obtain a quantitative comparison between them, allowing to determine the dominant power losses at the design stage. The calculation of the power losses of all the converter components is performed for both the boost and the buck converter.

Fig. 4.9 The lumped model for a metal-track or bondwire inductor, taking both the parasitic series resistance R_{Ls} and parasitic substrate capacitance C_{sub} into account

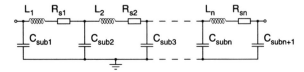

Note that the practical issues for the monolithic integration of the converter components and the effect on the parasitic elements are provided in Sect. 6.1.

The electrical models for the inductor, the capacitor, the switches, the buffers that drive these switches and the interconnect are discussed in the respective Sects. 4.2.1, 4.2.2, 4.2.3, 4.2.4 and 4.2.5.

4.2.1 Inductor

The model for the inductor that is used for the implementation in the mathematical steady-state model is shown in Fig. 4.9. This model is a lumped model which takes both the parasitic series resistance and parasitic substrate capacitance into account. More complex models for inductors exist [Cao03, Wu03, Hua06], which also take the inter-winding parasitic capacitance and substrate resistance into account. However, it is found that the overall accuracy of the steady-state model is not significantly influenced by introducing these additional parameters. The number of segments of the lumped inductor model is chosen equal to the number of bondwires for a bondwire inductor (see Sect. 6.1.1) and for an on-chip metal-track inductor (see Sect. 6.1.1) ten segments is found to yield sufficient accuracy.

Considerations on the Inductance

Any given electrical conductor has a certain self inductance L_{self}, which is proportional to its length and inverse proportional to its radius. For a straight conductor with a circular cross-section, surrounded by a dielectric medium with a relative permeability having a value of $\mu_r = 1$, L_{self} is determined by (4.38) [Gre74].[1]

$$L_{self} = \frac{\ell}{5}\left(\ln\left(\frac{2\ell}{r}\right) - \frac{3}{4} + \frac{r}{\ell}\right) \text{ [nH]} \qquad (4.38)$$

In (4.38) ℓ [mm] is the length of the conductor and r [mm] the radius of the perpendicular cross-section. For example when considering a bondwire with $r = 12.5$ μm and $\ell = 1$ mm, this yields an inductance of about $L_{self} = 0.87$ nH.

---

[1] Please note that the equations deduced in [Gre74] are obtained form [Gro62], which dates back to the year 1946, and that the latter work is in-turn based on the J.C. Maxwell's work [Max54], published in 1873.

4.2 Non-ideal Converter Components Models

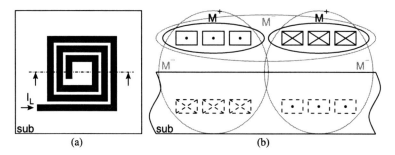

Fig. 4.10 (a) The top-view of a planar square spiral inductor above a conductive substrate and (b) the cross-sectional view with indication of the most significant mutual inductances

When two electrical conductors are in each other's neighborhood and if their perpendicular projection onto one another is not zero (not regarding their thickness), they will both experience a mutual inductance M. For two parallel conductors with a circular cross-section, surrounded by a dielectric medium with $\mu_r = 1$, M is calculated through (4.39) [Gre74].

$$M = \frac{\ell_{overlap}}{5} \left(\ln\left(\frac{\ell_{overlap}}{d} + \sqrt{1 + \left(\frac{\ell_{overlap}}{d}\right)^2}\right) \right.$$
$$\left. - \sqrt{1 + \left(\frac{d}{\ell_{overlap}}\right)^2} + \frac{d}{\ell_{overlap}} \right) \; [\text{nH}] \quad (4.39)$$

In (4.39) $\ell_{overlap}$ [mm] denotes the overlapping length of the two conductors and d [mm] is the pitch distance between the two center points of the conductors. The value of M will be positive when the respective currents through the two conductors flow in the same direction and vice versa. For example, when considering two parallel bondwires with $\ell_{overlap} = 1$ mm and $d = 0.1$ mm, this yields a mutual inductance of about $M = 0.59$ nH.

Apart from coupled inductors, the practical use of the mutual inductance between conductors is found in spiral inductors. The top-view of a planar square spiral inductor above a conductive substrate is illustrated in Fig. 4.10(a). The corresponding cross-section of this structure is shown in Fig. 4.10(b). When assuming a current I_L flows through the inductor, as shown in Fig. 4.10(a), the direction of I_L is denoted in Fig. 4.10(b) by means of the respective dots and crosses. It is observed that in this spiral inductor implementation, the windings in which I_L flows in the same direction are closer to one another, compared to those in which the currents flow in the opposite direction. Hence, the total positive mutual inductance M^+ will have a larger value than the total negative mutual inductance M^-. However, this statement is only generally valid when it is assumed that the inductor is surrounded by a non-conductive medium. When the inductor is placed above a conductive substrate, which is the case for monolithic inductors, the individual windings will be influenced with an additional negative mutual inductance. This situation is also visualized in Fig. 4.10(b). This effect can be intuitively modeled by an identical inductor, which is parallel to the physical inductor and at a distance to the latter equal to twice the distance from the conductive substrate. In other words, the conductive substrate acts as a mirror for the physical inductor, thereby introducing an additional M^-. The value of this additional M^- is inverse proportional to both the resistivity ρ of the conductive substrate and its distance to the inductor. This is also the reason why (low resistive) metal structures underneath the inductor, for instance used for a capacitor (see Sect. 6.1.1) should incorporate gaps which are perpendicular to the conductors, also referred to as a patterned ground shield [Yim02]. Obviously, the best performance of a monolithic inductor will be obtained by avoiding high-conductive structures underneath it. This is also the idea behind under-etched inductors [Til96].

Whether the total mutual inductance M of the monolithic spiral inductor is either positive or negative largely depends on the distance to the conductive substrate and its conductivity. The resulting inductance of the inductor is determined by the sum of L_{self} and M, as stated by (4.40).

$$L = \sum_{i=1}^{\#seg} L_{self\_i} + \sum_{i=1}^{\#seg}\sum_{j=1}^{\#seg} M_{i-j} - \sum_{i=1}^{\#seg} M_{i-i} \qquad (4.40)$$

Note that the last term in (4.40) is due to the fact that the mutual inductance of a conductor referred to itself M_{i-i} has no physical meaning.

Two types of monolithic hollow spiral inductors are considered in this work for the implementation of the inductor: bondwire inductors and metal-track inductors. Bondwire inductors can be modeled with a tolerance of about 10% through (4.38), (4.39) and (4.40), compared to 3-D finite element simulators [Qi00]. Moreover, the parasitic series resistance R_{Ls}, accounting also for the skin-effect, and the parasitic substrate capacitance can be easily modeled, as explained in the next section. For the case of metal-track monolithic inductors many specific models exist, such as [Cro96, Dan99, Moh99, Jen02]. However, for the purpose of gaining more accuracy both the bondwire inductors and metal-track inductors are fine-tuned using the 2-D field-solver FastHenry [Kam94]. The expected inductance of monolithic metal-track inductors can be expected to be in the range of 0.5–100 nH, with a maximal Q-factor up to 40 [Bur98]. This quality, or Q-factor, for inductors is defined by (4.41).

$$Q = \frac{2\pi f_{SW} L}{R_{Ls}} \qquad (4.41)$$

For bondwire inductors the value of the Q-factor can generally be expected to be higher, compared to monolithic metal-track inductors, because the mean distance of the conductors towards the substrate can be higher. This causes a lower value of M^- and thus a larger value of L. The practical issues on the design of monolithic inductors are discussed in Sect. 6.1.1.

Fig. 4.11 The perpendicular cross-sectional view of a conductor which is prone to the skin-effect

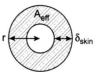

Parasitic Series Resistance

The series resistance of a conductor is determined by length ℓ, the area of its perpendicular cross-section $A_\varnothing$ and the resistivity[2] ρ of the material it consists of. For a round conductor, a bondwire for instance, the DC series resistance is given by (4.42).

$$R_{Ls} = \frac{\rho \ell}{A_\varnothing} = \frac{\rho \ell}{\pi r^2} \qquad (4.42)$$

When the current through the conductor is not DC the series resistance of the conductor will increase due to the skin-effect, whereby the electrons are forced towards the outer diameter of the conductor. This effect is illustrated in Fig. 4.11, which shows a perpendicular cross-sectional view of a conductor prone to the skin-effect. The skin-depth δ_{skin}, which the current will maintain, is inverse proportional to its frequency f and is calculated by means of (4.43).

$$\delta_{skin} = \sqrt{\frac{\rho}{\pi f \mu}} \qquad (4.43)$$

In (4.43) μ[3] is the permeability of the conductor material.

The actual series resistance increase can be calculated through the effective perpendicular cross-sectional area $A_{\varnothing\_eff}$, which is in-turn determined by δ_{skin}, yielding (4.44).

$$A_{\varnothing\_eff} = 2\pi r \delta_{skin} - \pi \delta_{skin}^2 \qquad (4.44)$$

The series resistance, taking the skin-effect into account, is finally found by substituting (4.44) into (4.42). For a gold bondwire with $r = 12.5$ μm, the series resistance per millimeter of ℓ is plotted as a function of f in Fig. 4.12, by the black curve. The gray curve shows the DC resistance. It shows that for bondwires the skin-effect only becomes significant at $f > 100$ MHz.[4]

For monolithic metal-track inductors the calculation of the skin-effect is significantly more complex because the perpendicular cross-section of the windings is rectangular. Therefore, this calculation is performed by means of FastHenry.

It is important to understand that, for both types of inductors, the series resistance should not be calculated for the switching frequency f_{SW} of the DC-DC converter

---

[2] For gold: $\rho_{Au} = 2.35 \cdot 10^{-8}$ Ωm @ 293 K.

[3] For gold: $\mu_{Au} = 4 \cdot \pi \cdot 10^{-7}$ F/m.

[4] The frequency f at which the skin-effect becomes significant is also inverse proportional to the radius r of the bondwire.

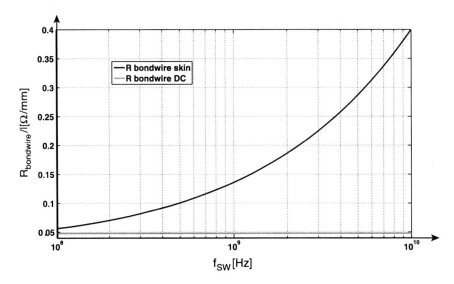

Fig. 4.12 The *black curve* shows the series resistance $R_{bondwire}$ per millimeter of length ℓ for a gold bondwire with $r = 12.5$ μm, as a function of frequency f and the *gray curve* denotes the DC value

when it is intended to operate in DCM. Indeed, this would result in an underestimation of R_{Ls}. Instead the frequency that is to be used for this calculation is given by (4.45).

$$f = \frac{1}{t_{on} + t_{off\_real}} \tag{4.45}$$

Eventually, R_{Ls} results in a certain power loss P_{RLs} in the DC-DC converter, which is calculated through (4.46).

$$P_{RLs} = I_{L\_RMS}^2 R_{Ls} = (I_{SW1\_RMS}^2 + I_{SW2\_RMS}^2) R_{Ls} \tag{4.46}$$

The theorem of superposition implies that the RMS current I_{L_RMS} through L in (4.46) can be replaced by the respective RMS currents I_{SW1_RMS} and I_{SW2_RMS} through SW_1 and SW_2. Equation (4.46) is valid for both a boost and a buck converter.

The piecewise linearized $i_{SW1}(t)$ and $i_{SW2}(t)$ in DCM yield the respective approximated expressions for I_{SW1_RMS} and I_{SW2_RMS}. These calculations are valid for both a boost and a buck converter and they are given by (4.47) and (4.48).

$$I_{SW1\_RMS} = \sqrt{\frac{1}{T}\int_0^T i_{SW1}^2(t)\,dt} \simeq I_{L\_max}\sqrt{\frac{t_{on}}{3(t_{on} + t_{off})}} \tag{4.47}$$

$$I_{SW2\_RMS} = \sqrt{\frac{1}{T}\int_0^T i_{SW2}^2(t)\,dt} \simeq I_{L\_max}\sqrt{\frac{t_{off\_real}}{3(t_{on} + t_{off})}} \tag{4.48}$$

4.2 Non-ideal Converter Components Models

The same calculation method, used for (4.47) and (4.48), can be used for the calculation of I_{SW1_RMS} and I_{SW2_RMS} in CCM, yielding (4.49) and (4.50).

$$I_{SW1\_RMS} = \sqrt{\frac{1}{T} \int_0^T i_{SW1}^2(t)\, dt}$$

$$\simeq \sqrt{\frac{t_{on}}{t_{on}+t_{off}} \left(I_{L\_min}^2 + I_{L\_min}(I_{L\_max} - I_{L\_min}) + \frac{(I_{L\_max} - I_{L\_min})^2}{3} \right)}$$

(4.49)

$$I_{SW2\_RMS} = \sqrt{\frac{1}{T} \int_0^T i_{SW2}^2(t)\, dt}$$

$$\simeq \sqrt{\frac{t_{off}}{t_{on}+t_{off}} \left(I_{L\_min}^2 + I_{L\_min}(I_{L\_max} - I_{L\_min}) + \frac{(I_{L\_max} - I_{L\_min})^2}{3} \right)}$$

(4.50)

Note that P_{RLs} is a resistive or Joule-loss and is therefore already taken into account in the respective second-order models of the boost and the buck converter.

Parasitic Substrate Capacitance

Apart from R_{Ls}, the parasitic substrate capacitance C_{sub} is a significant cause of power loss in L, which needs to be taken into account. The calculation of C_{sub} is conducted through (4.51).

$$C_{sub} = \frac{\epsilon_0 \epsilon_{r\_ox} A_L}{d_{ox}}$$

(4.51)

In (4.51) ϵ_0[5] is permittivity of vacuum, ϵ_{r_ox}[6] is the relative permittivity of the oxide, d_{ox} is the thickness of the oxide and A_L is the perpendicular projected area of the inductor windings onto the substrate for a monolithic metal-track inductor or the total area of the bonding pads of a bondwire inductor.

The electrical power P_C that is associated with the charging and discharging of capacitors is calculated by the product of the required energy, given by (2.17), and the f_{SW}, yielding (4.52).

$$P_C = f_{SW} E_{Uin \to RC} = f_{SW} C_{par} U_{in} \Delta U_C$$

(4.52)

In (4.52) ΔU_C denotes the voltage swing over the capacitor.

For a bondwire inductor, the power loss P_{L_Csub} caused by the total parasitic substrate capacitance of the bonding pads is calculated through (4.53).

$$P_{L\_Csub} = f_{SW} U_{in} C_{pad} \sum_{i=1}^{n-1} \left(i \frac{\Delta U_L}{n-1} \right) = \frac{n f_{SW} C_{pad} U_{in} \Delta U_L}{2}$$

(4.53)

---

[5]$\epsilon_0 \simeq 8.854188 \cdot 10^{-12}$ F/m.

[6]Depending on the used IC technology ϵ_{r_ox} varies between 2.7 and 7.

Fig. 4.13 The model for a capacitor, taking the parasitic series resistance R_{Cs}, the parasitic parallel resistance R_{Cp} and the parasitic series inductance L_{Cs} into account

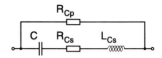

In (4.53) it C_{pad} is the parasitic substrate capacitance of one bonding pad, n denotes the number of bonding pads and ΔU_L denotes the voltage swing at the node where L connected to the switches. For a boost converter $\Delta U_L = U_{out}$ and for a buck converter $\Delta U_L = U_{in}$. Furthermore, it is assumed that ΔU_C of C_{sub_i}, as shown in the model in Fig. 4.9, varies in a linear fashion over the length of the inductor's windings, between zero and ΔU_L.

For a monolithic metal-track inductor P_{L_Csub} is calculated through (4.54).

$$P_{L\_Csub} = \frac{f_{SW} C_{sub} U_{in}}{\Delta U_L} \int_0^{\Delta U_L} u \, du = \frac{f_{SW} C_{sub} U_{in} \Delta U_L}{2} \qquad (4.54)$$

In (4.54) C_{sub} is the total parasitic substrate winding capacitance.

4.2.2 Capacitor

Figure 4.13 shows the equivalent circuit for modeling the output capacitor C. It takes the parasitic series impedance, consisting of the parasitic series inductance[7] L_{Cs} and the parasitic series resistance[8] R_{Cs}, and the parasitic parallel or leak resistance R_{Cp} into account. The effect of these parasitic elements on the operation of both the boost and buck converter is discussed in the following sections. In this discussion L_{Cs} is assumed to be small enough to be neglected.

Parasitic Series Resistance

The parasitic series resistance R_{Cs} of a monolithic capacitor is largely dependent on its geometry and lay-out and is due to the finite resistivity of the metal connections. The different methods for implementing monolithic capacitors and the effect of their lay-out on R_{Cs} are explained in Sect. 6.1.2. The effect of R_{Cs} is two-fold: it causes a power loss and it increases ΔU_{out}. This power loss P_{Rcs} is respectively determined by (4.55) and (4.56) for a boost and a buck converter.

$$P_{Rcs} = I_{SW2\_RMS}^2 R_{Cs} \qquad (4.55)$$

$$P_{Rcs} = I_{L\_RMS}^2 R_{Cs} = (I_{SW1\_RMS}^2 + I_{SW2\_RMS}^2) R_{Cs} \qquad (4.56)$$

---

[7] Also referred to as Equivalent Series Inductance (ESL).
[8] Also referred to as Equivalent Series Resistance (ESR).

Fig. 4.14 The output voltage $u_{out}(t)$ of a boost converter in (**a**) DCM and (**b**) CCM, as a function of time t. The *gray curves* are valid for $R_{Cs} = 0$ and the *black curves* for a finite value of R_{Cs}

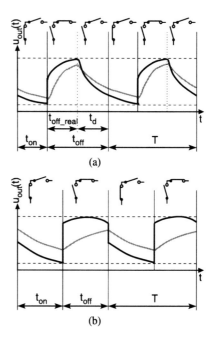

In (4.55) I_{SW1_RMS} and I_{SW2_RMS} for both CMs can be found in Sect. 4.2.1. As $P_{R_{Cs}}$ is a restive loss it is already taken into account in the second-order model, described in Sects. 4.1.2 and 4.1.4.

The effect of R_{Cs} on $u_{out}(t)$ for boost converter is illustrated in Fig. 4.14(a) for DCM and in Fig. 4.14(b) for CCM. The gray curves are valid for the case where $R_{Cs} = 0$ and the black curves are valid for a finite value of R_{Cs}. It is observed that steep transients occur at the transition between the charge and discharge phase in both CMs and that the effective amplitude of ΔU_{out} is increased. This increase is already approximated for the boost converter in Sect. 4.1.2. The steep transients are the result of the fact that the current towards the output capacitor is not continuous, which is explained in Sect. 3.2.1. As a consequence, the odd harmonics have a significantly larger amplitude, compared to the case where $R_{Cs} = 0$. This poses a potential problem in terms of Electro Magnetic Interference (EMI[9]) towards the circuitry of the load and/or towards other neighboring circuits. Obviously, these potential EMI issues depend on the intended application and are not further considered here.

For a buck converter the effect of R_{Cs} on $u_{out}(t)$ is illustrated for both DCM and CCM in Figs. 4.15(a) and 4.15(b), respectively. The gray curves are valid for the case where $R_{Cs} = 0$ and the black curves for a finite value of R_{Cs}. It can be seen that the amplitude of ΔU_{out} is increased and that no steep transients occur, as opposed to the boost converter. This increase of ΔU_{out} is approximated in Sect. 4.1.4. The

---

[9]Not to be confused with the record label that produced the Beatles.

Fig. 4.15 The output voltage $u_{out}(t)$ of a buck converter in (**a**) DCM and (**b**) CCM, as a function of time t. The *gray curves* are valid for $R_{Cs} = 0$ and the *black curves* for a finite value of R_{Cs}

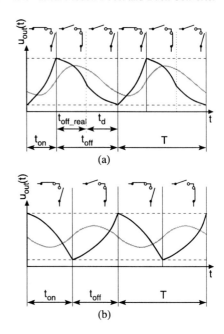

lack of steep transients in $u_{out}(t)$ is due to the fact that the current towards the output capacitor is continuous in a buck converter, as explained in Sect. 3.1.1. Nevertheless, the finite value of R_{Cs} does introduce discontinuities in $u_{out}(t)$. This is observed in its frequency spectrum through an increase of the amplitude of both the even and odd harmonics. The potential EMI issues are expected to pose fewer problems for the buck converter, compared to the boost converter, since this increase is less pronounced.

The increase of ΔU_{out} due to the finite value of R_{Cs} is disadvantageous for both the boost and buck converter. Indeed, to compensate for this effect f_{SW} and/or C is to be increased. An increase of f_{SW} will lead to increased switching-losses and thus a lower value of η_{SW} and an increase of C will lead to a larger chip area requirement. It is clear that this will result in a trade-off between A and η_{SW}.

Parasitic Parallel Resistance

Depending on the implementation, monolithic capacitors will tend to introduce a certain leakage current I_{C_leak}, which can be modeled by a parasitic parallel resistance R_{Cp}. This I_{C_leak} is linear proportional to the physical size (area) of the output capacitor and thus to its capacitance. The only effect of R_{Cp} is that it will introduce a power loss P_{Rcp}, which is calculated through (4.57).

$$P_{Rcp} = \frac{U^2_{out\_RMS}}{R_{Cp}} = \overline{U_{out} I_{C\_leak}} \qquad (4.57)$$

Fig. 4.16 The parallel circuit of two capacitors C_1 and C_2, with their respective parasitic series resistances R_1 and R_2, and the equivalent circuit with one capacitor $C_{eq}(f)$ and resistor $R_{eq}(f)$

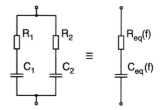

As P_{Rcp} is a resistive loss, it is already taken into account in the second-order model of the boost and the buck converter, which are explained in the respective Sects. 4.1.2 and 4.1.4.

On Parallel Capacitors

The output capacitor of a monolithic DC-DC converter is often implemented as the parallel combination of two different types of capacitors (see Chap. 6), in order to achieve a higher capacitance density. The implementations of this work often make use of the combination of MIM and MOS capacitors. The MIM capacitor is characterized by a lower capacitance density compared to the MOS capacitor, but the MIM capacitor has a lower parasitic series resistance compared to the MOS capacitor. In a DC-DC converter this combination will act as a single equivalent output capacitor, having an equivalent capacitance $C_{eq}(f)$ and an equivalent parasitic series resistance $R_{eq}(f)$, which are both frequency dependent. The idea behind this is illustrated in Fig. 4.16. It can be proven that the frequency dependent values of $C_{eq}(f)$ and $R_{eq}(f)$ are given by (4.58) and (4.59), respectively.

$$C_{eq}(f) = \frac{(C_1 + C_2)^2 + C_1^2 C_2^2 (R_1 + R_2)^2 (2\pi f)^2}{C_1 + C_2 + C_1 C_2 (C_1 R_1^2 + C_2 R_2^2)(2\pi f)^2} \quad (4.58)$$

$$R_{eq}(f) = \frac{C_1^2 R_1 + C_2^2 R_2 + C_1^2 C_2^2 R_1 R_2 (R_1 + R_2)(2\pi f)^2}{(C_1 + C_2)^2 + C_1^2 C_2^2 (R_1 + R_2)^2 (2\pi f)^2} \quad (4.59)$$

Figure 4.17 shows the graphical representation of (4.58) (upper graph) and (4.59) (lower graph), for the case where $R_1 < R_2$ and $C_1 < C_2$. The values of f_1 and f_2 are approximated by (4.60) and (4.61).

$$f_1 \simeq \frac{1}{2\pi R_1 C_1} \quad (4.60)$$

$$f_2 \simeq \frac{1}{2\pi R_2 C_2} \quad (4.61)$$

It can be seen that for moderately low frequencies $f < f_2$ the capacitance C_{eq} roughly equals the sum of C_1 and C_2 and that R_2 will be dominant in R_{eq}. For the case where $f > f_1$, R_{eq} equals about $R_1//R_2$ and C_{eq} is strongly decreased, but still higher than C_1. Thus, it is concluded that the capacitor with the largest capacitance and highest parasitic series resistance (MOS capacitor) is most effective

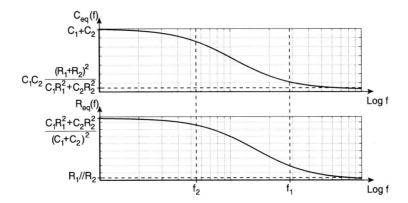

Fig. 4.17 The *upper graph* shows the equivalent capacitance $C_{eq}(f)$ and the *lower graph* shows the equivalent resistance $R_{eq}(f)$, both as a function of frequency f

at low frequencies and that the other capacitor (MIM capacitor) will be dominant at high frequencies. Obviously, the parasitic series resistance of both capacitors should be minimized in order to keep ΔU_{out} low. Notice that for the special case where $C_1 = C_2$ and $R_1 = R_2$ the value of C_{eq} is the sum of C_1 and C_2 and the value of R_{eq} equals $R_1/2 = R_2/2$. Moreover, for this case both C_{eq} and R_{eq} are frequency-independent. This property is used in the implementation of MOS capacitors, which is explained in Sect. 6.1.2.

4.2.3 Switches

The switches in the monolithic CMOS DC-DC converters, realized in this work, are all implemented as MOSFETs. This includes the freewheeling switches, denoted as SW_2 for both a boost and a buck converter (see Figs. 2.18 and 3.1), which could also be implemented as diodes. The reason for this is can be understood by comparing the power loss P_{Df} due to the forward voltage drop U_{Df} of a diode with the power loss P_{Ron} due to the on-resistance of a MOSFET, which are respectively defined by (4.62) and (4.63), for the static case.

$$P_{Df} = U_{Df} I_{ak} \tag{4.62}$$

$$P_{Ron} = R_{on} I_{ds}^2 \tag{4.63}$$

Figure 4.18 shows the graphical representation of P_{Df} and P_{Ron}, as a function of the current. It is observed that for a MOSFET with a certain on-resistance R_{on}, P_{Ron} will be lower than P_{Df} when the current is below a certain value. This threshold value for the current can be increased by lowering R_{on} of the MOSFET, while the curve of the diode is fixed by its constant forward voltage drop U_{Df} of about 0.6 V. Obviously, the dynamic losses, which are not taken into account in this comparison, are also determining for the current threshold point. For a MOSFET these dynamic

4.2 Non-ideal Converter Components Models

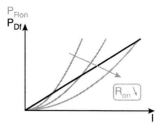

Fig. 4.18 The power loss P_{Df} of forward voltage drop of a diode (*black curve*) and the power loss P_{Ron} of the on-resistance of a MOSFET (*gray curve*), both as a function of the current I

losses mainly consist of parasitic capacitances (gate) and finite switching times, whereas for a diode the main dynamic losses are due to the reverse recovery and also the parasitic capacitances (junction). However, it turns out that for the designs conducted in this work the lowest power losses are obtained by means of MOSFET freewheeling switches. Note that this is partly due to the relatively low U_{out} of the converters in this work and that for high values of U_{out} a freewheeling diode might be the best choice.

On-Resistance

The first power loss which is caused by the MOSFET switches is related to their parasitic on-resistance. The first-order approximation of R_{on_n} for an n-MOSFET in the triode region, as illustrated in Fig. 1.18, is given by (4.64). For a p-MOSFET the expression for R_{on_p} is analogue.

$$R_{on\_n} = \frac{U_{ds}}{I_{ds}} = \frac{U_{ds}}{(1.5)} \simeq \frac{1}{\mu_n C_{ox} \frac{W_n}{L_n} \left(U_{gsn} - V_{Tn} - \frac{U_{dsn}}{2} \right)} \quad (4.64)$$

Because R_{on_n} is inverse proportional to the W_n/L_n ratio, the value for L_n will always be the minimum feature size of the given CMOS technology and the value of W_n is to be chosen large enough. It is clear that W_n is a design parameter that is to be determined in the design flow. For an improved matching between (4.64) and SPICE simulations the parameters of (4.64) are determined empirically, resulting in an improved accuracy.

The resulting power losses in SW_1 and SW_2, denoted as P_{Rsw1} and P_{Rsw2}, for both a boost and a buck converter are calculated through (4.65) and (4.66), respectively.

$$P_{Rsw1} = R_{SW1} I_{SW1\_RMS}^2 = R_{on} I_{SW1\_RMS}^2 \quad (4.65)$$

$$P_{Rsw2} = R_{SW2} I_{SW2\_RMS}^2 = R_{on} I_{SW2\_RMS}^2 \quad (4.66)$$

The expressions for I_{SW1_RMS} and I_{SW1_RMS}, for both a boost and buck converter and for both CMs, are calculated in Sect. 4.2.1. Note that P_{Rsw1} and P_{Rsw2} are resistive losses which are already taken into account in the respective second-order models for a boost and a buck converter, as discussed in Sects. 4.1.2 and 4.1.4.

Fig. 4.19 The parasitic capacitances in an n-MOSFET

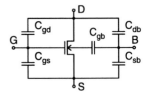

Parasitic Capacitances

The second power loss that is caused by the MOSFET switches is due to their parasitic capacitances. These different parasitic capacitances of an n-MOSFET, which are analogue to a p-MOSFET, are illustrated in Fig. 4.19. The parasitic capacitances of a MOSFET are namely:

- C_{gd}: The parasitic gate-drain capacitance.
- C_{gs}: The parasitic gate-source capacitance.
- C_{gb}: The parasitic gate-bulk capacitance.
- C_{db}: The parasitic drain-bulk capacitance.
- C_{sb}: The parasitic source-bulk capacitance.

The definition of these parasitic capacitances is given by (4.67), (4.68), (4.69), (4.70) and (4.71).

$$C_{gd} = \frac{\partial Q_g}{\partial U_{ds}} \qquad (4.67)$$

$$C_{gs} = C_{gg} + C_{gd} + C_{gb} = \frac{\partial Q_g}{\partial U_{gs}} + C_{gd} + C_{gb} \qquad (4.68)$$

$$C_{gb} = \frac{\partial Q_g}{\partial U_{sb}} \qquad (4.69)$$

$$C_{db} = \frac{\partial Q_d}{\partial U_{sb}} \qquad (4.70)$$

$$C_{sb} = \frac{\partial Q_s}{\partial U_{sb}} \qquad (4.71)$$

The resulting power losses from these parasitic capacitances are calculated through (4.52). In (4.52) U_{in} denotes the source voltage of which the parasitic capacitance is being charged, which can be either the in- or output voltage of the converter, and ΔU denotes the voltage swing over the parasitic capacitor.

Finite Switching Times

The third and last significant power loss in the MOSFET switches is caused by the finite switching times, needed to fully turn a MOSFET on and off. During these transients both a voltage $u_{SW}(t)$ over and a current $i_{SW}(t)$ through the MOSFETs exits,

4.2 Non-ideal Converter Components Models

resulting in non-ideal switching (see Fig. 1.3 for the concept of ideal switching). The power loss P_{tSW}, caused by the finite switching times, is calculated through (4.72).

$$P_{tSW} = f_{SW} \int_0^{t_{SW}} i_{SW}(t) u_{SW}(t) \, dt \qquad (4.72)$$

In (4.72) t_{SW} denotes the time during which the switching transients occur. When the switch is in steady-state, which can either be on or off, the power loss is either calculated through R_{on} or non-existent.

The rise- and fall-time of a MOSFET switch are mainly caused by the finite output resistance of the driver, used to charge and discharge the parasitic gate capacitance C_g of the MOSFET switch. The drivers for the MOSFET switches, used in the implementations in this work, consist of digital tapered buffers. For this type of driver the rise-time t'_r of the MOSFET switch is calculated by (4.73) [Rab03].

$$t'_r = -\ln\left(\frac{1}{2}\right) \frac{3}{4} \frac{U_{dd}}{\mu_n C_{ox} \frac{W_{p\_buff}}{L_{p\_buff}} \left((U_{dd} + V_{Tp})U_{dsatp} - \frac{U_{dsatp}^2}{2}\right)\left(1 - \frac{7}{9}\lambda_p U_{dd}\right)}$$
$$\cdot (C_g + 2C_{gdn} + 2C_{gdp} + C_{dbn} + C_{dbp}) \qquad (4.73)$$

All the parameters in (4.73) are associated with the n-MOSFET and p-MOSFET of the last stage of the digital tapered buffer, except for C_g which is the parasitic gate capacitance of the driven MOSFET switch. Furthermore, λ_p denotes the early voltage of a p-MOSFET and U_{dsatp} is the drain-source saturation voltage of a p-MOSFET. The calculation of the fall-time t'_f is analogue to (4.73).

In (4.73) it assumed that the last stage of the digital tapered buffer is driven by a signal with infinite steep flanks, which is not consistent with the reality. By means of (4.74) this effect is compensated for, yielding the eventual values for the rise- and fall-time, denoted as t_r and t_f.

$$t_{r/f} = \sqrt{t'^2_{r/f} + t^2_{flank}} \qquad (4.74)$$

In (4.74) t_{flank} is the mean rise- and fall-time of the signal that drives the last buffer stage. This parameter is CMOS technology dependent.[10]

In order to determine $i_{SW}(t)$ and $u_{SW}(t)$ during the switching transients for a boost converter, Fig. 4.20 is considered. The qualitative waveforms of $i_{SW1}(t)$, $u_{SW1}(t)$, $i_{SW2}(t)$ and $u_{SW2}(t)$ are illustrated, for both DCM and CCM, in the respective Figs. 4.20(a) and 4.20(b). The respective losses of both switches during t_r and t_f are calculated through (4.72), using the linear approximation of $i_{SW}(t)$ and $u_{SW}(t)$. For the boost converter in DCM, for which Fig. 4.20(a) is considered, this yields (4.75), (4.76) and (4.77).

---

[10]The approximate values of t_{flank} for a 180 nm CMOS, a 130 nm CMOS and a 90 nm CMOS process, respectively are 28.5 ps, 14 ps and 7.5 ps.

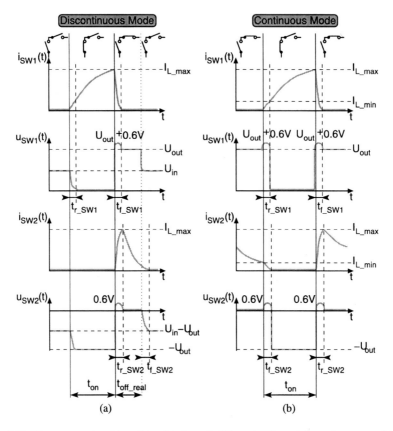

Fig. 4.20 The currents $i_{SW1}(t)$ and $i_{SW2}(t)$ through SW_1 and SW_2 and the voltages $u_{SW1}(t)$ and $u_{SW2}(t)$ over SW_1 and SW_2 for a boost converter in (**a**) DCM and (**b**) CCM

$$P_{tr\_SW1} = \frac{f_{SW} I_{L\_max} t_{r\_SW1}^2 U_{in}}{6 t_{on}} \quad (4.75)$$

$$P_{tf\_SW1} = \frac{f_{SW} I_{L\_max} t_{f\_SW1} (U_{Df} + U_{out})}{2} \quad (4.76)$$

$$P_{tr\_SW2} = \frac{f_{SW} I_{L\_max} t_{r\_SW2} U_{Df}}{2} \quad (4.77)$$

Note that the power loss P_{tf_SW2} during the fall-time of SW_2 is considered to be zero in DCM. The reason for this is the fact that SW_2 is turned off at the time where $i_{SW1}(t) = 0$, as can be seen in Fig. 4.20(a).

The switching losses due to the finite t_r and t_f of the boost converter in CCM, for which Fig. 4.20(b) is considered, is given by (4.78), (4.79), (4.80) and (4.81). In this case P_{tf_SW2} is not zero because SW_2 is turned off when $i_{SW2}(t)$ is still flowing.

4.2 Non-ideal Converter Components Models

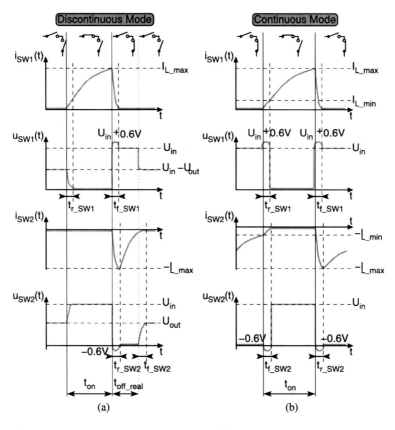

Fig. 4.21 The currents $i_{SW1}(t)$ and $i_{SW2}(t)$ through SW_1 and SW_2 and the voltages $u_{SW1}(t)$ and $u_{SW2}(t)$ over SW_1 and SW_2 for a buck converter in (a) DCM and (b) CCM

$$P_{tr\_SW1} = \frac{f_{SW} I_{L\_min} t_{r\_SW1} (U_{Df} + U_{out})}{2} \quad (4.78)$$

$$P_{tf\_SW1} = (4.76) \quad (4.79)$$

$$P_{tr\_SW2} = (4.77) \quad (4.80)$$

$$P_{tf\_SW2} = \frac{f_{SW} I_{L\_min} t_{f\_SW2} U_{Df}}{2} \quad (4.81)$$

For the buck converter the qualitative representation of $i_{SW1}(t)$, $u_{SW1}(t)$, $i_{SW2}(t)$ and $u_{SW2}(t)$ is shown in Figs. 4.21(a) and 4.21(b), for both DCM and CCM. Similar to the boost converter the combination of (4.72) together with the linearized functions for $i_{SW1}(t)$, $u_{SW1}(t)$, $i_{SW2}(t)$, $u_{SW2}(t)$ is used for the calculation of the respective rise- and fall power losses of the switches of the buck converter. When considering Fig. 4.21(a) for the buck converter in DCM, this yields (4.82), (4.83) and (4.84). Also, similar to the boost converter P_{tf_SW2} is assumed to be insignificant.

$$P_{tr\_SW1} = \frac{f_{SW}I_{L\_max}t_{r\_SW1}^2(U_{in} - U_{out})}{6t_{on}} \quad (4.82)$$

$$P_{tf\_SW1} = \frac{f_{SW}I_{L\_max}t_{f\_SW1}(U_{Df} + U_{in})}{2} \quad (4.83)$$

$$P_{tr\_SW2} = (4.77) \quad (4.84)$$

For the buck converter in CCM, illustrated in Fig. 4.21(b), the calculations of rise- and fall-time power losses yield (4.85), (4.86), (4.87) and (4.88).

$$P_{tr\_SW1} = \frac{f_{SW}I_{L\_min}t_{r\_SW1}(U_{Df} + U_{in})}{2} \quad (4.85)$$

$$P_{tf\_SW1} = (4.83) \quad (4.86)$$

$$P_{tr\_SW2} = (4.84) \quad (4.87)$$

$$P_{tf\_SW2} = (4.81) \quad (4.88)$$

In the waveforms of $u_{SW1}(t)$ and $u_{SW2}(t)$ of the boost and buck converter, shown in Figs. 4.20 and 4.21, an additional voltage of 0.6 V is observed during the switching transients. This voltage is caused by the forward voltage drop U_{Df} of the parasitic drain-bulk diode of the freewheeling MOSFET switches. This parasitic diode is forward biased during the transition between the charge and discharge phase and also during the transition between the discharge and the charge phase in CCM. Indeed, during these transitions both switches need to be off for a short period, avoiding a short-circuit of either the output capacitor in a boost converter or U_{in} in a buck converter. However, the current through the inductor cannot change instantly due to the law of Lenz (2.44) and will therefore keep on flowing through the parasitic drain-bulk diode during the switching transitions.

This bulk conduction is illustrated for both the boost and the buck converter in the respective Figs. 4.22(a) and 4.22(b). For the boost converter the freewheeling switch is a p-MOSFET and the bulk conduction will occur in the n-well, between the p^+-region of the drain and the n^+-region of the bulk. In the buck converter the freewheeling switch is an n-MOSFET and the bulk conduction will occur in the substrate, between the p^+-region of the bulk and the n^+-region of the drain. Clearly, this bulk conduction cannot be avoided, for sake of maintaining the correct operation of the DC-DC converter. Nevertheless, the bulk conduction introduces a serious risk for latch-up. Therefore, additional lay-out measures are to be taken in order to avoid this unwanted behavior, which will be discussed in Sect. 6.1.3.

4.2.4 Buffers

It is already mentioned in Sect. 4.2.3 that the gate of the MOSFET switches is driven by digital tapered buffers. The circuit of these buffers is shown in Fig. 4.23. Each stage of the buffer comprises a CMOS inverter, which in-turn consists of an n-MOSFET and a p-MOSFET. The output signal is non-inverted, with respect to the input signal, for an even number of stages. All the MOSFETs in the buffer

4.2 Non-ideal Converter Components Models

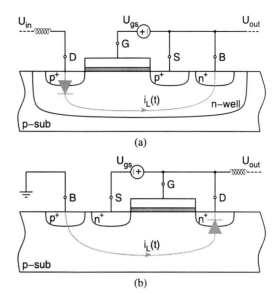

Fig. 4.22 The physical cross-sections (**a**) of the freewheeling p-MOSFET in a boost converter and (**b**) the freewheeling n-MOSFET in a buck converter. In both cross-sections the bulk current, which occurs at the transition between the charge and discharge phase, is shown

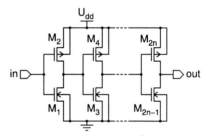

Fig. 4.23 The circuit of a digital tapered CMOS buffer with n-stages

have the minimum technology feature size for L. The p-MOSFETs in each inverter stage are given a larger W than the corresponding n-MOSFETs, such that $t_r = t_f$. This is important for the preservation of the pulse width of the buffered signal, as explained in Chap. 5. Each consecutive inverter stage is scaled to be larger than the previous stage, starting with a minimal sized inverter. The fixed scaling factor f_{scale} is designed for minimal propagation delay and is calculated through (4.89) [Rab03].

$$f_{scale} = \sqrt[n]{F} = \sqrt[n]{\frac{C_g}{C_{g\_min}}} \quad \text{with} \quad n = \ln(F) = \ln\left(\frac{C_g}{C_{g\_min}}\right) \qquad (4.89)$$

In (4.89) n denotes the number of stages, F is the global effective fan-out of the buffer, C_g is the parasitic gate capacitance of the driven MOSFET and C_{g_min} denotes the parasitic gate capacitance of a minimal sized inverter.

The first power loss associated with the tapered buffer is a dynamic loss caused by the parasitic capacitances of the MOSFETs. These parasitic capacitances are modeled, similar to the switches, by means of Fig. 4.19. The according total power loss P_{buff_cpar} is calculated through (4.52). The second power loss caused by the tapered buffer is due to the momentary drain-source short-circuit current I_{ds_i} that occurs in the individual buffer stages during the switching transients. This power loss P_{buff_short} is approximated by means of (4.90) [Vee84].

$$P_{buff\_short} = U_{dd} \sum_{i=1}^{n} \overline{I_{ds\_i}} \simeq U_{dd} \sum_{i=1}^{n} \frac{\mu C_{ox}}{12} \frac{W}{L} f_{SW} t_{r/f} (U_{dd} - 2V_t)^3 \qquad (4.90)$$

In (4.90) n denotes the number of stages of the tapered buffer and $\overline{I_{ds\_i}}$ is the mean drain-source short-circuit current of the ith buffer stage. Also, the following assumptions are made: $\mu_n = \mu_p = \mu$, $W_n = W_p = W$ and $t_r = t_f = t_{r/f}$.

It can be proven that by reducing the number of stages n of the tapered buffer, defined by (4.89), by one, a significant power reduction can be obtained. Moreover, by doing so the increase of t_r and t_f does not lead to a significant increase of the power loss, caused by the finite switching transients of the power switches of the DC-DC converter. Also, by eliminating the last stage the area requirement of the buffer can be significantly reduced. In the designs in this work this area decrease is typically about a factor two.

4.2.5 Interconnect

In the domain of DC-DC converters with off-chip components it is well known that the length of the interconnect of the current path of the DC-DC converter is to be minimized. When not designed and modeled carefully, this interconnect will have a significant negative impact on the performance of the DC-DC converter, in terms of η_{SW}, P_{out_max} and ΔU_{out}. These issues remain valid for monolithic DC-DC converters, where they can potentially pose severe problems. Therefore, special care has to be taken for the lay-out of both the converter components and the interconnect, which is elaborated upon in Chap. 6. Moreover, because of this fact, the design of monolithic DC-DC converters is an iterative process, involving a constant feedback between the actual design, simulations and the lay-out.

The main types of on-chip interconnect consists of on-chip metal-tracks and bondwires.[11] The calculation of the series resistance of the latter, including the frequency dependent skin-effect, is discussed in Sect. 4.2.1. On-chip metal track connections have a rectangular cross-section, as shown in Fig. 4.24, of which the thickness d is determined by the CMOS technology and the metal layer number. In standard CMOS technologies these metal tracks consist of aluminum[12] or copper.[13] Because of the fixed d the total parasitic series resistance R_{track} is calculated through the parameter called the square-resistance $R_\square$, by means of (4.91).

$$R_{track} = R_\square \frac{L_{track}}{W_{track}} \quad \text{with} \quad R_\square = \frac{\rho}{d} \qquad (4.91)$$

---

[11] Other types of interconnect, such as flip-chip and passive-die, or chip-stacking, are not considered in this work.

[12] For aluminum: $\rho_{Al} = 2.7 \cdot 10^{-8}$ Ωm @ 293 K.

[13] For copper: $\rho_{Cu} = 1.7 \cdot 10^{-8}$ Ωm @ 293 K.

Fig. 4.24 A perspective view of a square metal-track conductor, with the definition of its width W_{track}, its length L_{track} and its thickness d

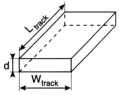

In (4.91) L_{track} and W_{track} are the respective length and width of the metal-track, as defined in Fig. 4.24. Obviously, (4.91) is valid for moderately low-frequencies ($f < 100$ MHz), where the skin-effect is not yet significant. For higher frequencies the series resistance is simulated FastHenry.

Another factor that contributes to the series resistances of on-chip interconnect consists of the vias that provide the connection between two consecutive metal layers, which is shown in Fig. 1.21. These vias consist of tungsten[14] or aluminum and they usually have, technology determined, fixed dimensions. Thus, the parasitic resistance R_{via_tot}, introduced by the vias between two metal-tracks, is calculated through (4.92).

$$R_{via\_tot} = \frac{R_{via}}{\#via} \quad (4.92)$$

In (4.92) R_{via} is the resistance of a single via and $\#via$ denotes the number of vias.

Input & Output

The interconnect at the in- and output of monolithic DC-DC converters comprises both bondwires, for the connection with the package or PCB, and on-chip metal-tracks. The combined parasitic series resistance R_{in} of the bondwires and metal-tracks at the input of the converter introduces an on-chip input voltage ripple ΔU_{in} and a power loss P_{Rin}, which is calculated through (4.93) for a boost and a buck converter.

$$P_{Rin} = R_{Uin} I_{in\_RMS}^2$$
$$\text{with} \begin{cases} \text{Boost: } I_{in\_RMS}^2 = I_{SW1\_RMS}^2 \\ \text{Buck: } I_{in\_RMS}^2 = I_{SW1\_RMS}^2 + I_{SW2\_RMS}^2 \end{cases} \quad (4.93)$$

The currents I_{SW1_RMS} and I_{SW2_RMS} are provided in Sect. 4.2.1, both for a boost and a buck converter. Note that the effect of R_{in} is already taken into account in the second-order models for the boost and the buck converter, which are explained in the respective Sects. 4.1.2 and 4.1.4.

As explained in Sect. 4.2.1, bondwires introduce a certain parasitic inductance. At the input of a monolithic DC-DC converter this parasitic inductance L_{in} causes,

---

[14] For tungsten: $\rho_W = 5.5 \cdot 10^{-8}$ Ωm @ 293 K.

Fig. 4.25 (a) The model for the parasitic input resistance R_{in} and inductance L_{in}, with an on-chip decouple capacitor C_{dec} and its parasitic series resistance R_{Cdec}. (b) The equivalent impedance circuit of this model

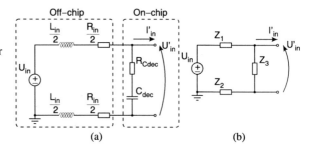

in addition to R_{in}, the on-chip input voltage ripple ΔU_{in} to increase even more. The equivalent circuit of the interconnect at the input of the converter is shown in Fig. 4.25(a). An input decouple capacitor C_{dec}, which is necessary in some cases to reduce ΔU_{in}, is also added. Together with C_{dec}, its parasitic series resistance R_{Cdec} is taken into account, as it influences the effectiveness of C_{dec}. Note that other types of input decoupling, for instance RLC-decoupling, are also possible [Ing97]. However, these alternatives are not practical for DC-DC converters because they introduce an additional parasitic input resistance. Also, in addition to the on-chip decoupling, off-chip decoupling is used to avoid excessive voltage swing due to long PCB-tracks. This will be discussed in Sect. 6.2.

For the calculation of ΔU_{in} the equivalent impedance circuit of Fig. 4.25(b) is considered. The result of this calculation and its translation to the circuit of Fig. 4.25(a) is given by (4.94), both for a boost and a buck converter.

$$\Delta U_{in} = U'_{in\_max} - U'_{in\_min} = \Delta I_{in}(Z_3//(Z_1+Z_2))$$
$$= \frac{R_{Cdec}R_{in} + 2\pi f R_{Cdec}L_{in}}{R_{Cdec} + 2\pi f R_{in}C_{dec} + (2\pi f)^2 L_{in}C_{dec}}$$

with
$$\begin{cases} \text{Boost: } \Delta I_{in} = I'_{in\_max} - I'_{in\_min} = I_{L\_max} \\ \text{Buck: } \Delta I_{in} = I'_{in\_max} - I'_{in\_min} = I_{L\_max} - I_{L\_min} \end{cases} \quad (4.94)$$

In (4.94) f is defined by (4.45), which is not equal to f_{SW} in DCM. Furthermore, it is observed that ΔU_{in} is linear proportional to the current ripple ΔI_{in} at the input

Table 4.1 The parameters used for the calculation example of the input decouple capacitor of a DC-DC converter, shown in Fig. 4.26

Parameter	Value
L_{in}	2 nH
R_{in}	0.2 Ω
R_{Cdec_1}	0.05 Ω
R_{Cdec_2}	0.2 Ω
R_{Cdec_3}	0.5 Ω
f	100 MHz
ΔI_{in}	0.2 A

4.2 Non-ideal Converter Components Models

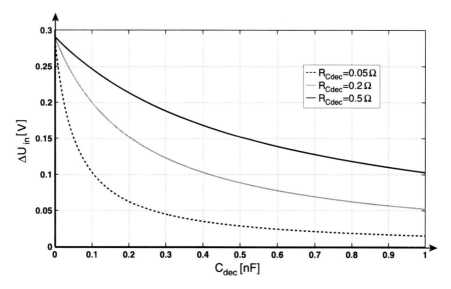

Fig. 4.26 The on-chip input voltage ripple ΔU_{in} as a function of the capacitance of the decouple capacitor C_{dec}, for three different values of the parasitic series resistance R_{Cdec} of the decouple capacitor. The parameters for which this plot is valid are given in Table 4.1

of the DC-DC converter. Compared to the boost converter, this ΔI_{in} is smaller for the buck converter, as explained in Chap. 3.

In order to gain a better insight into the effect of C_{dec} and R_{Cdec} on ΔU_{in}, realistic numerical data for an example is provided in Table 4.1. This data is used in the graph of Fig. 4.26, which shows ΔU_{in} as a function of the capacitance C_{dec}, for three different values of R_{Cdec}. It is observed that the value of ΔU_{in} decreases upon increasing values of C_{dec}. Also, the value of ΔU_{in} is lower for lower values of R_{Cdec}. The case where $R_{Cdec} = 0.2\ \Omega$ corresponds to a MIM capacitor and the case where $R_{Cdec} = 0.5\ \Omega$ corresponds to a MOS capacitor, which is verified in Sect. 6.1.2. For example, when a value of $\Delta U_{in} = 0.1$ V is required, the capacitance of the MOS capacitor will have to be approximately a factor 2.5 larger than for a MIM capacitor. Hence, whether to use a MIM and/or a MOS capacitor depends on the capacitance density and the maximum operating voltage.

Note that ΔU_{in} also has a negative impact on ΔU_{out}, by increasing the latter. However, this effect is complex to model because it depends on many variables, including the output capacitance of the converter and various parasitics. Through simulations, which are confirmed by measurements, it is observed that this effect will be negligible, providing $\Delta U_{in} < 10\%\ U_{in}$.

At the output of the converter metal-track interconnect and bondwires (for the case of off-chip loads) also introduce a parasitic series resistance R_{out}. The power loss P_{Rout} caused by R_{out} is calculated through (4.95).

$$P_{Rout} = R_{out} I_{out\_RMS}^2 = R_{out} \frac{U_{out\_RMS}^2}{R_L} \tag{4.95}$$

Obviously, the effect of R_{out} becomes more significant at higher values for P_{out}, which is also noticed through the measurements in Chap. 6.

On-Chip Interconnect

The on-chip interconnect with metal-tracks introduce additional parasitic resistances, capacitances and inductances between the DC-DC converter components. It is assumed that the parasitic inductances of these connections are minimal and cause no significant effects. The parasitic series resistances and capacitances of the on-chip metal connections, on the other hand, are considered to be potentially significant and thus added to the already identified parasitic series resistances and capacitances.

It is clear that the additional parasitic series resistances and capacitances due to interconnect are largely dependent on the chip lay-out of the monolithic DC-DC converter. Therefore, a periodical feedback between the design and lay-out is necessary to obtain an optimal design.

4.3 Temperature Effects

A final effect which is to be taken into account for the boost and buck converter steady-state model is the influence of the chip die temperature on the performance characteristics. Any given DC-DC converter has a power conversion efficiency η_{SW} which is lower than 100%, causing the dissipation of a certain power P_{diss}, as calculated by (1.3). As a result of P_{diss} the temperature of the converter components will rise to a certain amount above the ambient temperature. In DC-DC converters with external components the temperature increase will differ for the individual converter components, requiring a customized model [Zar08]. For monolithic DC-DC converters, on the other hand, it is reasonable to assume that this temperature increase will be uniformly distributed over the entire chip die, because the thermal conductivity[15] λ_{th} of silicon is fairly high. Therefore, the temperature increase ΔT above the ambient temperature of a monolithic DC-DC converter chip can be first-order approximated by (4.96).

$$\Delta T = \gamma P_{diss} \qquad (4.96)$$

In (4.96) γ is the thermal resistance of the chip die to the ambient. When the chip die is directly mounted onto a PCB (FR4), which is the case for the majority of measurement setups which are discussed in Chap. 6, the value of γ can easily be in the order 250 K/W [Tak00]. This implies that the temperature increase of the chip

---

[15] For silicon and copper the thermal conductivity is respectively $\lambda_{th\_Si} = 150$ W/(m · K) and $\lambda_{th\_Cu} = 390$ W/(m · K).

4.3 Temperature Effects

die will become significant at values of P_{diss} in the order of 100 mW, which is well in the range of the designs presented in this work (see Chap. 6).

In Sects. 4.3.1 and 4.3.2 the effect of ΔT on the performance of the inductor and the MOSFET switches is discussed. The actual implementation into the second-order model for the boost and the buck converter is discussed in Sect. 4.4.2.

4.3.1 Inductor

The metal-tracks or bondwires of the on-chip inductor have a parasitic series resistance R_{Ls}, as explained in Sect. 4.2.1. This R_{Ls} has a positive temperature coefficient (PTC), implying that its value will increase upon an increasing temperature. For a certain temperature increase ΔT above the ambient temperature, R_{Ls} is calculated by means of (4.97).

$$R_{Ls@T+\Delta T} = R_{Ls@T}(1 + \alpha \Delta T) \qquad (4.97)$$

In (4.97) $R_{Ls@T}$ is the parasitic series resistance of the inductor at a certain known temperature T, $R_{Ls@T+\Delta T}$ is the parasitic series resistance of the inductor at a temperature $T + \Delta T$ and α is the resistance temperature coefficient[16] of the metal of the windings.

For a metal-track inductor it can be intuitively seen that the temperature of the windings will be quite evenly distributed and approximately equal to the chip die temperature. Whereas for a bondwire inductor, a temperature gradient in the bondwires exists, which can be approximated by a parabola [Nob00]. However, for sake of simplicity it is assumed that the average bondwire temperature is approximately equal to the chip die temperature.

It is clear that the increase of the chip die temperature ΔT, due to P_{loss}, will result in a higher value of R_{Ls}. This will in-turn cause P_{RLs}, and therefore also P_{loss}, to increase. Obviously, this effect will become more significant at high values of P_{loss}, which is associated with high values of P_{out}.

4.3.2 Switches

For the MOSFET switches the temperature effects are considered both in the on- and in the off-state. In the on-state the on-resistance will be prone to a PTC-effect and in the off-state the MOSFET switches will introduce a leakage current, which increases with their temperature. These two phenomena will be discussed in the following sections, along with their effect on the performance of the DC-DC converter.

---

[16] For gold, aluminum and copper: $\alpha_{Au} \simeq \alpha_{Al} \simeq \alpha_{Cu} = 4 \cdot 10^{-3}$ 1/K.

On-Resistance

Similar to the metal of an inductor, the induced channel of a MOSFET is prone to a PTC-effect. The nature of this PTC-effect is mainly due to the fact that V_t will increase and μ will decrease upon increasing temperature, but this is further beyond the scope of this work. Although specialized electrothermal SPICE-models, which include self-heating, are described in the literature [Zar10], a simpler and straightforward empirical approach is preferred in this work. More concrete, the PTC-effect of R_{on} is first-order approximated through (4.98), which is similar to (4.97).

$$R_{on@T+\Delta T} = R_{on@T}(1 + \alpha_{MOS}\Delta T) \qquad (4.98)$$

In (4.98) the on-resistance temperature coefficient α_{MOS} of the MOSFET is determined through SPICE simulations. It is assumed to be equal for both an n-MOSFET and a p-MOSFET and independent of both ΔT and W. As R_{on} also comprises the parasitic series resistance of the drain and source connections, as explained in Sect. 4.2.5, this is also included in the value of α_{MOS}. For the CMOS technologies used in this work, explained in Chap. 6, α_{MOS} is in the order of $3.5 \cdot 10^{-3}$ 1/K.

Eventually, the result of the PTC-effect of R_{on} is similar to that of R_{Ls} of the inductor, which is discussed in Sect. 4.3.1.

Leakage Current

The drain-source leakage current I_{ds_leak} of a MOSFET increases upon increasing temperature. For the CMOS technologies used in this work I_{ds_leak} at room temperature 300 K is in the order of a few µA and the increase of I_{ds_leak} at a temperature of 400 K is about two orders of magnitude. When assuming that the DC-DC converter will operate at near-room temperature, the power loss due to I_{ds_leak} at low values of P_{out} will be in the order of a few µW. When keeping in mind that the minimum targeted P_{out} in this work is in the order of a few mW, this power loss will not be significant. At higher chip die temperatures (400 K) these losses can add up to a few hundred µW, which is still not significant because in that case the total dissipated power loss P_{diss} will be in the order of hundreds of mW. For these reasons the power loss due to the MOSFET leakage current is not taken into account in the models for the boost and the buck converter.

4.4 The Final Model Flow

The most significant resistive power losses of the boost and the buck converter are taken into account in their respective second-order models, which are explained in Sect. 4.1. In addition the most significant dynamic losses are identified and explained in Sect. 4.2. These dynamic losses are however not yet taken into account in the second-order models for the boost and the buck converter. Therefore, the

4.4 The Final Model Flow

integration of these dynamic losses into the second-order models is explained in Sect. 4.4.1. This will lead to the final calculation of U_{out_RMS} and η_{SW}.

The influence of the chip die temperature onto the power loss in the inductor and the switches, as explained in Sect. 4.3, is also taken into account for the final model. The method for doing this is discussed in Sect. 4.4.2. Finally, important general considerations on the design of monolithic DC-DC converters, which are deduced by means of the mathematical steady-state model, are discussed in Sect. 4.4.3. Note that all the complex symbolic calculations, performed until this point, are performed by means of Mathematica®. The final model, used for the numerical calculations, is programmed in MatLAB®.

4.4.1 Inserting the Dynamic Losses

A number of the additional losses, which are discussed and modeled in Sect. 4.2, will cause less power to be transferred to the output of the converter or they will cause an additional power loss at the output of the converter. From this point of view, these losses can be modeled as an additional load resistance R_{loss} at the output of the converter. Hence, the real load R'_L, which the converter sees at its output, is calculated through (4.99).

$$R'_L = R_L // R_{loss} \tag{4.99}$$

This artificial additional load resistance will result in a decreased real RMS output voltage, denoted as U'_{out_RMS}, compared to the model which only accounts for the resistive losses, for the same input parameters. This U'_{out_RMS} is calculated through the power balance at the output of the converter. This means that the real output power P'_{out} of the converter is equal to the output power P_{out}, calculated with merely the resistive power losses, reduced with the dynamic losses P_{loss} at the output. This is formally translated into (4.100).

$$\begin{aligned} U'_{out\_RMS} &= \sqrt{R'_L P'_{out}} = \sqrt{R'_L (P_{out} - P_{loss})} \\ &= \sqrt{R'_L (P_{out} - P_{L\_sub} - P_{SW\_C} - P_{SW\_t} - P_{buff\_C} - P_{buff\_short} - P_{Rout})} \end{aligned} \tag{4.100}$$

In (4.100) P_{loss} consists of the following losses, valid for both the boost and the buck converter unless noted otherwise:

- P_{L_sub}: The parasitic substrate capacitance of the inductor.
- P_{SW_C}: The parasitic capacitances of SW_2 in both the boost converter and the buck converter.
- P_{SW_t}: The finite switching times of SW_1 and SW_2.
- P_{buff_C}: The parasitic capacitances in the buffer which drives SW_2, because in the designs of this work this buffer is powered through the output of the converter.

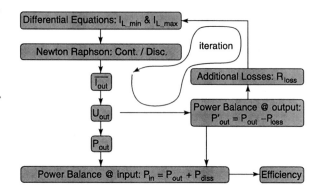

Fig. 4.27 The flow-chart of the model flow for the boost and the buck converter, starting from the differential equations and taking all the significant resistive and dynamic losses into account, except for the temperature effects

- P_{buff_short}: The short-circuit current of the buffer which drives SW_2, for the same reason as P_{buff_C}.
- P_{Rout}: The parasitic series resistance of the interconnect at the output of the converter. This is the only resistive loss not yet taken into account into the second-order model for the boost and the buck converter.

All the individual power losses in P_{loss} are a function of U'_{out_RMS}. This implies that the resulting value of R_{loss} will depend on both P_{loss} and U'_{out_RMS}, as stated by (4.101).

$$R_{loss} = \frac{U'^2_{out\_RMS}}{P_{loss}} = \frac{U'^2_{out\_RMS}}{P_{L\_sub} + P_{SW\_C} + P_{SW\_t} + P_{buff\_C} + P_{buff\_short} + P_{Rout}} \quad (4.101)$$

Because of the mutual dependency between U'_{out_RMS} and R_{loss}, an iteration is required for the final calculation of U'_{out_RMS}. In order to understand this, the entire model flow is to be considered, which is illustrated by the flow-chart of Fig. 4.27. This model flow, which is analogue for the boost and the buck converter, comprises the following steps:

1. The basic second-order differential equations of the converter are used for the calculation of I_{L_min} and I_{L_max}.
2. t_{off_real} is iteratively calculated by means of Newton-Raphson, when $t_{off\_real} < t_{off}$ the converter is operating in DCM, otherwise it is operating in CCM.
3. Through the charge balance of the output capacitor, when the converter is in steady-state operation, $\overline{I_{out}}$ is calculated.
4. $\overline{U_{out}}$ is calculated through the output RC network for the boost converter and directly out of $\overline{I_{out}}$ for the buck converter.
5. The additional losses P_{loss}, which are not included in the differential equations, are calculated by means of the power balance at the output of the converter. These losses are dependent on U_{out_RMS}.
6. The combination of U_{out_RMS} and P_{loss} yields the additional load resistance R_{loss}.
7. R_{loss} is used to recalculate U'_{out_RMS}, from step 1 onwards. This iteration is performed until the last calculated value of U'_{out_RMS} differs less than 0.1%

4.4 The Final Model Flow

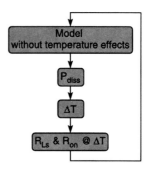

Fig. 4.28 The flow-chart showing the additional flow to take the temperature and self-heating effects into account for the model of the boost and the buck converter

from the previous one. This process typically requires five iterations, which is still acceptable for the total calculation time.

8. P_{out} is determined by means of the final value of U'_{out_RMS}.
9. The power-balance at the input of the converter is calculated to obtain P_{in}.
10. Finally, η_{SW} is determined through P_{in} and P_{out}.

The last two steps of the model flow are mathematically translated into (4.102).

$$\eta_{SW} = \frac{P'_{out}}{P_{in}} = \frac{P'_{out}}{P_{in}} = \frac{U'^2_{out\_RMS}}{R_L(P_{loss} - P_{RLs} - P_{Rcs} - P_{Rcp} - P_{Rsw} - P_{Rin})} \quad (4.102)$$

For the validation of this model through measurements the reader is referred to Chap. 6, where a maximal error of η_{SW} of about 4% is observed. Note that this model calculates all the important output parameters of the boost and the buck converter in steady-state, with a comparable accuracy compared to SPICE simulations. Moreover, the simulation time using this model is decreased approximately by a factor 30, compared to SPICE.

4.4.2 Inserting the Temperature Effects

The effect of the temperature and the self-heating is not yet taken into account in the model for the boost and the buck converter, which is discussed in Sect. 4.4.1. The additional flow required to achieve this is shown in the flow-chart of Fig. 4.28. This straightforward flow comprises the following steps:

1. The entire model for the boost and the buck converter, without the temperature and the self-hearing effects, is calculated for the standard temperature of $T = 300$ K. This yields the total dissipated power P_{diss} of the converter.
2. P_{diss} results in a temperature increase ΔT of the entire converter.
3. This ΔT in turn results into an increase of both R_{Ls} of the inductor and R_{on} of the two switches.
4. The model for the boost and buck converter is calculated a second time, for the new values of R_{Ls} and R_{on}. This yields the output parameters of the converter, accounting for the temperature and self-heating effects.

The results obtained with the added temperature and self-heating effects are validated by means of measurements in Chap. 6.

4.4.3 Reflections on Design

The main questions to be answered when designing monolithic DC-DC converters is the required chip die area, which determines the cost of the system, and the realistically achievable power conversion efficiency, which emphasizes the practical use. Prior research has attempted to answer these questions, without achieving realistic results [Kur02, Mus05b, Sch06], because of the simple models (see Sect. 2.3.3) that are being used. Moreover, none of this prior research, including [Kar06], leads to the basic qualitative trade-offs for monolithic DC-DC converters, which are crucial for their feasibility analysis and optimal design. It should not be underestimated that, due to the numerous input and output parameters, resulting into a multidimensional design space, the design of monolithic DC-DC converters is quite complex. This is exactly where the mathematical model, derived in the previous sections, plays a key role.

It is understood that the optimal design point is dependent on various CMOS IC technology parameters, of which many vary with the technology node (see Table 1.20). Examples of such varying parameters and their potential effect on the performance of a monolithic DC-DC converter are:

- The MOSFET transistor parameters, such as the minimum feature size (gate length L) and the oxide thickness t_{ox}. These parameters determine both the static and dynamic behavior of the MOSFET switches, altering the optimal R_{on} and f_{SW}. In addition, t_{ox} is inverse proportional to the capacitance density of MOS capacitors and will therefore significantly affect the achievable power density of the DC-DC converter.
- The number of metal layers, their individual thickness and resistivity. This has a large influence on the achievable Q-factor of metal-track inductors and R_{Cs} of monolithic capacitors. For the DC-DC converter this will have implications on η_{SW} and ΔU_{out}.
- Whether or not MIM capacitors are available in the given technology and their capacitance density and maximum operating voltage. This will also have a significant effect on the power density P_{out}/A and ΔU_{out} of the converter.

Due to these CMOS technology dependent parameters, a quantitative formulation of the design trade-offs for various different technology nodes is not considered useful. The large amount of available IC technologies, with their possible options and extensions, and their rapid evolution (see Sect. 1.3.1), demands for a more qualitative approach. Therefore, a discussion on qualitative design trade-off's, which are generally valid for monolithic DC-DC converters, is provided in this section. These trade-offs are obtained through the previously deduced mathematical model in combination with concrete practical know-how, acquired through numerous designs and implementations, in various CMOS technologies (see Chap. 6), in this work.

4.4 The Final Model Flow

For the determination of the design trade-offs a number of assumptions and requirements have to be formulated for the in- and output parameters of the DC-DC converter. Note that they are valid for both the boost and the buck converter. For the input parameters of the converter the following assumptions and requirements apply:

- U_{in}: The input voltage is kept constant.
- R_L: The load resistance is kept constant.
- f_{SW}: The switching frequency is varied such that the requirements for the output parameters are met.
- δ: The duty-cycle is varied such that the requirements for the output parameters are met.
- W_{SW1} & W_{SW2}: The width W of the switches is designed such that the condition $P_{Rsw} = P_{SW\_c} + P_{SW\_t}$ is fulfilled.
- A_L: The inductor area is proportional to $L^{2/3}$, for a constant track width [Cro96].
- R_{Ls}: The parasitic series resistance of the inductor is proportional to $L^{1/2}$, for a constant track width.
- C_{L_Csub}: The parasitic substrate capacitance of the inductor is linear proportional to A_L.
- A_C: The capacitor area is linear proportional to C.
- Other parasitics: Realistic values for other parasitics are chosen and kept constant, unless noted otherwise.

Analogue to the input parameters, the output parameters are to meet with the following assumptions and requirements:

- U_{out}: The output voltage is kept constant.
- ΔU_{out_max}: The output voltage ripple is kept smaller than $10\% \cdot U_{out}$.
- η_{SW}: Is maximized.

It is noted that the focus of these trade-offs is on the DC-DC converter only, regardless of the control strategy. For this reason P_{out} is assumed to be constant, because its impact on η_{SW} and ΔU_{out} is dependent on the type of control strategy, which is discussed in Chap. 5.

The resulting deduced qualitative design trade-offs, valid for both the boost and the buck converter, are graphically illustrated in Fig. 4.29. The clarification of each of these eight basic trade-offs, and their impact on other parameters, is as follows:

1. $f_{SW} \rightleftharpoons L$: Figure 4.29(a) shows f_{SW} as a function of L, for different values of C. A larger L requires a lower f_{SW}. This is due to the fact that more energy E_m can be stored in the inductor per switching cycle, as a higher L allows for a longer t_{on}, which is explained in Sect. 2.3.1. At high values of L, f_{SW} evolves towards a finite asymptote, due to the loss caused by the increasing R_{Ls} and C_{L_sub} of the inductor. At low values of L, f_{SW} becomes theoretically infinitely high, due to the increasing switching losses and the limited magnetic energy E_m stored in L. At higher values of C a lower f_{SW} suffices, for the same ΔU_{out}.
2. $\eta_{SW} \rightleftharpoons L$: Figure 4.29(b) shows η_{SW} as a function of L, for different values of C. At the extreme values for L, η_{SW} is not optimal. For low values of L this is due

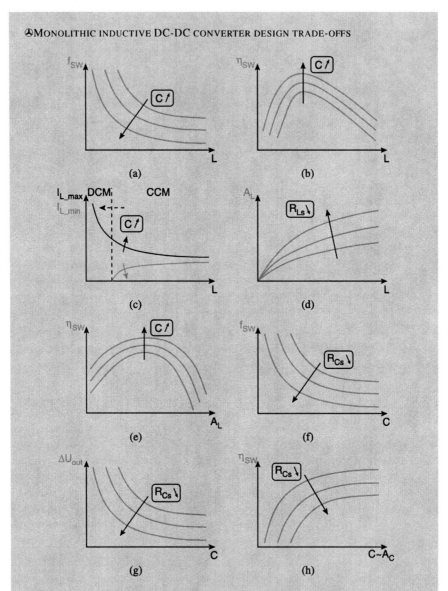

Fig. 4.29 The qualitative design trade-offs for monolithic DC-DC converters: (a) f_{SW} as a function of L for different values of C, (b) η_{SW} as a function of L for different values of C, (c) I_{L_max} and I_{L_min} as a function of L for different values of C, (d) A_L as a function of L, for different values of R_{Ls}, (e) η_{SW} as a function of A_L for different values of C, (f) f_{SW} as a function of C for different values of R_{Cs}, (g) ΔU_{out} as a function of C for different values of R_{Cs} and (h) η_{SW} as a function of $C \sim A_C$, for different values of R_{Cs}

to the high switching losses, associated with high values of f_{SW}. Whereas at high values of L the losses due to R_{Ls} and C_{L_sub} become dominant. Higher values of C are associated with higher η_{SW}, for the same L and ΔU_{out}, as a lower f_{SW} is required.

3. I_{L_min} & $I_{L\_max} \rightleftharpoons L$: Figure 4.29(c) shows I_{L_min} and I_{L_min} as a function of L, for different values of C. High values of L are associated with a longer t_{on} and t_{off} and vice versa. Thus, for high values of L the converter will tend to operate in CCM, whereas low values of L will lead to DCM. Higher values of C will allow a higher ΔI_L for the same ΔU_{out}.

4. $A_L \rightleftharpoons L$: Figure 4.29(d) shows A_L as a function of L, for different values of R_{Ls}. This curve follows from the assumption that $A_L \sim L^{2/3}$, for a constant winding width. Higher winding widths yield a lower R_{Ls} and a higher A_L, which is in-turn proportional to C_{L_sub}.

5. $\eta_{SW} \rightleftharpoons A_L$: Figure 4.29(e) shows η_{SW} as a function of A_L, for different values of C. The optimum value of η_{SW} is obtained for a certain value of A_L, such that $P_{RLs} = P_{L\_Csub}$. Higher values of C result in a higher value of η_{SW}, for the same ΔU_{out}, because optimal f_{SW} is lower.

6. $f_{SW} \rightleftharpoons C$: Figure 4.29(f) shows f_{SW} as a function of C, for different values of R_{Cs}. Lower values of C require a higher f_{SW} and vice versa, for the same ΔU_{out}. For a lower value of R_{Cs}, a lower value of C suffices for a certain ΔU_{out}.

7. $\Delta U_{out} \rightleftharpoons C$: Figure 4.29(g) shows ΔU_{out} as a function of C, for different values of R_{Cs}. A lower value of C yields a higher value of ΔU_{out} and vice versa, for the same optimal value of η_{SW}. A lower value of R_{Cs} requires a lower C, for the same ΔU_{out}.

8. $\eta_{SW} \rightleftharpoons C \sim A_C$: Figure 4.29(h) shows η_{SW} as a function of C, which is proportional to A_C, for different values of C. A larger value of C yields a higher value of η_{SW}, which saturates at a certain values due to the finite, non-zero losses in the other converter components. For a lower value of R_{Cs}, a lower value of C suffices to obtain the same η_{SW}.

Monolithic DC-DC converters in standard CMOS tend to have limited values of L and C (in the order of nH and nF), because of the limited available chip area. This inevitably leads to high values of f_{SW} and also in most cases to DCM operation. Therefore, the above mentioned trade-offs often result in a trade-off between: $\eta_{SW} \rightleftharpoons \Delta U_{out} \rightleftharpoons A$. The implications on real designs are illustrated with various design examples, in various CMOS technologies, in Chap. 6.

4.5 Conclusions

An accurate mathematical steady-state design model, for both a monolithic boost and buck converter, is deduced in this chapter. This model takes all the significant resistive, dynamic and self-heating related power losses into account. By means of the input parameters and the circuit parameters of the DC-DC converter, all the output parameters are calculated, including the important η_{SW}. The flow for building this model comprises the following four steps, which are elaborated upon in the respective Sects. 4.1, 4.2, 4.3 and 4.4:

1. *Second-order model*: When only considering the significant resistive losses, the boost and buck converter's respective charge and discharge circuits are described by (second-order) differential equations. These equations are used to obtain a closed-form expression for the calculation of U_{out}.
2. *Modeling the dynamic power losses*: Each converter component is modeled through an equivalent circuit, enabling the determination of the additional non-resistive power losses.
3. *Temperature effects*: The combined resistive and dynamic losses add up to a certain power dissipation P_{diss}. Due to the finite, non-zero thermal resistance γ form the chip die to the ambient, a certain amount of self-heating is caused by P_{diss}. The associated temperature increase ΔT in-turn results in an increase of R_{on} and R_{Ls}, which causes η_{SW} to decrease.
4. *The final model flow*: The dynamic losses are added to the second-order model by modeling them into an additional parasitic load resistance. The effect on the output parameters is calculated through an iteration. The temperature effects are added afterwards, such that the effect of both the resistive and the dynamic losses is taken into account for the self-heating.

Apart from the influence of the non-ideal converter components on the power losses, their influence on other important output parameters, such as ΔU_{in} and ΔU_{out}, is also discussed in Sect. 4.2.

Finally, this chapter is concluded with some considerations on important general qualitative design trade-offs for monolithic DC-DC converters, which are obtained through the model. The most important conclusion is that monolithic inductive DC-DC converters are limited by the fundamental trade-off between:

- η_{SW}: The power conversion efficiency.
- ΔU_{out}: The output voltage ripple.
- A: The chip die area.

It is clear that the performance of the DC-DC converter will also be dependent on the CMOS technology parameters, which is illustrated by means of the practical implementations in Chap. 6.

Chapter 5
Control Systems

The DC-DC power converter stage needs drive signals to enable the switches, which are generated by the control system. This control system is responsible for the second important task of the DC-DC converter, apart from achieving a certain voltage conversion: Regulating the output voltage to the desired level. This includes providing sufficient immunity against load and line variations. For the purpose of monolithic DC-DC converters a new set of challenges emerge, both in terms of the control strategy and the basic design of the control system. This is due to the requirement of high switching frequencies and short switching times, needed to guarantee the optimal performance of the monolithic DC-DC converter.

The fact that both the power stage and the control system of an inductive monolithic DC-DC converter tend to operate at extreme conditions, compared to converters that use external components, requires a hands-on approach. Indeed, the known control theory for analyzing and modeling the DC-DC control systems is up to this point no longer of practical use. This is partly due to the high switching frequencies and also because most of the known theory is intended for standard PWM and PFM control strategies. Therefore, the novel control strategies, based on the well-known PFM method, proposed in this chapter are discussed from a practical point of view. These discussions include the principle of the control strategies and the circuits used in the practical chip implementations of Chap. 6.

In this chapter the two conventional control strategies PWM and PFM are discussed and compared to each other in Sect. 5.1. The Constant On/Off-time (COOT) control strategy, together with implementation examples for single-phase, single-output and multi-output converters, is explained in Sect. 5.2. The Semi-Constant On/Off-time (SCOOT) control strategy, together with a implementation examples for multi-phase and multiple-output converters, is explained in Sect. 5.3. The Feed-forward Semi-Constant On/Off-Time (F^2-SCOOT) control strategy, together with an implementation example, is discussed in Sect. 5.4. The aspect of start-up, in combination with some start-up circuit implementations, is discussed in Sect. 5.5. Finally, the chapter is concluded in Sect. 5.6.

Fig. 5.1 The concept of a control system for an inductive DC-DC converter

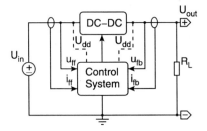

5.1 Inductive Type Converter Control Strategies

Controlling the output voltage of an inductive DC-DC converter is virtually always performed by means of a feedback/feed-forward mechanism. Most applications require a constant output voltage, which remains stable under varying load and line conditions. This is measured through the *line regulation* and *load regulation*, which are calculated by means of (5.2) and (5.1), respectively.

$$\text{Load Regulation} = \frac{1}{U_{out}} \frac{dU_{out}}{dI_{out}} \left[\frac{\%}{A}\right] \quad \text{with} \quad U_{in} = C^{st} \quad (5.1)$$

$$\text{Line Regulation} = \frac{1}{U_{in}} \frac{dU_{out}}{dU_{in}} \left[\frac{\%}{V}\right] \quad \text{with} \quad I_{out} = C^{st} \quad (5.2)$$

The concept of a control system, performing this task, is shown in Fig. 5.1. It physically controls the switches of the DC-DC converter, based on the information obtained from either the input voltage, the output voltage, the input (inductor) current or the output (inductor) current. Combinations of these inputs are also possible. The power supply U_{dd} for the control system can either be the input or the output voltage, possibly preceded by a dedicated (linear) voltage converter. The control systems in this work will be directly supplied by the input voltage (boost converter) or the output voltage (buck converter).

Control mechanisms for inductive DC-DC converter can essentially be divided into two categories: Pulse Width Modulation (PWM) and Pulse Frequency Modulation (PFM). PWM is the most widely used option for non fully-integrated DC-DC converters. The basic principle and an implementation example are provided in Sect. 5.1.1. Nevertheless, it will be proven that variations on PFM are the better choice for monolithic DC-DC converters. The principle of PFM is explained in Sect. 5.1.2. A comparison between PWM and PFM control schemes and their effect on power conversion efficiency, output voltage ripple and load regulation, is discussed in Sect. 5.1.3.

5.1.1 Pulse Width Modulation

PWM is a well known and widely used method for controlling inductive DC-DC converters with off-chip components. Modeling the small-signal behavior of such

5.1 Inductive Type Converter Control Strategies

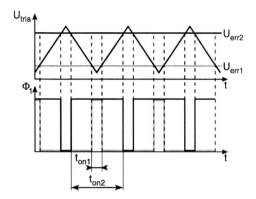

Fig. 5.2 The concept of Pulse Width Modulation (PWM) signal Φ_1 generation by means of comparing a triangular waveform U_{tria} to an error-voltage U_{err}

a non-linear system is performed through the state-space averaging method, which can be used both for CCM [Wes73] and DCM [Mak91]. These small-signal models are used to simulate the dynamic behavior, frequency response and stability of a DC-DC converter with a PWM control loop. Many extensions and adaptations of these basic models are described in the literature, for example to model multi-phase converters [Qiu06], to introduce component parasitics [Dav07] and for using current-mode feedback [Dav09]. However, these models are only usable until about one decade below f_{SW} [Wu98], making them unsuitable for high-frequency monolithic DC-DC converters, due to a lack of accuracy. Therefore, these models are omitted in this work. The PWM control system design is performed through transistor-level SPICE simulations, in addition to the stability and dynamic behavior characterization.

In the following two sections the concept of PWM and a practical implementation for a monolithic boost converter are discussed.

The Concept

Figure 5.2 illustrates the concept of generating a PWM signal Φ_1 by means of comparing triangular waveform U_{tria}, having a constant frequency equal to f_{SW}, to an error-voltage U_{err}. The error-voltage is generated by subtracting U_{out} from a reference voltage U_{ref} (see next section). When U_{err} has a larger amplitude than U_{tria}, Φ_1 is logic high and vice versa. By doing so it can be observed that t_{on} becomes larger with an increasing value of U_{err}. Thus, it follows that the duty-cycle δ is proportional to the amplitude of U_{err}. It can intuitively be understood U_{err} should not be allowed to rise or fall beyond the amplitude of U_{tria}, as this can lead to instability.

The PWM control method is suited for both CMs. However, for CCM and for $\delta > 50\%$ the stability of the converter can be compromised due to subharmonic oscillations, which can be modeled [Ki98, Pap04]. Nevertheless, it is demonstrated that these subharmonic oscillations are avoided by adding a compensation ramp to the control loop [Dei78]. Also, it is known that under certain circumstances in CCM chaos can emerge in a PWM control loop, which can be predicted to some extend

⌬SUBHARMONIC OSCILLATIONS IN PWM CONVERTERS

The operation of the DC-DC converter is considered normal when a constant value of t_{on}, and thus also δ, is maintained in steady-state operation. When the PWM controlled DC-DC converter operates in CCM and when $\delta > 50\%$, subharmonic oscillations can occur. The time-domain illustration of this phenomenon is shown in Fig. 5.3.

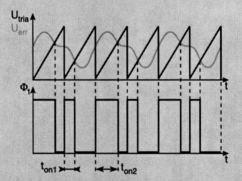

Fig. 5.3 The basic principle of subharmonic oscillations in a DC-DC converter with a PWM control loop

It is observed that in the case of subharmonic oscillation a periodic sequence of different $t_{on1} \neq t_{on2}$, and therefore also different $\delta_1 \neq \delta_2$, occur. This results in a dramatically increased ΔU_{out}, of which the fundamental frequency is lower than f_{SW}.

[Ham88, Li07]. Because of the fact that the PWM implementation in this work is designed to operate in DCM, subharmonic oscillations and chaos are not an issue. Therefore, these effects will not be considered here.

Implementation Boost Converter

For the purpose of controlling U_{out} of a monolithic boost converter in a 180 nm 1.8 V CMOS technology [Wen07], a DCM PWM control system is designed and implemented. The measurements of this converter are discussed in Sect. 6.3. The block diagram of this control system is illustrated in Fig. 5.4. The operation is explained by means of its three sub-blocks:

1. *PWM*: This block generates the PWM-signal for the control of the output voltage. First, an error-voltage U_{err} is generated by applying both a fraction (1/2) of output voltage U'_{out} and the reference voltage U_{ref}, which is used to set U_{out} to the desired level, to an integrator. This integrator consists of an OPAMP with negative feedback through a capacitor C_f. Next, U_{err} is level-shifted through the resistive divider R_{f1}–R_{f2}. The resulting signal is compared to a triangular

5.1 Inductive Type Converter Control Strategies

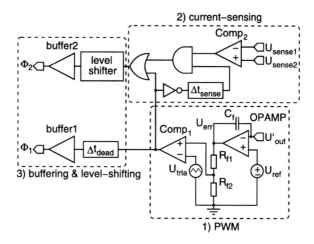

Fig. 5.4 The block diagram of the PWM control system implementation of a fully-integrated boost converter [Wen07]

waveform voltage U_{tria}, by means of a comparator $Comp_1$, yielding the PWM signal analogue to Fig. 5.2. U_{tria} also determines the constant $f_{SW} = 100$ MHz of the converter. The PWM signal is used to drive the current-sensing block and the buffering & level-shifting block.

2. *Current-sensing*: Because this control system is designed for DCM, the current through SW_2 (see Fig. 2.18(a)) is to be measured in order to determine the moment in time t_{off_real} to open SW_2. This is achieved by measuring a fraction of the voltage $U_{sense2} - U_{sense1}$ over SW_2 means of the comparator $Comp_2$. More precisely, $U_{sense2} - U_{sense1}$ is deduced from two resistive dividers, each between a node of SW_2 and GND, which are sized such that $U_{sense1} - U_{sense2}$ becomes zero when $i_{SW2}(t)$ is still slightly positive. This is done to compensate for the respective delays, introduced by $Comp_2$, the level-shifter and the buffer2. At the moment when SW_1 is opened the current-sensing block is overruled by the signal from the PWM block and SW_2 is closed automatically. After the time-delay $\Delta t_{sense} = 1.2$ ns has elapsed the actual current-sensing is enabled, ensuring SW_2 is opened when $i_{SW2}(t)$ becomes zero.

3. *Buffering & level-shifting*: The respective signals from the PWM and current-sensing block are used to control SW_1 and SW_2 of the boost converter. First, the signal from the PWM block is delayed through a time-delay Δt_{dead} of 350 ps and buffered through buffer1, yielding the active-high Φ_1 to drive SW_1. Δt_{dead} compensates for the delay of the level-shifter and the buffer2, ensuring the right timing between SW_1 and SW_2. Secondly, the signal from the current-sensing block is level-shifted from $U_{in} - U_{GND}$ to $U_{out} - U_{in}$ and buffered through buffer2, yielding the active-low Φ_2 to drive SW_2.

The OPAMP is implemented as a symmetrical cascoded Operational Transconductance Amplifier (OTA), with a current-loaded common emitter output stage, of which the circuit illustrated in Fig. 5.5. The DC-gain of 36 dB, the bandwidth (BW) of 200 kHz and the gain-bandwidth (GBW) of 8.4 MHz are empirically determined

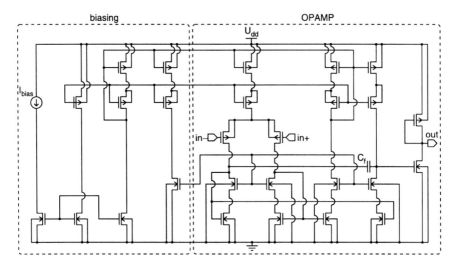

Fig. 5.5 The circuit of a symmetrical cascoded OTA with a current-loaded common emitter output stage

Fig. 5.6 The circuit of a comparator

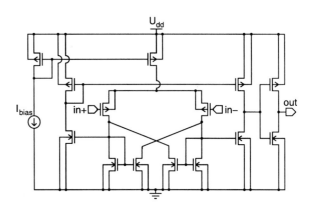

to ensure stable operation of the boost converter with its feedback loop, for a P_{out}-range of 25 mW to 150 mW.

Figure 5.6 shows the circuit used for the two comparators $Comp_1$ and $Comp_2$. The cross-coupled section introduces positive feedback, which is responsible for the fast clipping of the comparator. The rise- and fall-time are kept below a value of 200 ps, in order to ensure a sufficient fast response.

The circuit used for the time-delays Δt_{sense} and Δt_{dead} is shown in Fig. 5.7. It consists of a chain of minimal-sized digital inverters, loaded with a MOS capacitor. Δt_{sense} is implemented by means of multiple identical, series connected delay-cells, improving the duty-cycle preservation and increasing its immunity against mismatch.

5.1 Inductive Type Converter Control Strategies

Fig. 5.7 The circuit of a time-delay

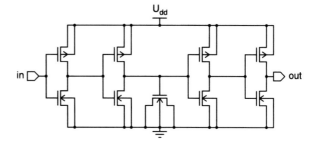

Fig. 5.8 The circuit of a level-shifter [Ser05]

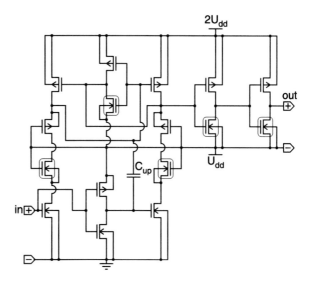

Figure 5.8 shows the circuit of the level-shifter [Ser05]. It converts the input voltage, varying between U_{in} and GND, to the output voltage, which varies between U_{out} and U_{in}. Capacitor C_{up} enables faster clipping of the cross-coupled pair, by means of charge-coupling. The resulting time-delay is minimized to a value of 210 ps.

Finally, the buffers are implemented as digital tapered buffers (see Fig. 4.23). Each buffer is designed to introduce minimal delay and consists of seven stages, with a scaling factor of 2.6.

5.1.2 Pulse Frequency Modulation

Similar to PWM, Pulse Frequency Modulation[1] (PFM) is also a well known method for controlling DC-DC converters, which originates from the 1960's [Sch64]. As

---

[1]The PFM control technique for DC-DC converters is also referred to as *ripple-based control*, *hysteric control*, *bang-bang control* and *one-shot control*.

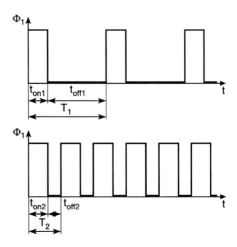

Fig. 5.9 The concept of Pulse Frequency Modulation (PFM), with a constant on-time t_{on}. The *upper graph* shows the timing for low load, low frequency operation and the *lower graph* shows the timing for high load, high frequency operation

such, models for both small and large signal behavior are described in the literature [Hon00, Sun02]. However, analogue to models for PWM, they lack accuracy at high switching frequencies (>100 MHz), making them unreliable for monolithic DC-DC converters. Consequentially, the design and analysis of PFM control systems is performed by means of SPICE transient simulations.

The concept of PFM, comprising a constant t_{on}, is illustrated in Fig. 5.9. The upper graph illustrates the timing for low load, low frequency operation and the lower graph shows the timing for high load, high frequency operation. Both the low and the high load operation use the same on-time $t_{on1} = t_{on2}$. As a consequence, when U_{in} and U_{out} are assumed constant, the regulation of U_{out} under varying loads is achieved by changing f_{SW}. This is in contrast to PWM, where f_{SW} is constant and the regulation of U_{out} under varying loads is achieved by altering t_{on}. Also, similar to PWM, PFM does not suffer any instability issues in DCM [Red09].

Apart from a constant t_{on}, other PFM control techniques exist [Red09], such as constant t_{off} control and hysteric control. All the control systems used in this work, except for one (see Sect. 5.1.1), are based on novel variations on the PFM control method.

5.1.3 Pulse Width Modulation vs. Pulse Frequency Modulation

In the previous Sects. 5.1.1 and 5.1.2 the basic concept of both PWM and PFM is explained. The main difference is the fact that PWM uses t_{on} to regulate U_{out}, while PFM uses f_{SW}. The impact of this difference on important specifications of DC-DC converters, for the purpose of monolithic integration, is discussed in the next sections.

5.1 Inductive Type Converter Control Strategies

Fig. 5.10 The power conversion efficiencies η_{SW_PFM} and η_{SW_PWM} of a PFM (constant t_{on}) and a PWM controlled DC-DC (*gray curve*) converter, as a function of the output power P_{out}. The *solid black curve* and the *dashed black curve* denote η_{SW_PFM} for equal switching frequencies $f_{SW\_PFM} = f_{SW\_PWM}$ at the maximal output power P_{out_max} and at the minimal output power P_{out_max}, respectively

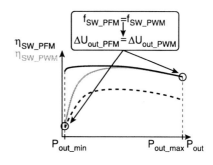

Power Conversion Efficiency

Figure 5.10 shows η_{SW_PFM} and η_{SW_PWM} of both a PFM (constant t_{on}) and a PWM controlled DC-DC converter, as a function of P_{out}. These trends are deduced by means of the mathematical model for the boost and the buck converter, discussed in Chap. 4. Realistic values, which are valid for monolithic DC-DC converters, are assumed for the input parameters.

When considering the gray curve for a PWM control system, it is observed that η_{SW_PWM} tends to drop towards low values of P_{out}. This is due to the constant f_{SW}, which causes the switching losses to become dominant at low values of P_{out}. At high values of P_{out} η_{SW_PWM} also tends to drop, which is due to the fact that the conduction losses become dominant. For values P_{out} in the order of a few hundred mW this effect is even more pronounced, due to the temperature effects (see Sect. 4.3).

For a PFM control system two different cases are plotted. The first case is denoted by the solid black curve, where η_{SW_PFM} is plotted for $f_{SW\_PFM} = f_{SW\_PWM}$ at P_{out_max}. This implies that $\eta_{SW\_PFM} = \eta_{SW\_PWM}$ at P_{out_max}. At low values of P_{out}, η_{SW_PFM} stays more constant and drops significantly slower. The reason is the variable f_{SW}, which decreases as P_{out} decreases. Hence, the switching losses are less dominant for PFM compared to PWM, at low values of P_{out}. At high values of P_{out} the conduction losses are dominant for both the PFM and PWM control technique, yielding comparable η_{SW}. The second case for the PFM control system is denoted by the dashed black curve, showing η_{SW_PFM} for $f_{SW\_PFM} = f_{SW\_PWM}$ at P_{out_min}. Again, this implies that $\eta_{SW\_PFM} = \eta_{SW\_PWM}$ at P_{out_min}. Due to the high value of f_{SW_PFM} at P_{out_min}, a low value for η_{SW_PFM} is obtained. This value increases somewhat for increasing values of P_{out}, however it never exceeds the value of η_{SW_PWM}. Obviously this is due to the massive switching losses, combined with the conduction losses and temperature effects at high values of P_{out}.

It is concluded that PFM controls system can obtain a higher overall η_{SW_PFM}, caused by an increase at the low end of the P_{out} range, compared to PWM control systems. This statement is only true when $f_{SW\_PFM} \leqslant f_{SW\_PWM}$. For constant t_{on} PFM control systems, this statement is also translated into the fact that the value of

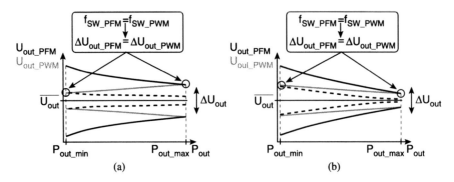

Fig. 5.11 The boundaries of the output voltages U_{out_PFM} and U_{out_PWM} of a PFM and a PWM (*gray curve*) controlled DC-DC converter as a function of the output power P_{out}, (**a**) for a boost converter and (**b**) for a buck converter in DCM. The *solid black curve* and the *dashed black curve* denote the boundary of U_{out_PFM} for equal switching frequencies $f_{SW\_PFM} = f_{SW\_PWM}$ at the maximal output power P_{out_max} and at the minimal output power P_{out_max}, respectively

t_{on} is to be chosen large enough. Indeed, for a constant t_{on}, and when assuming U_{in} and U_{out} are also constant and the converter is operating in DCM, a PFM control system can be regarded as a system that allows the DC-DC converter to transfer packets with a fixed amount of energy towards the output. Ideally, f_{SW} would therefore be linear proportional to P_{out}. This conclusion also follows from the equations for $k(\delta)$ in DCM, for both a boost and a buck converter, given by (2.72) and (3.2).

Output Voltage Ripple

Apart from the dependency of the output parameter ΔU_{out} upon the value of C, R_{Cs}, f_{SW} and δ, as explained in Sect. 4.4.3, ΔU_{out} also depends on P_{out} and the type of control system being used. Both these parameters should be taken into account in the design of a monolithic DC-DC converter, in order to make sure the specification for ΔU_{out} is met. Moreover, as explained in the previous section, the choice of the type of control system has significant impact on η_{SW}. This leads to an additional trade-off between ΔU_{out}, η_{SW} and A.

In order to gain a better insight into this trade-off Fig. 5.11 is considered. Both graphs show the boundaries of U_{out} for both a PWM and a PFM control system as a function of P_{out}, for a boost converter in Fig. 5.11(a) and for a buck converter in Fig. 5.11(b). The distance between the boundaries is equal to ΔU_{out}. These graphs are qualitative trends, valid for DCM and deduced with the model from Chap. 4. First, the case for a boost converter is considered in Fig. 5.11(a). The gray lines indicate the boundaries of U_{out} for a PWM controlled DC-DC converter. It is observed that ΔU_{out} is proportional to P_{out}. The solid black lines indicate the boundaries of U_{out} for a PFM control system, of which $f_{SW\_PFM} = f_{SW\_PWM}$ at P_{out_max}. Hence, $\Delta U_{out\_PFM} = \Delta U_{out\_PWM}$ at P_{out_max}, as indicated. It is observed that, as opposed to ΔU_{out_PWM}, ΔU_{out_PFM} is inverse proportional to P_{out}.

5.1 Inductive Type Converter Control Strategies

Consequentially, $\Delta U_{out\_PFM} > \Delta U_{out\_PWM}$, except at P_{out_max}. The dashed black line denotes the boundaries of U_{out} for a PFM controlled DC-DC converter, where $f_{SW\_PFM} = f_{SW\_PWM}$ at P_{min}. Therefore, in this case $\Delta U_{out\_PFM} = \Delta U_{out\_PWM}$ at P_{out_min}. Obviously, the inverse proportional relation between ΔU_{out_PFM} and P_{out} is maintained, resulting in the fact that $\Delta U_{out\_PFM} < \Delta U_{out\_PWM}$. It is understood that, depending on the choice of the range of f_{SW_PFM}, the value of ΔU_{out_PFM} is larger or smaller, compared to ΔU_{out_PWM}. Moreover, a higher f_{SW_PFM} range will result in a smaller overall ΔU_{out_PFM}. However, the f_{SW_PFM} range is inverse proportional to η_{SW_PFM}, which can be observed in Fig. 5.10. In other words, this observation leads to a trade-off between ΔU_{out_PFM} and η_{SW_PFM}, for the boost converter. This trade-off indirectly implies a second trade-off between η_{SW_PFM} and the required chip area A. Indeed, for a larger value of η_{SW_PFM}, a larger output capacitance C is required and thus a larger A.

A similar trade-off can be deduced for the buck converter, by examining Fig. 5.11(b). The difference between the boost and the buck converter is found in the fact that ΔU_{out_PWM} is inverse proportional to P_{out} for the buck converter, rather than being proportional to P_{out}. This is due to the nature of the output filter, which consists of both the inductor and the output capacitor in a buck converter. As a result, the difference between ΔU_{out_PFM} and ΔU_{out_PWM} is smaller for the case where $f_{SW\_PFM} = f_{SW\_PWM}$ at P_{out_max} (solid black line), compared to the boost converter. Nevertheless, the conclusion is analogue to that of a boost converter.

Load Regulation

The load regulation of a voltage converter is given by (5.1) and is used to measure the variation of U_{out} upon a varying I_{out}, normalized over the nominal U_{out}. The implementation example of a PWM control system, shown in Fig. 5.4, indicates that U_{out} is fed into an integrator, which acts as a low-pass filter in the frequency-domain. This implies that $\overline{U_{out}}$ is the actual parameter being used in the feedback-loop. Therefore, regardless of the type of converter, an ideal PWM control system will regulate $\overline{U_{out}}$ to the desired constant value. This is illustrated in the respective Figs. 5.12(a) and 5.12(b), which show $\overline{U_{out}}$ (black curves) and the boundaries of U_{out} (gray curves) for a boost and a buck converter. It is observed that the progress of ΔU_{out} as a function of P_{out}, which is proportional for boost converter and inverse proportional for a buck converter, has no influence on the ideal behavior of the PWM controlled DC-DC converter. Obviously, in reality the load regulation of a PWM controlled DC-DC converter will have a finite, non-zero value. This is due to the non-ideal control system components, such as the finite gain of the error amplifier, the delay in the feedback-loop, the ripple in the feedback signal… An in-depth discussion of these phenomena and their effect on the regulation of the converter are omitted in this work.

In contrast to PWM control systems, PFM control systems in general do not regulate upon the value of $\overline{U_{out}}$, they regulate upon the value of U_{out_min}. This is illustrated in Fig. 5.12(c), which shows $\overline{U_{out}}$ (black curves) and the boundaries of

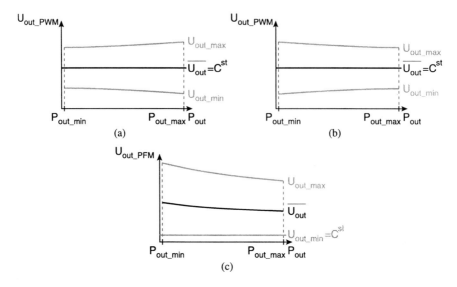

Fig. 5.12 The boundaries of the output voltages U_{out} (*gray curves*) of (**a**) a PWM controlled boost converter, (**b**) a PWM controlled buck converter and (**c**) a PFM controlled boost or boost converter. All the graphs are valid for DCM. The *black curves* denote the mean output voltage $\overline{U_{out}}$

U_{out} (gray curves) as a function of P_{out}, for an ideal PFM controlled boost or buck DC-DC converter. It is observed that U_{out_min} is kept constant and because ΔU_{out} varies inverse proportional to P_{out}, $\overline{U_{out}}$ is also inverse proportional to P_{out}. For an ideal PFM controlled monolithic DC-DC boost or buck converter this will result in a load regulation in the order of 150‰/mA, depending inverse proportionally upon the value of ΔU_{out}. In real PFM controlled DC-DC converters the load regulation will somewhat degrade due to additional factors, such as: the delay of the feedback-loop, the offset of the comparator… These additional effects are taken into account through transistor level SPICE simulations, rather than being the subject of a theoretical discussion.

In- and Output Noise (EMI)

Switched-mode power supplies have a tendency to introduce EMI at both the supply and the output. In Sect. 4.2.2 it is stated that both the finite output capacitance and ESR introduce mostly odd harmonics at the output of a boost converter and odd/even harmonics at the output of a buck converter. For fixed f_{SW} PWM controlled DC-DC converters this EMI is quite well defined and predictable, whereas for a PFM controlled DC-DC converter f_{SW} varies proportionally to P_{out} and is therefore less well defined. The input of the DC-DC converter introduces similar EMI, of which the suppression mainly depends on the capacitance of the input decoupling.

Which system proves to be the optimal choice for sake of minimizing the influence of EMI on neighboring systems and circuits largely depends on the application.

5.2 Constant On/Off-Time: COOT

For instance a PWM controlled DC-DC converter can be beneficial if f_{SW} is higher than the signal band of the application. In other cases PWM can pose problems because of the fact that the energy of the EMI is densely concentrated at f_{SW}. From this point of view a PFM controlled DC-DC converter can be advantageous, since the energy of the EMI is spread in the frequency-domain. As a result the average amplitude of the EMI will be lower compared to a PWM control system, but it will also occupy a larger frequency band.

In the end it can be concluded that much will depend on the specifications demanded by the application, which is not considered for sake of generality.

In Sect. 5.1.3 it is deduced that PFM controlled DC-DC converters can obtain a larger overall η_{SW}, compared to PWM controlled converters. This comes at a chip die area penalty, when ΔU_{out} is to be kept at the same maximal value. Nevertheless, for monolithic DC-DC converters the PFM method is preferred, due to the intrinsic lower values of η_{SW} which are achievable. Although many variants of PFM control systems exists [Red09], none of them are particularly optimized for monolithic DC-DC converters. For this purpose, the Constant On/Off-Time (COOT) control technique for monolithic DC-DC converters is introduced [Wen08a].

The basic concept of a COOT control system is explained in Sect. 5.2.1. Two practical implementations of a single-phase, single-output COOT control system are discussed in Sect. 5.2.2. The practical implementation of a single-phase, two-output SIMO COOT control system is discussed in Sect. 5.2.3.

5.2.1 The COOT Concept

The basic COOT concept can be deduced from the fundamental (differential) equations of the inductive DC-DC converter. In this section this deduction is performed for both the boost and the buck converter in DCM. For CCM an analogue method may be used, which is trivial.

First, the ideal boost DC-DC converter is considered, of which the fundamental calculations are performed in Sect. 2.3.3. During the charging of the inductor, $i_L(t)$ ramps up in a linear fashion during t_{on}. This yields I_{L_max} according to (2.67). When assuming the values of L, t_{on} and U_{in} are constant, a constant value for I_{L_max} is obtained, given by (5.3).

$$\frac{di_L(t)}{dt} = \frac{U_{in}}{L} \xrightarrow{DCM} I_{L\_max} = t_{on}\frac{U_{in}}{L} = C^{st} \blacksquare$$
$$\text{with} \quad L = C^{st};\ U_{in} = C^{st};\ t_{on} = C^{st} \qquad (5.3)$$

A similar argumentation can be used for the discharging of the inductor, of which the ideal ramp of $i_L(t)$ is calculated through (2.67). When also assuming U_{out} is constant, a constant t_{off_real} is in turn obtained, as stated by (5.4).

$$\frac{di_L(t)}{dt} = \frac{U_{in} - U_{out}}{L} \xrightarrow{DCM} t_{off\_real} = I_{L\_max} \frac{L}{U_{out} - U_{in}} = t_{on} \frac{U_{in}}{U_{out} - U_{in}}$$
$$= C^{st} \blacksquare$$

with $\quad L = C^{st};\ U_{in} = C^{st};\ U_{out} = C^{st};\ t_{on} = C^{st}$ \hfill (5.4)

Therefore, both t_{on} and t_{off_real} can be kept constant, providing that U_{in} and U_{out} are also kept constant. Through the fundamental differential equations of the boost converter, which are given in Sect. 4.1.1, it can be proven that the COOT principle is always valid, independently of the value of the load R_L. In other words, when t_{on}, U_{in} and U_{out} are constant, t_{off_real} can also be kept constant, regardless of the value of R_L.

The deduction for a buck converter is analogue to that of a boost converter. During the charging of the inductor, the course of $i_L(t)$ is determined by the constant values of L, t_{on}, U_{in} and U_{out}. As a result, a constant value for I_{L_max} is obtained, given by (5.5).

$$\frac{di_L(t)}{dt} = \frac{U_{in} - U_{out}}{L} \xrightarrow{DCM} I_{L\_max} = t_{on} \frac{U_{in} - U_{out}}{L} = C^{st} \blacksquare$$

with $\quad L = C^{st};\ U_{in} = C^{st};\ U_{out} = C^{st};\ t_{on} = C^{st}$ \hfill (5.5)

This constant I_{L_max} in turn leads to a constant value of off_real, as stated by (5.6).

$$\frac{di_L(t)}{dt} = -\frac{U_{out}}{L} \xrightarrow{DCM} t_{off\_real} = I_{L\_max} \frac{L}{U_{out}} = t_{on} \frac{U_{in} - U_{out}}{U_{out}} = C^{st} \blacksquare$$

with $\quad L = C^{st};\ U_{in} = C^{st};\ U_{out} = C^{st};\ t_{on} = C^{st}$ \hfill (5.6)

This leads to the same conclusion, made for the boost converter: when t_{on}, U_{in} and U_{out} are constant, t_{off_real} can also be kept constant, regardless of the value of R_L. The independency of the COOT principle on the value of R_L, for a buck converter, can be demonstrated through the fundamental differential equations of the buck converter, which are provided in Sect. 4.1.3.

The concept of COOT timing is illustrated in Fig. 5.13, for both a low and high load condition. The upper graph shows the COOT timing for a low load condition. For low loads, implying a high value for R_L, the idle time t_{idle1} between two consecutive switching cycles is large. As a consequence the switching frequency f_{SW1} has a low value. For a high load condition the opposite takes place: t_{idle2} becomes small and therefore f_{SW2} becomes large. Obviously, the conditions for COOT time are $t_{on1} = t_{on2}$, $t_{off\_real1} = t_{off\_real2}$, $U_{in} = C^{st}$ and $U_{out} = C^{st}$. In essence, COOT timing can be understood by considering the fact that during each switching cycle a fixed amount of energy is delivered to the output of the converter. Thus, U_{out} can be kept constant under varying load conditions by merely changing t_{idle}, which is equivalent to adapting f_{SW}. In this way, the number of fixed energy packets that are transferred to the output per unit of time can be controlled, which is equivalent to controlling P_{out}.

One of the main circuit-level advantages of the COOT technique is the fact that no current-sensing of the freewheeling switch is required. This current-sensing

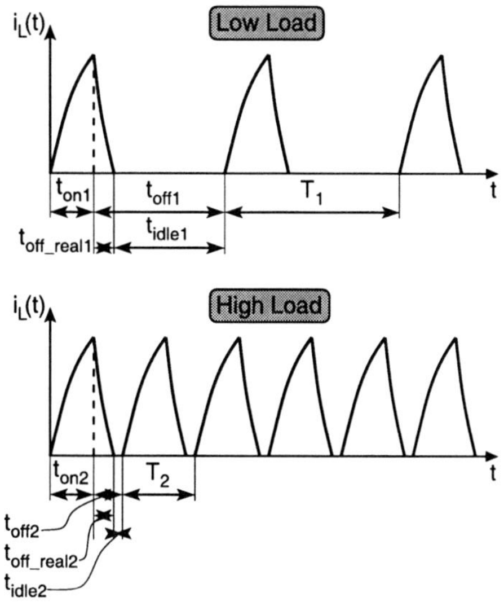

5.2 Constant On/Off-Time: COOT

Fig. 5.13 The basic concept of a Constant On/Off-Time (COOT) control system, illustrated by means of the current $i_L(t)$ through the inductor as a function of time t. The *upper* and *lower graphs* show the respective timing for low and high load operation

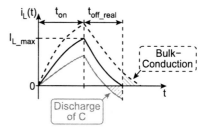

Fig. 5.14 The current $i_L(t)$ trough the inductor as a function of time t, for a single switching cycle of a COOT controlled DC-DC converter. The *solid black curve* denotes the nominal timing, whereas the *dashed black line* and the *solid gray curve* denote the respective case where bulk-conduction and a short-circuit of $i_L(t)$ occur

proves to be a problem for the correct timing at values of f_{SW} in the order of 100 MHz and more. This is confirmed by the practical implementation of the PWM boost control system implementation, described in Sect. 5.1.1. For a buck converter this current-sensing poses even more problems, as the voltage on switching node of the freewheeling switch drops below the *GND* voltage (see Fig. 4.21).

The absence of current-sensing of the freewheeling switch comes at a cost when the value of U_{in} and/or U_{out} is not constant. This is illustrated in Fig. 5.14, where $i_L(t)$ is shown for a single switching cycle of a COOT controlled DC-DC converter. The solid black curve denotes the nominal case, where both U_{in} and U_{out} have their nominal value. This results in the fact that $i_L(t)$ becomes exactly zero when t_{off_real} has elapsed, during the discharge phase of the inductor. When U_{in} becomes larger for a boost converter, or when U_{in} becomes larger and/or U_{out} becomes lower for a buck converter, the value of I_{L_max} will increase. Hence, during the discharge phase of the inductor, the time for $i_L(t)$ to reach zero will also increase. Because of the fixed t_{off_real}, the freewheeling switch will be opened before $i_L(t)$ reaches zero and bulk conduction will occur. This situation is illustrated by the dashed black curve. The opposite situation can also occur, resulting in the partially discharging of the output capacitor into the supply source for a boost converter and to *GND* for a buck converter. This case is denoted by the solid gray curve. It is clear that both these cases are unwanted as they result in increased power losses and thus a lower value of η_{SW}. This effect becomes less dominant for increasing values of the voltage conversion ratio k. A quantitative study of this phenomenon is omitted, as it is observed through measurements in Chap. 6.

The discussion on the COOT control system concept is concluded with an overview of the benefits and drawbacks:

- ✔ No current-sensing is required for the freewheeling switch.
- ✔ Good performance of η_{SW} at low loads, due to adaptive f_{SW}.
- ✔ U_{out} is always stable in DCM.
- ✔ Low dependency on deep-submicron CMOS variability, due to the possibility of the implementation by means of mostly digital building blocks.
- ✘ Mismatch of U_{in} and/or U_{in} results in lower η_{SW}, due to the optimization for near-constant values of U_{in} and U_{in}.
- ✘ Load regulation is theoretically non-zero and dependent on ΔU_{out}.

5.2.2 Single-Phase, Single-Output Implementations

Two practical implementations of the COOT control technique are discussed in the following sections [Wen08b, Wen08a]. Both serve the purpose of controlling a single-phase, single-output monolithic buck converter. For the measurement results of these converters, the reader is referred to the respective Sects. 6.4.1 and 6.4.2.

Version 1

Figure 5.15 shows the block diagram of the practical implementation of a COOT control system [Wen08b], used to control a single-phase, single-output, monolithic buck converter, in a 180 nm 1.8 V CMOS technology. This control system enables U_{out} to be regulated to a constant value, under varying load and line conditions. It allows the converter to operate in synchronous DCM. The control system consists of five building blocks, having the following functionality:

1. *Comparator*: The comparator block compares a fraction U'_{out} of U_{out} to a reference voltage U_{ref}. The desired U_{out} is controlled through the value of U_{ref}. When $U'_{out} < U_{ref}$ the comparator block outputs an active-high signal and vice versa.
2. *Busy-detector*: This block holds the signal from the comparator block until the converter has ended its entire switching cycle. This is monitored through two signals. First, the signal of SW_1 is monitored, preventing a new switching cycle to commence when the converter is in the charging phase of the inductor. Second, after the charging phase has ended, a minimal off-time t_{off_min} is to be waited

5.2 Constant On/Off-Time: COOT

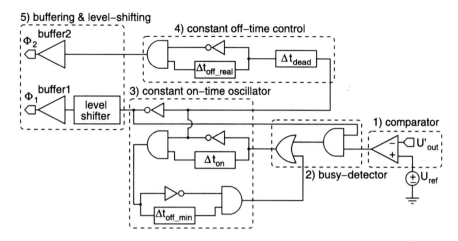

Fig. 5.15 The block diagram of the COOT control system implementation for a single-phase, single-output, fully-integrated buck DC-DC converter, using a bondwire inductor [Wen08b]

until a new switching cycle may commence. This block is simple, but crucial, as it avoids faulty timing, resulting in reduced performance and even chaos and malfunction of the converter.

3. *Constant on-time oscillator*: The oscillator consists of two ring-coupled monostable multi-vibrators, which are triggered by a falling-edge signal and output an active-high pulse. These active-high pulses have a respective constant duration of t_{on} of 4.4 ns and t_{off_min} of 2.7 ns. The feedback of the two monostable multi-vibrators is gated by the busy-detector, allowing it to be halted. The constant t_{on} pulse is used to drive SW_1 through an active-low pulse. The constant t_{off_min} pulse determines the minimal t_{off}, assuring the converter operates in DCM. t_{off_min} may also be used to limit the maximal P_{out} of the converter, which is not performed in this implementation.

4. *Constant off-time control*: When the active-low pulse of the drive signal of SW_1 has ended, thereby opening SW_1, SW_2 needs to be closed during a constant t_{off_real}. This is achieved by feeding the inverted active-low drive signal of SW_1 into a time-delay Δt_{dead} and consecutively into a mono-stable multi-vibrator. Δt_{dead} provides a dead-time of 490 ps between the opening of SW_1 and the closing of SW_2. It also compensates for the additional time-delay introduced by the level-shifter and buffer1. The mono-stable multi-vibrator is triggered by a falling-edge signal and outputs an active-high pulse, having a constant duration t_{off_real} of 2 ns. This active-high pulse is used to drive SW_2.

5. *Buffering & level-shifting*: The active-low signal from the constant on-time oscillator, to drive SW_1, is first level-shifted from U_{out}–GND to U_{in}–U_{out}. Afterwards it is buffered through buffer1, yielding the active-high Φ_1 signal to drive the gate of SW_1. The active-high signal from the constant off-time control is directly buffered through buffer2, yielding the signal Φ_2 to drive the gate of SW_2.

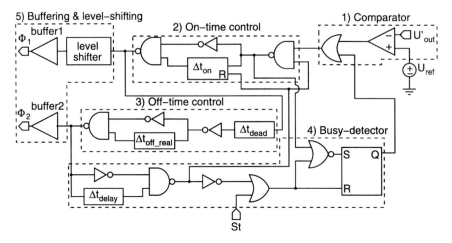

Fig. 5.16 The block diagram of the COOT control system implementation for a single-phase, single-output, fully-integrated buck DC-DC converter, using a metal-track inductor [Wen08a]

The circuit of the comparator is shown in Fig. 5.6. The circuit of the time-delays is shown in Fig. 5.7. For Δt_{on}, Δt_{off_min} and Δt_{off_real} multiple time-delay cells are cascaded for a better preservation of the duty-cycle and higher immunity against mismatch. The circuit of the level-shifter is shown in Fig. 5.8. Finally, the buffers are implemented as digital tapered inverters. Buffer1 and buffer2 consist of eight and seven scaled stages, respectively. Both buffers have a scaling factor of 2.6.

Version 2

Figure 5.16 shows the block diagram of a second practical implementation of a COOT control system [Wen08a], used to control a single-phase, single-output, monolithic buck converter, in a 130 nm 1.2 V CMOS technology. This control system enables U_{out} to be regulated to a constant value, under varying load and line conditions. It allows the converter to operate in synchronous DCM. The control system consists of five building blocks, having the following functionality:

1. *Comparator*: The comparator block compares a fraction U'_{out} of U_{out} to a reference voltage U_{ref}. The desired value of U_{out} is controlled through the value of U_{ref}. When the value of U'_{out} drops below the value of U_{ref} and when the busy-detector outputs an active-low signal to the comparator block, a rising-edge signal is generated.
2. *Constant on-time control*: When the comparator block outputs a rising-edge signal, indicating that U_{out} is too low, and when the busy-detector outputs a logic-high signal, the mono-stable multi-vibrator is activated. This mono-stable multi-vibrator generates an active-low pulse with a constant duration t_{on} of 2 ns. This pulse is used to drive SW_1. The time-delay Δt_{on} in the mono-stable multi-

5.2 Constant On/Off-Time: COOT

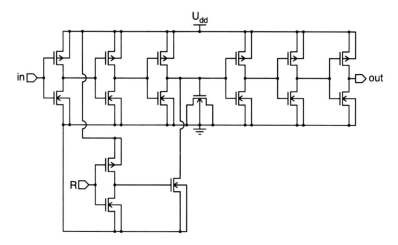

Fig. 5.17 The circuit of a time-delay with external reset functionality

vibrator is reset after each switching cycle, enabling a new switching cycle to commence more rapidly.

3. *Constant off-time control*: When the active-low pulse of the constant on-time control outputs a rising-edge signal, indicating that SW_1 is opened, the off-time control is activated. First, the output signal of the constant on-time control is fed into a constant time-delay Δt_{dead} of 150 ps. This Δt_{dead} provides a dead-time between the opening of SW_1 and the closing of SW_2, thereby also compensating for the delays introduced by the level-shifter and buffer2. The delayed rising-edge signal triggers a mono-stable multi-vibrator, which outputs an active-high pulse with a constant duration t_{off} of 750 ps. This active-high pulse drives SW_2.

4. *Busy-detector*: The busy-detector serves two purposes. First, it ensures that no new switching cycle is started before the previous one has ended. This avoids faulty timing, leading to decreased performance or malfunction of the converter. This is achieved by monitoring the output of the constant off-time control. When a new switching cycle is started the SR-flipflop is set and it is reset when the switching cycle has ended. The second function of the busy-detector is to allow switching cycles to follow one another more rapidly, when U_{out} has not yet reached the desired value. This is achieved by directly triggering the on-time control, when $U'_{out} < U_{ref}$. The internal mono-stable multi-vibrator serves the purpose of resetting the SR-flipflop. The duration of this reset t_{delay} is equal to 150 ps. At startup the SR-flipflop is reset through the signal St, which originates from the start-up circuit (see Sect. 5.5).

5. *Buffering & level-shifting*: The output signal from the on-time control is first level-shifted from U_{out}–GND to U_{in}–U_{out}. Afterwards it is buffered through buffer1, yielding the active-low signal Φ_1 to drive the gate of SW_1. The output signal of the off-time control is directly buffered through buffer2, yielding the active-high signal Φ_2 to drive the gate of SW_2.

The circuit of the comparator is shown in Fig. 5.6. For Δt_{off}, Δt_{dead} and Δt_{delay} the time-delay topology of Fig. 5.7 is used. For Δt_{on} a time-delay with external reset functionality is used, as illustrated in Fig. 5.17. The active-low reset signal R enables a faster discharge of the MOS capacitor, enabling consecutive charge cycles at a higher repetition rate. Δt_{on} and Δt_{off} are implemented by using multiple cascaded time-delay cells, yielding an improved preservation of the duty-cycle and higher immunity against mismatch. The circuit of the level-shifter is shown in Fig. 5.8. Finally, the buffers are implemented as digital tapered inverters. Buffer1 and buffer2 consist of seven and six scaled stages, respectively. Both buffers have a scaling factor of 2.6.

Compared to version 1 of the single-phase, single output COOT control system, this control system has two advantages. First, the consecutive switching cycles are enabled to follow each other faster. This enables a higher f_{SW}, which allows the converter to operate nearly at the boundary of DCM and CCM. As a result, the achievable P_{out_max} is larger. The second advantage is that this implementation uses only two relative large time-delays, compared to three for version 1. This results in a lower total transistor count and a slightly reduced area of the control system. The disadvantage of this control system is that P_{out_max} cannot be limited, which is possible with version 1.

5.2.3 Single-Phase, Two-Output SIMO Implementation

In this section a practical implementation of the COOT control technique for a two-output boost converter is discussed. This implementation comprises a COOT control system for a single-phase two-output SIMO boost converter in a 130 nm 1.2 V CMOS technology, of which the circuit of the converter stage is shown in Fig. 3.48. Note that the implementation of the entire converter is discussed in Sect. 6.3.2.

The COOT control system is based on dedicated switching cycles per output in DCM, as this provides a higher value η_{SW} compared to a shared switching cycle scheme. In monolithic DC-DC converters this fact is due to two reasons. First, I_{L_max} should be minimized to guarantee the lowest losses, as explained in Sect. 2.3.1. The second reason is the low values for t_{on} and t_{off_real}, which are in the order of ns. This causes the timing of the power switches to be critical, which would merely be worsened by a shared switching cycle.

The practical implementation of a COOT control system, is shown in Fig. 5.18. This control system enables U_{out} to be regulated to a constant value, under varying load and line conditions. It allows the converter to operate in synchronous DCM, with dedicated switching cycles per output. The control system consists of five building blocks, having the following functionality:

1. *Comparator*: The comparator block compares a fraction U'_{out1} of U_{out1} and a fraction U'_{out2} of U_{out2} to two respective reference voltages U_{ref1} and U_{ref2}. The desired U_{out1} and U_{out2} of the converter are controlled through the respective values of U_{ref1} and U_{ref2}. The respective comparators output a logic-high signal to

5.2 Constant On/Off-Time: COOT

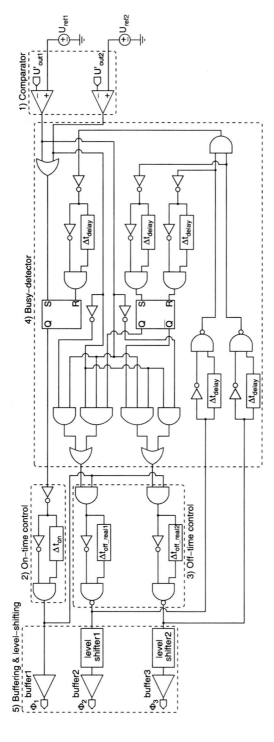

Fig. 5.18 The block diagram of the COOT control system implementation for a single-phase, two-output, SIMO, fully-integrated buck DC-DC converter, using a metal-track inductor

the busy-detector when $U'_{out1/2}$ is lower than $U_{ref1/2}$, indicating that the converter is to be enabled to charge the relevant output(s).

2. *Constant on-time control*: The mono-stable multi-vibrator is triggered through a rising-edge signal from the busy-detector, thereby generating an active-high output pulse with a duration t_{on} of 3 ns. This signal is used to drive SW_1, thereby charging the inductor.

3. *Constant off-time control*: When SW_1 is opened, either SW_2 or SW_3 is to be closed simultaneously, which is decided by the busy-detector. For this purpose, the busy-detector enables one of the two mono-stable multi-vibrators, which are both triggered by a falling-edge signal, to be triggered. Thus, either the mono-stable multi-vibrator of SW_2 or the one of SW_3 is triggered, which in turn will generate an active-low pulse. This active-low pulse has a duration of t_{off1} of 1.2 ns, or t_{off2} of 1 ns. The respective active-low pulses are used for driving SW_2 and SW_3. A dead-time delay is not required, because this is intrinsically provided by the steering of the switches, as explained in Sect. 6.3.2.

4. *Busy-detector*: This block serves three purposes. First, it will start a new switching cycle when the converter has completely finished the previous one and when $U'_{out1} < U_{ref1}$ and/or $U'_{out2} < U_{ref2}$. This is done by outputting a rising-edge signal to the on-time control. The second function is to avoid a new switching cycle to commence until the previous one is completely ended. This is achieved by setting an SR-flipflop when a new switching cycle is started, preventing another one to be started. When the switching cycle has ended the SR-flipflop is reset again, allowing for a consecutive switching cycle to be started, if required. Finally, the busy-detector is responsible for determining the priority of the charging of the outputs. When both $U'_{out1} < U_{ref1}$ and $U'_{out2} < U_{ref2}$, the busy detector will toggle between output one and two. When either $U'_{out1} < U_{ref1}$ or $U'_{out2} < U_{ref2}$ the busy detector will give priority to the output that needs to be charged. This is done until this output is at the desired voltage level or until the other output also needs charging, at which point the busy-detector will toggle between the two outputs. To achieve this functionality, the busy-detector saves the last output that was charged in the state of an SR-flipflop. The actual selection of the outputs is done by delivering two select signals to the off-time control.

5. *Buffering & level-shifting*: The output signal from the on-time control is directly buffered between U_{in}–GND, providing an active-high signal Φ_1 for driving the gate of SW_1. The output signal of the off-time control with a duration of t_{off1} is level-shifted from U_{in}–GND to U_{out2}–U_{in}. Afterwards, it is buffered by buffer2, yielding the active-low signal Φ_2 to drive the gate of SW_2. The other output signal, with a duration of t_{off2}, is level-shifted from U_{in}–GND to U_{out2}–U_{out1}. Subsequently, it is buffered by buffer3, yielding the active-low signal Φ_3 to drive the gate of SW_3.

The circuit topology used for the comparators is the similar to that used in the other implementations and is shown in Fig. 5.6.

For Δt_{delay}, which is 150 ps, the time-delay topology of Fig. 5.7 is used. For Δt_{on}, Δt_{off1} and Δt_{off2} an auto-reset time-delay, realized by connecting the reset R

5.2 Constant On/Off-Time: COOT

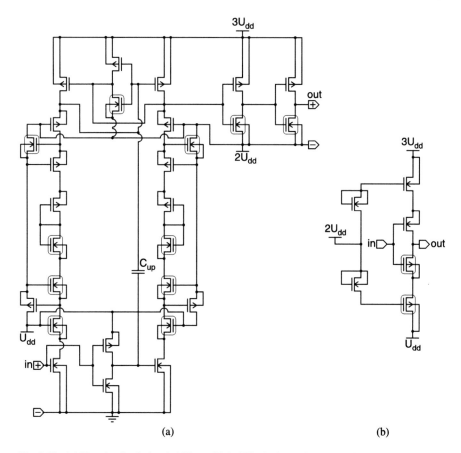

Fig. 5.19 (a) The circuit of a level-shifter, which shifts the input from U_{dd}–GND to $3 \cdot U_{dd}$–$2 \cdot U_{dd}$ and (b) the circuit of a modified inverter, of which the in- and output may vary between $3 \cdot U_{dd}$ and U_{dd}

input with the input *in* in the time-delay topology shown in Fig. 5.17, is used. These are also implemented by means of multiple cascaded auto-reset time-delay cells, for minimizing the duty-cycle variation and providing immunity against mismatch.

The circuit of level-shifter2 is shown in Fig. 5.19(a), which is based on the circuit shown in Fig. 5.8. It shifts the input signal, varying between U_{dd}–GND, to the output signal, which varies between $2 \cdot U_{dd}$–$3 \cdot U_{dd}$.

The circuit of level-shifter1 is in-turn based on the level-shifter from Fig. 5.19(a), were the high-side cross-coupled inverter is modified. The circuit of this modified inverter is shown in Fig. 5.19(b). Both its in- and output voltage may vary between U_{dd}–$3 \cdot U_{dd}$. This modification enables level-shifter1 to shift the input signal from U_{dd}–GND to $3 \cdot U_{dd}$–U_{dd}. All the triple-well n-MOSFETs are denoted with a rectangle in Figs. 5.19(a) and 5.8(b). For more information on these level-shifters, the reader is referred to [Ser07].

⑤SIMO/SMOC DCM CONTROL MODES

For the purpose of controlling SIMO or SMOC DC-DC converters two main methods can be identified [Kwo09]: a dedicated switching cycle per output scheme, shown for a boost and a buck converter in the respective Figs. 5.20(a) and 5.20(b), and a shared switching cycle per output scheme, shown for a boost and a buck converter in the respective Figs. 5.20(c) and 5.20(d). The illustrations of Fig. 5.20 are valid for PWM or PFM operation, in DCM. Also, the illustrations are valid for two-output converters, which can evidently be extrapolated to multiple-output converters.

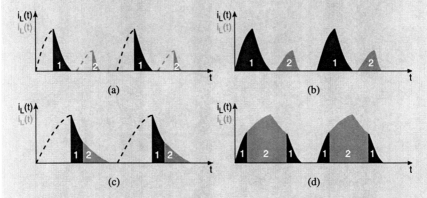

Fig. 5.20 The current $i_L(t)$ through the inductor as a function of time t, for different two-output SIMO/SMOC DC-DC converter switching schemes in DCM: a dedicated switching cycle scheme for (**a**) a boost and (**b**) a buck converter, a shared switching scheme for (**c**) a boost and (**d**) a buck converter

For a boost converter the inductor is charged first (dashed curves) and afterwards the energy is divided over the outputs (solid, filled curves). Whereas for a buck converter the outputs are charged both during the charge and discharge phase (solid, filled curves). Both switching schemes allow for variations in the priority of the outputs and also in the sequence in which the outputs are powered. For monolithic converters, the dedicated switching scheme is preferred, due to the following reasons:

- No freewheeling switch is required parallel with the inductor [Le07].
- The value of I_{L_max} can be lower.
- The effective f_{SW} can be lower.
- The timing of the power switches is less critical.

Due to these facts, dedicated switching schemes may achieve higher values of η_{SW} in monolithic DC-DC converters. However, this comes at the cost of a higher value of ΔU_{out}.

Buffer1 and buffer3 are implemented as standard digital tapered inverters. Buffer2 is implemented as a modified digital tapered buffer, of which the individual stages consist of the modified inverter, shown in Fig. 5.8(b). Buffer1, buffer2 and buffer3 have a scaling factor of 2.6 and consist of respectively six, seven and six stages.

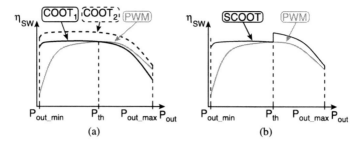

Fig. 5.21 (a) The power conversion efficiency η_{SW} as a function of the output power P_{out} for a PWM (*gray curve*) and two COOT controlled DC-DC converters (*solid* and *dashed black curves*). (b) The power conversion efficiency η_{SW} as a function of the output power P_{out} for PWM (*gray curve*) and a SCOOT controlled DC-DC converter

5.3 Semi-Constant On/Off-Time: SCOOT

The COOT control scheme, discussed in Sect. 5.2, is the PFM variant for monolithic inductive single-phase DC-DC converters. When a higher output power is required, at a minimized chip die area en maximal power conversion efficiency, the step towards multi-phase converters is self-evident (see Sect. 3.5.1). In order to obtain these maximal performance parameters, adaptations are have to be made to the COOT control scheme, as it is optimized for single-phase converters. The solution to tackle this problem is provided by the novel control technique called Semi-Constant On/Off-Time (SCOOT) timing [Wen09b]. This control technique supports monolithic inductive multi-phase converters, providing them with an increased high load power conversion efficiency, compared to COOT timing.

The basic concept of a SCOOT control scheme is explained in Sect. 5.3.1. Two practical implementations of SCOOT control systems are discussed in Sect. 5.3.2, comprising a single-output and a two-output version.

5.3.1 The SCOOT Concept

The basic concept of the SCOOT control scheme is based on the COOT control scheme, which is explained in Sect. 5.2.1. The limiting factor of the COOT technique can be understood by examining Fig. 5.11, where it can be seen that for a PFM control scheme, in general, ΔU_{out} is inverse proportional to P_{out}. This is due to the fact that value of f_{SW} increases upon increasing values of P_{out}, when assuming U_{in} and U_{out} are constant. Hence, the worst-case value of ΔU_{out} for a PFM controlled DC-DC converter occurs at P_{out_min}. For a COOT implementation this implies that the value of t_{on} has to be chosen sufficiently small, such that the specification of ΔU_{out} is met at P_{min}. Usually, this will not correspond with the value of t_{on} for which the maximal η_{SW} is obtained, especially at high values of P_{out}.

The SCOOT control scheme overcomes the limited η_{SW} issue, posed by the COOT timing scheme. This is explained by means of Fig. 5.21(a), which illustrates

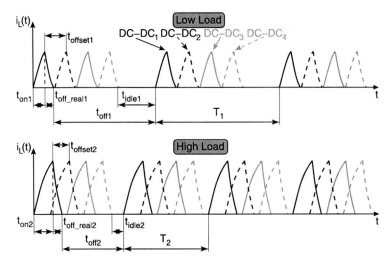

Fig. 5.22 The basic concept of a four-phase Semi-Constant On/Off-Time (SCOOT) control system, illustrated by means of the currents $i_L(t)$ through the respective inductors as a function of time t. The *upper* and *lower graphs* show the respective timing for low and high load operation

η_{SW} as a function of P_{out} for a PWM (gray curve) and two COOT (dashed and solid black curve) controlled DC-DC converters. The corresponding t_{on1} and t_{off_real1} are smaller for the COOT$_1$ case, compared to the COOT$_2$ case. For the COOT$_1$ case t_{on1} and t_{off_real1} are chosen sufficiently small, such that the specification for ΔU_{out} is met at P_{out_min}. For the COOT$_2$ case t_{on2} and t_{off_real2} are chosen larger, such that the specification for ΔU_{out} is met at a certain threshold output power P_{out_th}. As a result the overall η_{SW} for the COOT$_1$ case is lower compared to the COOT$_2$ case, due to the higher associated f_{SW} at a given value of P_{out} for the COOT$_1$ case. The combination of the COOT$_1$ and the COOT$_2$ case results in SCOOT timing, which is illustrated in Fig. 5.21(b). Note that in both Figs. 5.21(a) and 5.21(b) the PWM curve is merely added for comparison. In the literature similar concepts also exist for PWM, called maximal efficiency tracking [AH09]. However, these concepts are optimized for non fully-integrated DC-DC converters and are therefore not optimized for monolithic DC-DC converters.

It is already mentioned that the SCOOT timing scheme will be used to control multi-phase DC-DC converters. This is not a strict requirement, as the SCOOT timing scheme might also be used for single-phase DC-DC converters. Nevertheless, P_{out_max} is usually lower for single-phase converters and the effect of SCOOT timing is more pronounced at high values of P_{out}.

An example of the SCOOT timing scheme for a four-phase converter is illustrated in Fig. 5.22, which shows $i_L(t)$ for each phase as a function of time. First, the upper graph is considered, which is valid for low loads. During one switching cycle all four converters are consecutively enabled, with a certain constant offset-time $t_{offset1}$ between each converter. Each individual converter is operated with a constant t_{on1} and t_{off_real1}, complying with COOT timing. The idle-time t_{idle} between consecutive

5.3 Semi-Constant On/Off-Time: SCOOT

switching cycles controls the amount of P_{out} being delivered to the output of the converter. Thus, U_{out} of the converter is kept to the desired value under varying load conditions by means of varying f_{SW}, at which the full switching cycles are repeated. For the low load timing case t_{on1} and t_{off_real1} are chosen sufficiently small, keeping $i_L(t)$ limited and ensuring a low value of ΔU_{out}. In addition $t_{offset1}$ is made large enough, such that the total current delivered to the output is effectively distributed in the time-domain. This will also contribute to a sufficiently low value of ΔU_{out}. The lower graph is valid for the high load case. At high load operation the values of t_{on2} and t_{off_real2} are made larger than for low load operation. In this way, the amount of energy delivered to the output of the converter per switching cycle is increased, resulting in a lower required f_{SW}. To further increase this energy per switching cycle $t_{offset2}$ is made smaller than $t_{offset1}$, allowing the consecutive switching cycles of the individual converters to follow one another faster. Similar to low load operation, P_{out} is regulated through the adaptation of t_{idle}, which is equivalent to f_{SW}.

Unlike for the fixed f_{SW} PWM timing, the interleaving of multi-phase SCOOT timing cannot be performed in a symmetrical fashion. This is due to the causality principle, which implies that it is not possible to predict when the next switching cycle will occur, since the load is unknown. As a result, the positive effect of multi-phase interleaving on ΔU_{out} for SCOOT timing will be smaller compared to PWM timing.

To conclude the discussion on the SCOOT control system concept, the benefits and drawbacks are listed:

✔ No current-sensing required for the freewheeling switch.
✔ Good performance for η_{SW} at both low and high load operation, due to the combination of the adaptive f_{SW} and the optimized t_{on} and t_{off_real} at both low and high load operation.
✔ In DCM U_{out} is always stable.
✔ The implementation requires mostly digital building blocks, making it ideally suited for implementation in deep-submicron CMOS processes.
✔ Is well-suited for multi-phase converters, due to the adaptive interleaving and the η_{SW} benefit at high P_{out}.
✗ Mismatch of U_{in} and/or U_{out} results in lower η_{SW}, similar to COOT timing.
✗ Load regulation is theoretically non-zero and dependent on ΔU_{out}.
✗ Perfect symmetrical interleaving of the phases in a multi-phase implementation cannot be achieved, resulting in a larger ΔU_{out} than is achievable with PWM.

5.3.2 Multi-phase Implementations

Two practical implementations of the SCOOT control technique are discussed in the next sections [Wen09b]. The first version controls a four-phase, single-output buck converter and the second version controls a four-phase, two-output SMOC buck converter. The entire realization on silicon of these converters is discussed in the respective Sects. 6.4.3 and 6.4.4.

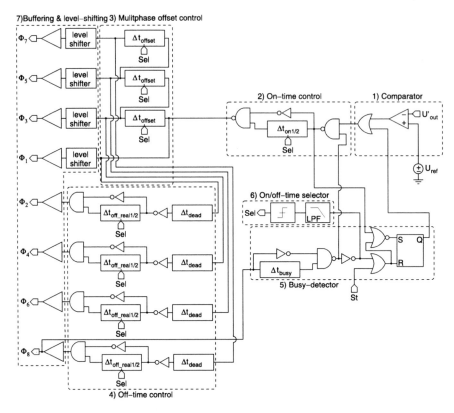

Fig. 5.23 The block diagram of the SCOOT control system implementation for a four-phase, single-output, fully-integrated buck DC-DC converter, using metal-track inductors [Wen09b]

Version 1: Multi-phase, Single-Output

Figure 5.23 shows the block diagram of the practical implementation of the SCOOT control system [Wen09b], acquired to control a four-phase, single-output, monolithic buck converter, in a 130 nm 1.2 V CMOS technology. This control system enables U_{out} of the converter to be regulated to a constant value, under varying load and line conditions. In order to achieve this, the converter is operated in synchronous DCM. The basic operation of the SCOOT control system is explained by means of its seven building blocks:

1. *Comparator*: This block compares a fraction of U_{out} of the converter U'_{out} to a reference voltage U_{ref}, setting U_{out} to the desired value. The converter is enabled to start switching, whereby the comparator block outputs a rising-edge signal, if two conditions are both met: 1) $U'_{out} < U_{ref}$ and 2) the converter has ended its entire four-phase switching cycle, as indicated by the busy-detector.
2. *On-time control*: The on-time control consists of a two-input mono-stable multi-vibrator, which may be triggered by two events: 1) a rising-edge signal from the

5.3 Semi-Constant On/Off-Time: SCOOT

comparator block, if the busy-detector outputs a logic high signal or 2) a rising-edge signal from the busy-detector, if the comparator block outputs a logic-high signal. Both cases indicate that the converter has ended a complete switching cycle and that U_{out} is below the desired value, indicated by U_{ref}. When the monostable multi-vibrator is triggered, the on-time control outputs an active-low pulse. This pulse has a duration of either t_{on1} of 0.8 ns, or t_{on2} of 1.6 ns. This duration depends on the state of the *sel* signal, which is determined by the on/off-time selector block.

3. *Multi-phase offset control*: This block serves two purposes. The first purpose is to generate three additional phase-shifted signals, starting from the t_{on1}/t_{on2} pulse from the on-time control block. This is achieved through three time-delays, having a fixed-duration of either $t_{offset1}$ of 1.5 ns or $t_{offset2}$ of 0.75 ns, depending on the state of the signal *sel*. By doing so, the duration of the on-time pulse should not be affected.

4. *Off-time control*: The off-time control has two main functions. First, the four phase-shifted signals from the multi-phase offset control are delayed with a dead-time $t_{dead} = 0.15$ ns, preventing the high- and low-side switches to be turned on simultaneously. Secondly, these four phase-shifted signals are used to initiate the generation of four active-high pulses, with a duration of either t_{off_real1} of 0.3 ns or t_{off_real2} of 0.6 ns. These off-time pulses are generated separately with individual mono-stable multi-vibrators. These are triggered on a rising-edge, indicating that the corresponding high-side switch is opened. The reason for generating these off-time pulses separately is because of the fact that the timing of these pulses, with regard to the on-time pulses, is critical. If this timing is not correct the low-side switches might be enabled too soon/late, causing increased losses through short-circuit or bulk-diode conduction.

5. *Busy-detector*: The busy-detector detects when the last converter has ended its switching cycle and postpones new switching cycles of the first converter to commence until this is fulfilled. Hence, faulty timing, leading to a higher ΔU_{out} or chaos in the switching scheme, is avoided. This is achieved by setting an SR-flipflop when a new switching cycle is started, of which the output is used to hold the output signal of the comparator block. When the last converter has ended its switching cycle, the SR-flipflop is reset by a mono-stable multi-vibrator. If at this point U_{out} is higher than the desired value the converter will be idle, otherwise a new switching cycle is automatically started.

6. *On/off-time selector*: This block is used to decide which t_{on}, t_{off_real} and t_{offset} is to be used, depending on the load of the converter. For this purpose, P_{out} is measured indirectly through f_{SW} of the converter, which is derived from the mono-stable multi-vibrator of the busy-detector block. As this signal incorporates a fixed pulse-length t_{busy} of 150 ps, its mean value is linear proportional to f_{SW} and in-turn to the delivered P_{out}. This mean value is derived from a low-pass filter with a cutoff-frequency of 8.7 MHz. The output of the filter is connected to a schmitt-trigger, which outputs the *Sel* signal to select the delay values. When triggered to active-high, the delays are switched from t_{on1}, t_{off_real1} and $t_{offset1}$ to t_{on2}, t_{off_real2} and $t_{offset2}$.

Fig. 5.24 The circuit of the selectable time-delay

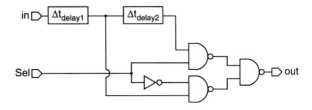

Fig. 5.25 The circuit of the low-pass RC filter

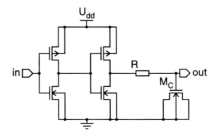

7. *Buffering & level-shifting*: The four phase-offset t_{on} pulses from the multi-phase offset-control are first level-shifted from U_{out}–GND to U_{out}–U_{in}. Afterwards, they are buffered, yielding four active-low signals Φ_1, Φ_3, Φ_5 and Φ_7 to drive the gates of the high-side switches. The four phase-offset t_{off_real} pulses are directly buffered, yielding the active-high signals Φ_2, Φ_4, Φ_6 and Φ_8 to drive the gates of the low-side switches.

The circuit topology used for the comparators is the similar to that used in the other implementations and is shown in Fig. 5.6.

For Δt_{busy}, which is 150 ps, the time-delay topology of Fig. 5.7 is used. The selectable time-delays are implemented as shown in Fig. 5.24. When the *Sel* signal is active-low the time-delay between in- and output is equal to Δt_{delay1} and when the *Sel* is active-high the time-delay between in- and output is equal to Δt_{delay1} + Δt_{delay2}. For $\Delta t_{on1/2}$, $\Delta t_{off1/2}$ a cascade of auto-reset time-delays are used, of which the topology is based on Fig. 5.17.

The circuit of the low-pass RC filter is shown in Fig. 5.25. The total resistance of the RC filter is formed by the series combination of the resistance of resistor R and the output resistance of the second inverter. The capacitor is implemented as a MOS capacitor, formed by the n-MOSFET M_C. The cutoff-frequency of the low-pass RC filter is 8.7 MHz.

For the schmitt-trigger the circuit of Fig. 5.26 is used. The threshold levels are mainly determined by the sizing of the n-MOSFET and the p-MOSFET of the first inverter.

The circuit of the level-shifter is shown in Fig. 5.8. It shifts the input signal, varying between U_{dd}–GND, to the output signal, which varies between $2 \cdot U_{dd}$–U_{dd}.

The buffers are standard digital tapered inverters, having a scaling factor of 2.6. The buffers for the high-side switches contain eight stages, whereas the buffers for the low-side switches contain seven stages.

Fig. 5.26 The circuit of the schmitt-trigger

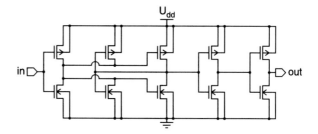

Version 2: Multi-phase, Dual-Output SMOC

Figure 5.27 shows the block diagram of the second practical realization of the SCOOT control system. In this case the control system is used to control a four-phase, two-output SMOC, monolithic buck converter, in a 90 nm 1.2 V CMOS technology. The concept of the novel SMOC topology, for multiple-output converters, is explained in Sect. 3.5.3. This control system enables both U_{out1} and U_{out2} of the converter to be regulated to a constant value, under varying load and line conditions. The converter is operated in synchronous DCM. The basic operation of the SCOOT control system is explained by means of its seven building blocks:

1. *Comparator*: This block compares a fraction of U_{out1} and U_{out2} of the converter, denoted as U'_{out1} and U'_{out2}, to the reference voltages U_{ref1} and U_{ref2}, respectively. The reference voltages determine the value of U_{out1} and U_{out2}. When $U'_{out1} < U_{ref1}$ and/or $U'_{out2} < U_{ref2}$, the comparator(s) will output a logic-high signal(s) to the busy-detector, indicating that at least one of the output voltages of the converter is too low.
2. *On-time control*: When one or both the output voltages of the converter is too low, the busy detector will output a falling-edge signal to one of the mono-stable multi-vibrators of the on-time control. By doing so the on-time control outputs an on-time pulse signal $\Phi_{1/1}$, used to drive the high-side switch. The duration of this signal is dependent on the output of the converter, which needs to be charged and the state of the *Sel1* or *Sel2* signal. For the first output the on-time pulse has a duration of either t_{on1a} of 0.6 ns or t_{on1b} of 0.8 ns. The second output has an on-time pulse duration of either t_{on2a} of 0.6 ns or t_{on2b} of 1.4 ns.
3. *Off-time control*: When the high-side switch is opened, the off-time control block is activated by either the t_{on1} or t_{on2} pulse. The signals from the on-time control are first delayed through the dead-time delays Δt_{dead}, avoiding simultaneous conduction of the high- and low-side switch. Afterwards, the corresponding t_{off_real1} or t_{off_real2} mono-stable multi-vibrator is activated to drive the low-side switch of the first converter stage, by means of signal $\Phi_{2/1}$. When the first output of the converter is to be charged, the off-time pulse has a duration of either t_{off_real1a} of 0.8 ns or t_{off_real1b} of 1 ns. The second output of the converter is charged with an off-time pulse, having a duration of either t_{off_real2a} of 0.4 ns or t_{off_real2b} of 0.6 ns. This block is repeated four times in total, providing the correct timing of all of the four converter stages.

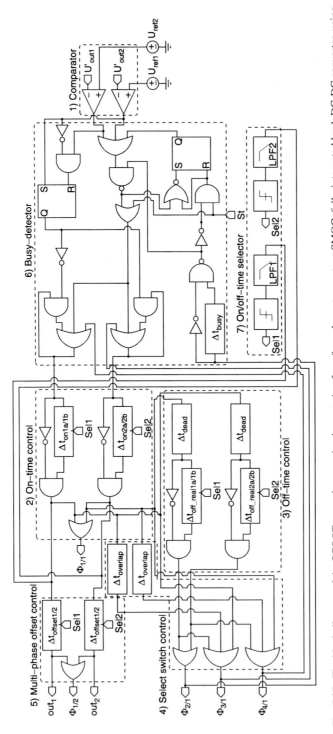

Fig. 5.27 The block diagram of the SCOOT control system implementation for a four-phase, two-output SMOC, fully-integrated buck DC-DC converter, using metal-track inductors

5.3 Semi-Constant On/Off-Time: SCOOT

4. *Select switch control*: The output select switches of the converter need to be closed during the time which the high- and low-side switch of the converter are closed. Thus, these signals $\Phi_{3/1}$ and $\Phi_{4/1}$, used for driving the selector switches for each of the two outputs, are the combination of the output signals $\Phi_{1/1}$ and $\Phi_{2/1}$ from the on- and off-time control. The overlap-delays $\Delta t_{overlap}$ are necessary to fill the gap of the dead-time between the opening and closing of the high- and low-side switches. Therefore, $\Delta t_{overlap}$ needs to be slightly larger than Δt_{dead}. This block is also repeated four times in total, for each converter stage.
5. *Multi-phase offset control*: The individual on-time pulses of t_{on1} and t_{on2} are delayed through selectable offset time-delays Δt_{offset}, providing the interleaving between the four phases of the power converter. The offset values of the time-delays are the same for both the outputs, which is 1 ns for $t_{offset1}$ and 0.6 ns for $t_{offset2}$. This block is repeated three times in total, providing the interleaving for each of the four converter stages.
6. *Busy-detector*: The first task of the busy-detector is to prevent a new switching cycle from commencing until the previous one is completely finished. This is achieved by setting an SR-flipflop until the switching cycle is finished. The second task is to decide when which output of the converter is to be charged. When both outputs are to be charged, the busy-detector will toggle through both outputs, giving them equal priority. When only one output is to be charged, it will receive full power from the converter, until it is either at its desired voltage level or until the other output needs to be powered as well. For this purpose, the state of the last powered output is saved in another SR-flipflop. The third task is to immediately commence a new switching cycle if one or both the outputs are not at the desired voltage level at the end of a switching cycle. This ensures fast follow-up of the switching cycles. Note that the busy-detector is to be enabled by the start-up circuit of the converter, through signal *St*. The start-up circuit will be discussed in Sect. 5.5.2.
7. *On/off-time selector*: The *Sel1* and *Sel2* signals for the selection of the on-times, the off-times and the offset-times are generated by this block. The respective *Sel1* and *Sel2* signals are dependent on the respective output powers P_{out1} and P_{out2} of the two outputs of the converter. When P_{out1} and P_{out2} are higher than their respective threshold values, the corresponding *Sel* signal will be logic-high and vice versa. For this purpose the two on-time signals from the on-time control are used to determine P_{out1} and P_{out2}, by feeding them consecutively into a low-pass filter and a schmitt-trigger.

The circuit topology used for the comparators is the similar to that used in the other implementations and is shown in Fig. 5.6.

For Δt_{busy}, which is 150 ps, the time-delay topology of Fig. 5.7 is used. The selectable time-delays for the on-time, the off-time and the offset-time are implemented as shown in Fig. 5.24. For $\Delta t_{on1/2}$, $\Delta t_{off1/2}$ a cascade of auto-reset time-delays are used, of which the topology is based on Fig. 5.17.

The on/off time selector is implemented similar as in version 1 of the SCOOT control system. The circuit of the RC low-pass filter is shown in Fig. 5.25 and the circuit of the schmitt-trigger similar to the one of Fig. 5.26.

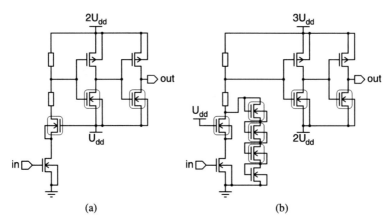

Fig. 5.28 Two circuit implementations of a level-shifter: (**a**) to shift the input from U_{dd}–GND to $2 \cdot U_{dd}$–U_{dd} and (**b**) to shift the input from U_{dd}–GND to $3 \cdot U_{dd}$–$2 \cdot U_{dd}$

The circuits of the level-shifters are shown in Fig. 5.28. The level-shifter of Fig. 5.28(a) shifts the input from U_{dd}–GND to $2 \cdot U_{dd}$–U_{dd}. This is done by switching a resistive divider on and off, yielding a level-shifted signal. The signal is then buffered through two inverters, also restoring the levels between $2 \cdot U_{dd}$–U_{dd}. The switch is implemented as a stacked MOSFET switch, effectively distributing the voltage of $2 \cdot U_{dd}$ over the individual n-MOSFETs. The level-shifter of Fig. 5.28(b) shifts the input from U_{dd}–GND to $3 \cdot U_{dd}$–$2 \cdot U_{dd}$. The level-shifting is also achieved by switching a resistive divider on and off. In this implementation the resistive divider is continuously loaded with a chain of diode-connected n-MOSFETs, preventing the voltage over the stacked MOSFET switch to rise above $2 \cdot U_{dd}$. The functionality of the diode loaded resistive divider is similar as that of linear shunt voltage converter, as explained in Sect. 2.1.2. The disadvantage of these level-shifter implementations, compared to that of Fig. 5.8, is that they draw a constant bias current for the resistive divider, when the input signal is logic-high. For this implementation this is not an issue as the converter, for which this control system is designed, is intended for achieving high values of P_{out} (ca. 1.2 W).

These level-shifters are not explicitly shown in Fig. 5.27, nevertheless they are required for the driving of the stacked transistor MOSFET implementations of the switches, which is explained in Sect. 6.4.4.

The buffers are also not explicitly shown in Fig. 5.27, but obviously they are required to drive the switches of the converter. They are implemented as standard digital tapered inverters, having a scaling factor of 2.6. The buffers for the high-side switches and the select switches for the second output contain nine stages, whereas the buffers for the low-side switches and the select switches for the first output contain eight stages.

5.4 Feed-Forward Semi-Constant On/Off-Time: F^2-SCOOT

In Sect. 5.1.3 it is concluded that the PFM control scheme is preferred for controlling monolithic inductive DC-DC converters, due to its many advantages over PWM. The proposed practical timing implementations of the PFM control scheme: COOT and SCOOT timing (see Sects. 5.2 and 5.3) are relatively straightforward to implement and well suited for full-integration. Both from a theoretical and practical point of view, these timing schemes are promising for achieving the best performance of the monolithic DC-DC converters. However, they are not well suited for applications that need to cope with a high variation of U_{in} and U_{out}. The variation of U_{out} in most practical cases is negligible, as it is usually required to be constant. The variation of U_{in}, on the other hand, can be quite large in some typical applications. The technical answer to solve this shortcoming is the Feed-Forward Semi-Constant On/Off-Time (F^{2}SCOOT) timing scheme.

The basic concept of the F^{2}SCOOT timing scheme is explained in Sect. 5.4.1. A practical implementation of an F^{2}SCOOT control system is discussed in Sect. 5.4.2, which is used to control a variant of the two-output SMOC converter.

5.4.1 The F^2-SCOOT Concept

The concept of the F^{2}SCOOT timing scheme can be understood by examining (5.5) for the buck converter. When U_{in} increases, I_{L_max} will also increase proportionally, providing t_{on} is constant, and vice versa. Thus, for a constant t_{on} the maximal η_{SW} design point is not maintained, which will result in a lower overall η_{SW}. This is illustrated by the gray curve in Fig. 5.29(a), which shows η_{SW} as a function of U_{in}, for a buck converter with a constant P_{out}. For this case the value of t_{on} is to be designed sufficiently small, such I_{L_max} that at U_{in_max}, and in turn ΔU_{out}, is kept within the converter's specifications. As a consequence I_{L_max} will be lower at lower values of U_{in}, resulting in less transferred energy per switching cycle. Therefore, f_{SW} will increase, causing η_{SW} to decrease due to the associated switching losses. Obviously, when I_{L_max} increases upon increasing U_{in}, t_{off_real} will also increase according to (5.6). In other words, the COOT concept will no longer be applicable when U_{in} varies over a wide range.

To cope with this problem t_{on} will need to be adaptive upon the value of U_{in}. Ideally, I_{L_max} should be maintained at a constant value. Through (5.5) it is understood that t_{on} needs to be varied inverse proportional to U_{in} in order to keep I_{L_max}, at a constant value of U_{out}. In this ideal case t_{off_real} would remain constant. However, at high values of U_{in}, the low value of t_{on} would cause f_{SW} to increase dramatically, resulting in strongly decreased η_{SW}. Therefore, in a real situation t_{on} will need to be varied upon U_{in} as indicated in Fig. 5.29(b), rather than linear. For the assumption of U_{out} being constant, this implies that I_{L_max} will still slightly increase upon increasing values of U_{in}, causing t_{off_real} also to slightly increase. This feed-forward concept of U_{in}, combined with the feedback concept of U_{out} of COOT timing (see

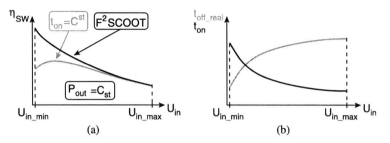

Fig. 5.29 (a) The power conversion efficiency η_{SW} as a function of the input voltage U_{in}, at constant output power P_{out} for a buck converter. The *gray curve* denotes a constant on-time control scheme, the *black curve* denotes an F^{2}SCOOT control scheme. (b) The on-time t_{on} and the real off-time t_{off_real} of an F^{2}SCOOT control scheme, as a function of the input voltage U_{in}

Sect. 5.2), yields the novel F^{2}SCOOT timing concept. F^{2}SCOOT timing results in an increased η_{SW}, compared to a constant t_{on} timing scheme, as illustrated by the black curve in Fig. 5.29(a).

Note that the trends in the graphs of Figs. 5.29(a) and 5.29(b) are deduced by means of the mathematical model for the buck converter, which is discussed in Chap. 4, and that a similar discussion and conclusion can be made for a boost converter.

The idea of altering t_{on} upon varying values of U_{in} is also described for non fully-integrated DC-DC converters in the literature, but the implementations differ from the F^{2}SCOOT timing concept. In [Kaz99] a PWM variant for a boost converter is described, which only uses feed-forward of U_{in}. Clearly, this is not ideal for controlling the load regulation. In [Sah07] a PFM variant for a buck converter is described, which, unlike the F^{2}SCOOT timing concept, requires current-sensing of the low-side switch.

To conclude the discussion on the F^{2}SCOOT timing concept, the benefits and drawbacks are listed:

✔ No current-sensing of the freewheeling switch is required.
✔ As opposed to COOT and SCOOT timing, F^{2}SCOOT timing is suited for applications where U_{in} may vary over a wide range.
✔ Increased η_{SW} performance, compared to PWM, at varying loads. This is due to the varying f_{SW}, similar to the COOT timing concept.
✔ Increased η_{SW} performance, compared to constant t_{on}, at varying U_{in}.
✔ U_{out} is always stable in DCM.
✔ The implementation is feasible using mostly digital building blocks, making it ideally suited for deep-submicron CMOS technologies.
✘ Mismatch of U_{out} still results in a decreased η_{SW}.
✘ The load regulation is theoretically non-zero and dependent on ΔU_{out}.

5.4.2 Single-Phase, Two-Output Implementation

The practical implementation of the F^{2}SCOOT control system is used to control a single-phase, two-output SMOC, monolithic buck converter, in a 0.35 μm 3.3 V/80 V high-voltage CMOS technology. The F^{2}SCOOT control system allows U_{out} to remain constant, for a very wide U_{in} range of the converter and under varying load conditions. Furthermore, the converter is operated in synchronous DCM. The concept of the novel SMOC topology for multiple-output converters is explained in Sect. 3.5.3. In this case only one output is fed by the DC-DC converter and the other one is derived from the switched output, by means of a linear series converter. The discussion in this section is limited to the F^{2}SCOOT control system implementation. The entire converter, including the linear converter, and its measurements results is discussed in Sect. 6.4.5.

The implementation part of the F^{2}SCOOT used for the feedback of U_{in}, is similar to the second implementation of the COOT control system, of which the circuit is shown in Fig. 5.16 and the principle of operation is explained in Sect. 5.2.2.

In order to make t_{on} and t_{off_real} adaptive and dependent upon U_{in}, the time-delays Δt_{on} and Δt_{off_real} of the circuit of Fig. 5.16 are replaced by those of the circuit in Fig. 5.30. In this circuit t_{on_in}, t_{on_out}, t_{off_in} and t_{off_out} are the respective in- and outputs of the adaptive t_{on} and t_{off_real}. As such, the circuit of Fig. 5.30 provides the feed-forward of U_{in}, in addition to the feedback of U_{out}. The basic principle of the feed-forward circuit is explained by means of its four building blocks:

1. *ADC*: The Analog to Digital Converter (ADC) block generates a 4-bit binary code out of the value of U_{in}. To achieve this, U_{in} is first decreased through the resistive divider R_1–R_2, yielding U'_{in}. It is also filtered through the low-pass RC filter, consisting of C and $R_1//R_2$, to suppress ΔU_{in} and other input related noise. The resulting voltage is compared with the output of the Digital to Analog Converter (DAC), by means of the comparator. When $U'_{in} > U_{DA}$ the comparator will output a logic-high signal and vice versa. The output of the comparator is fed into the Up/Down (U/D) input of the 4-bit binary counter, which will count up when the comparator outputs a logic-high signal. The output of the counter is in turn fed into the input of the DAC. Thus, when $U'_{in} > U_{DA}$ the counter will count up until $U'_{in} > U_{DA}$ and vice versa, effectively converting U'_{in} into a 4-bit binary code. The clock frequency at which this process takes place is $f_{SW}/16$, as determined by the drive signal for the high-side switch SW_1.
2. *Hold*: The 4-bit binary code from the ADC block is stored into four edge-triggered D-flipflops, which are made transparent at the end of each switching cycle of the converter. This prevents the value of Δt_{on} and Δt_{off_real} to be altered during a switching cycle, potentially causing faulty timing.
3. *On-time selector*: The 4-bit binary code from the hold block is consecutively fed into a 16 × 1 MUltipleXer (MUX), which selects the appropriate Δt_{on}. As explained in Sect. 5.4.1, the resulting value of t_{on} is inverse proportional to the value of U_{in}.
4. *Off-time selector*: This block works similar to the on-time selector block, except for the fact that the selected Δt_{off_real} is proportional to U_{in}.

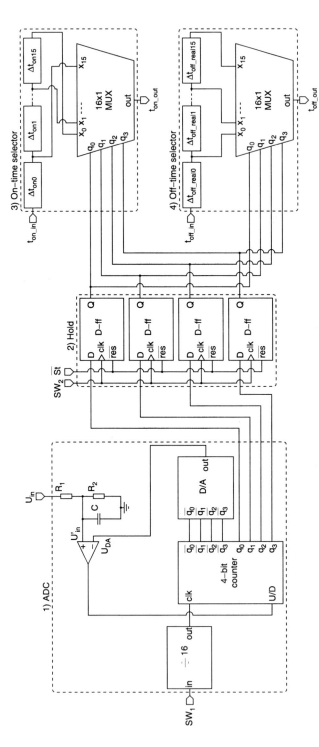

Fig. 5.30 The block diagram of the F^{2}SCOOT control system implementation of the feed-forward control loop

5.4 Feed-Forward Semi-Constant On/Off-Time: F²-SCOOT

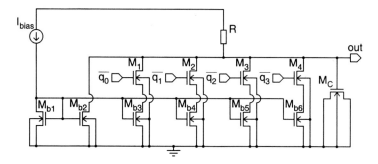

Fig. 5.31 The circuit of a 4-bit binary Digital to Analog Converter (DAC), using binary weighted current sources

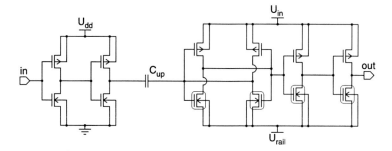

Fig. 5.32 The circuit of a high voltage ratio level-shifter

The ÷16 clock divider is implemented as a cascade of four D-flipflops. The 4-bit counter is implemented as a 4-bit binary ripple counter, which saturates at its minimal/maximal value. The comparator is similar to that in Fig. 5.6. The time delays of Δt_{on} and Δt_{off_real} are implemented as auto-reset time delays, based on the circuit of Fig. 5.17. The 16 × 1 MUXs consist of selectable transmission gates, with complementary switches.

The circuit implementation of the 4-bit binary DAC is shown in Fig. 5.31. It consists of four binary weighted current sources and a bias current source. The sum of these currents, depending on the 4-bit inverted input signal $\overline{q}$, flows through the resistor, thereby causing a voltage drop over it. This voltage drop, referred to *GND*, is the output, which is in-turn low-pass filtered by a MOS capacitor for improved Power Supply Rejection Ratio (PSRR).

The COOT feedback, shown in Fig. 5.16, uses the same components, except for the level-shifter. The circuit of the level-shifter, used in this implementation, is illustrated in Fig. 5.32. Two inverters act as buffers for the input signal and drive the capacitor C_{up} of 150 fF. The other side of the capacitor is connected to a latch. The actual level-shifting is performed through charge-coupling between the input buffer and the output latch. The output latch is in turn buffered through two inverters.

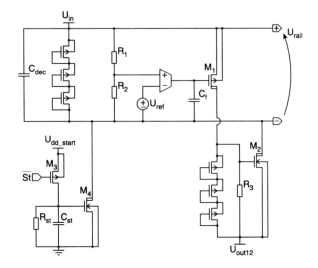

Fig. 5.33 The circuit of a rail-shifter for generating a fixed offset voltage U_{rail}, which is referred to U_{in}, together with its start-up circuit

This implementation enables the input to be shifted from U_{dd}–GND to U_{in}–U_{rail}. The voltage difference between the level-shifted signals is only restricted by the technology supply voltage, rather than by the topology itself, as opposed to the level-shifters of Figs. 5.8 and 5.19.

The buffers are implemented as digital tapered buffers, with a scaling factor of 2.6. Both the high- and low-side switch buffers contain nine stages.

The gate of the high-side switch is to be switched between U_{in}, turning it off, and U_{rail}, turning it on. This U_{rail} has a fixed offset referred to U_{in}, of 3.3 V. Because of the fact that this converter is designed to cope with a high range of U_{in}, U_{rail} cannot have a constant value, referred to GND. Therefore, for the purpose of generating U_{rail}, the rail-shifter circuit of Fig. 5.33 is acquired. In essence the rail-shifter is a linear shunt voltage converter, which generates U_{rail} by the pull-down transistor M_2. To achieve this, U_{rail} is first lowered through the resistive divider R_1–R_2. The resulting voltage is then compared with the reference voltage U_{ref}, by means of the OTA. The output of the OTA drives a level-shifter, consisting of M_1, R_3 and protection diodes, which in turn drives the gate of the pull-down transistor M_2. Transistor M_2 pulls U_{rail} towards the 12 V output U_{out12}, rather than to GND, to increase η_{lin}. The decouple capacitor C_{dec} of 500 pF minimizes the voltage ripple of U_{rail} to about 10% of its value. This is necessary due to the fast charging and discharging of the large parasitic gate-capacitance of the high-side switch. During the start-up phase of the converter $U_{rail} = U_{in}$ and the rail-shifter will not be able to perform its task. Therefore, a start-up circuit is added, pulling U_{rail} down until the OTA of the rail-shifter is able to operate normally. This is achieved by providing an active-low start pulse $\overline{St}$ to M_3, which causes C_{St} to be charged and in turn M_4 to pull U_{rail} towards GND. During this phase the protection diodes between U_{in} and U_{rail} prevent the value of $U_{in} - U_{rail}$ from becoming too large. After the start-up pulse, C_{St} is discharged through R_{St} and M_4 is

5.5 Start-up

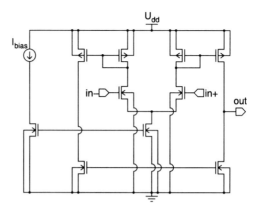

Fig. 5.34 The circuit of a basic symmetrical Operational Transconductance Amplifier (OTA)

turned off again. At this point the rail-shifter can perform its task in the normal fashion.

Note that the supply voltage U_{dd_start} for the start-up circuit of the rail-shifter is in turn generated by the start-up circuit of the DC-DC converter itself, which is discussed in Sect. 5.5.2.

The circuit which is used for the OTA in the rail-shifter is a basic symmetrical OTA, as illustrated in Fig. 5.34. This OTA has a GBW of 2 GHz and a phase-margin of 90°, which is achieved by adding the additional output capacitor C_f, parallel to the parasitic gate-capacitance of M_1 in the rail-shifter.

5.5 Start-up

Most of the practical implementations of monolithic converters in this work use U_{out} of the DC-DC converter to power (a part of) the control system, including the level-shifters and the buffers for driving the power switches. Consequentially, the control system is not able to perform its task when the output(s) of the DC-DC converter are not yet powered to the right voltage level, causing a start-up issue. Alternatively, dedicated linear voltage converters may be used for powering (a part of) the control system [Haz07], when possible. Nevertheless, this is to be avoided due to their low associated η_{lin}. Moreover, many of the proposed control system implementations require a start signal, allowing them to initialize the state of flipflops and eventually to enable the converter to start switching.

For these purposes a start-up circuit is required in many converter implementations, enabling U_{out} of the DC-DC converter to be raised to a sufficient level during the initialization phase. The possible concepts for realizing start-up circuits are discussed in Sect. 5.5.1. A discussion on the practical implementations of start-up circuits, which are used in coexistence with the previously discussed control systems, is provided in Sect. 5.5.2.

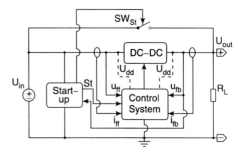

Fig. 5.35 The start-up method concept, suited for DC-DC step-down converters

5.5.1 The Concept

The start-up circuit has the task of raising the output voltage(s), which are evidently zero when the converter is turned off, of the DC-DC converter to a sufficient level during its initialization. By doing so, the main control system is powered, enabling it to commence its task. In general, this function can be achieved in two ways. The first way is to provide a second, simplified control system, which enables the converter to start switching at the initialization phase [Leu05]. This start-up control system may be powered directly by U_{in}, or through a linear voltage converter or a basic charge-pump. This method is suited for both DC-DC step-up and DC-DC step-down converters. For sake of simplicity, it is not used in the practical implementations in this work.

The second method to achieve start-up is only suited for DC-DC step-down converters and its concept is illustrated in Fig. 5.35. During the initialization phase the input of the converter is connected with its output, until U_{out} has reached a sufficient level for the control system to take over. If necessary, a start signal St may be provided by the start-up circuit to the control system, enabling it. To achieve these tasks, the start-up circuits needs to be powered solely by U_{in}, unlike the control system. Also, the start-up circuit needs to sense U_{out}, in order for it to be disabled when the level of U_{out} is sufficiently high.

In Fig. 5.35 it can be observed that U_{in} and U_{out} are connected through a dedicated switch SW_{St}. Alternatively, in a buck converter this can also be done by using the high-side switch. However, because the high-side switch typically has a low on-resistance, this may cause a large current peak during start-up. Such a current peak, of which the amplitude can easily be in the order of a few Ampère, could potentially cause damage to the switch and/or interference with neighboring circuits. To avoid this, a dedicated switch is used in the practical implementations in this work, allowing a better control of the start-up current.

5.5.2 Implementations

Figure 5.36 shows the start-up circuit, which is used in various practical implementations in this work. The first task of the start-up circuit is to provide power to itself.

Fig. 5.36 The start-up circuit, used in various practical realizations in this work

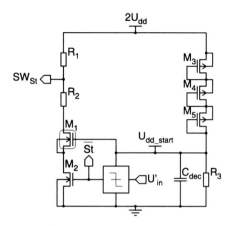

This is achieved by means of a linear shunt converter (see Sect. 2.1.2), consisting of M_3, M_4, M_5, R_3 and C_{dec}. The output voltage of this linear shunt converter is U_{dd_start}, of which the value is ideally equal to the nominal technology supply voltage U_{dd}. Obviously, the number of diode-connected p-MOSFETs is dependent on the value of U_{in}, which is equal to $2 \cdot U_{dd}$ in this example. U_{dd_start} is acquired to power the schmitt-trigger and to bias the stacked transistor M_1.

The second task of the start-up circuit is to close the start-up switch SW_{st}, shown in Fig. 5.35, until U_{out} of the converter has reached a sufficient level. To achieve this, a fraction of the output voltage of the converter U'_{out} is fed into the schmitt-trigger, which outputs a logic-high signal when $U'_{out} < U_{th\_L}$. This activates the switch formed by the stacked n-MOSFETs M_1 and M_2, pulling the resistive divider R_1–R_2 towards GND. As a result, the voltage on node SW_{st} is lowered to U_{dd}, activating the p-MOSFET start-up switch. When $U'_{out} > U_{th\_H}$, the schmitt-trigger will output a logic-low signal and the start-up switch is opened. Because $U_{th\_L} < U_{th\_H}$ a safety barrier is intrinsically present, avoiding the start-up circuit to be activated unintended by ΔU_{out}, for instance.

Note that this start-up circuit is also suited for $U_{in} > 2 \cdot U_{dd}$, providing the stacked switch is implemented as shown in Fig. 5.28(b). This start-up circuit is also used in the multiple-output SMOC converter implementation, discussed in Sect. 6.4.4, where a dedicated start-up circuit per output is used and control logic is added to determine when both outputs have a sufficient voltage level.

5.6 Conclusions

A constant output voltage of a DC-DC converter, under varying load and line conditions, is obtained by means of a feedback mechanism, possibly also supplemented by a feed-forward loop. The two main feedback switching schemes are PWM and PFM. In Sect. 5.1 it is deduced that PFM is the preferred method for monolithic DC-DC converters, as it is capable of achieving a higher overall η_{SW}, especially at low values of P_{out}. It is also known that PFM control systems contain less analog building blocks, making them ideal for implementation in deep-submicron CMOS technologies. The drawbacks of a PFM control scheme are: a higher ΔU_{out} at low load operation and the theoretically non-zero load regulation.

From this knowledge three novel PFM switching schemes, which are optimized for monolithic DC-DC converters, are proposed:

1. *COOT*: (see Sect. 5.2) Constant On/Off-Time is well suited for single-phase, single- and multiple-output converters. It uses both a constant t_{on} and t_{off_real}, thereby transferring fixed amounts of energy to the output of the converter. This results in straightforward implementations, optimized for a limited variation of U_{in} and U_{out}.
2. *SCOOT*: (see Sect. 5.3) Semi-Constant On/Off-Time is well suited for multi-phase, single- and multiple-output converters. Two or more t_{on} and t_{off_real} pairs are used, in addition with an adaptive offset time between the consecutive phases of the converter. Their values all depend on the demanded P_{out}. This approach results in an improved high load η_{SW}, compared to COOT timing. SCOOT timing is also optimized for a limited variation of U_{in} and U_{out}.
3. F^2-*SCOOT*: (see Sect. 5.4) Feed-Forward Semi-Constant On/Off-Time is well suited for single-phase, single- and multiple-output converters, requiring a large U_{in} range. This timing scheme is based on COOT timing, whereby the constant t_{on} and t_{off_real} pairs are dependent on the value of U_{in}. As a result, the overall η_{SW} is drastically increased and may be maintained over a wide range of U_{in}.

All three proposed PFM control schemes incorporate the lack of needing current-sensing of the freewheeling switch, an optimized overall η_{SW} performance and a stable output voltage in DCM. Note that combinations of these three control schemes are possible.

Multiple practical implementations of both the PWM and the proposed PFM control schemes are also provided throughout this chapter. These implementations are used in the monolithic DC-DC converter realizations, which are discussed in Chap. 6.

Finally, the concept of start-up circuits, together with an implementation example, is discussed in Sect. 5.5. These start-up circuits are required in most of the monolithic DC-DC converter implementations of this work. This is due to the fact that the control system (or a part of it) is supplied by the output(s) of the converter. As such, additional lossy linear voltage converters or charge-pumps are avoided.

Chapter 6
Implementations

The goal of this work is to realize monolithic inductive DC-DC converters, having both a maximal overall power conversion efficiency and a maximal power density. In Chap. 3 it is concluded that the boost and the buck converter, and their multiphase, SIMO and SMOC variants, are the most promising topologies to achieve this goal. The mathematical steady-state design model for the boost and the buck converter, described in Chap. 4, leads to the important design trade-offs associated with monolithic integration. As such, it enables the designer to find the optimal design point for the intended application or specifications. Furthermore, the discussion on control systems in Chap. 5, points out that this optimal design point is a function of the output power and the in- and output voltage. Therefore, novel control schemes, together with practical circuit implementations, are proposed to increase the average performance of a monolithic DC-DC converter over a wider range of output powers and input voltages.

The combined knowledge of both inductive DC-DC converter and control techniques and systems, leads to the various practical chip realizations, which are described in this chapter. Thereby, the hands-on approach from the previous Chap. 5 is continued, providing the designer the essential feeling of the various practical implementation and measurements issues. At the same time the reader is provided with the idea of what (and what not) is to be expected from monolithic inductive DC-DC converters in various standard CMOS technologies, performance wise.

The practical implementation possibilities, involving the essential components of the DC-DC converter's power stage: inductors, capacitors and switches, are discussed in Sect. 6.1. Comments on the main measurement principles and setups are provided in Sect. 6.2. The various practical implementations of monolithic inductive boost and buck converters are discussed in the respective Sects. 6.3 and 6.4. In order to situate this work, a side-by-side comparison of the measurements of the implementations in this work and the implementations described in the literature, is performed in Sect. 6.5. Finally, this chapter is concluded in Sect. 6.6.

6.1 Monolithic Converter Components

The essential building blocks of inductive DC-DC converters, namely inductors, capacitors and switches, incorporate non-ideal characteristics, as discussed in Sect. 4.2. As a consequence, power losses are introduced, which in turn result in a decreased power conversion efficiency η_{SW}. Therefore, it is clear that care has to be taken in the lay-out of the converter components, in addition to their design. For this reason, the next sections provide some basic considerations on the lay-out of monolithic inductive converter components, which will yield an increased η_{SW}.

For the inductor, the bondwire and the metal-track implementations are discussed in Sect. 6.1.1. The implementation of a capacitor by means of a MIM, MOM and MOS capacitor is explained in Sect. 6.1.2. The implementation of the power switches, with the emphasis on waffle-shaped structures, is provided in Sect. 6.1.3.

6.1.1 Inductor

Basically, there are two ways to realize on-chip inductors: with or without extra processing. The process of fabricating inductors by means of additional processing steps is also referred to as micro-machining. Many techniques have been demonstrated in the literature to improve the specifications of on-chip inductors. A first way to achieve this is to apply ferro-magnetic core materials to metal-track inductors [Gar07] or even bondwire inductors [Lu10]. These techniques increase the self inductance of the inductor, for the same number of windings. As a drawback, core losses are introduced, due to eddy-currents. A second way to increase the total inductance of on-chip inductors is to increase the mutual inductance. This may be achieved by bar and meander inductors [Ahn96], in an attempt to mimic solenoid inductors. However, this usually results in a higher parasitic series resistance, due to the use of multiple vias between metal layers. A third way is to decrease the parasitic series resistance by using an additional thick metal-film processing step on top the chip [Per04]. In this process a relatively thick, compared to on-chip metal layers, metal-film is deposited. This kind of inductors do not provide a significant advantage, as they are still prone to eddy-current losses in the conductive silicon chip substrate. A fourth method, and presumably the most effective one, decreases these eddy-current losses in the substrate by means of under-etching [Til96, Wu09]. Note that due to the higher costs of additional processing steps, micro-machined inductors are not further considered in this work. Nevertheless, this does not mean that their application in monolithic inductive DC-DC converters is to be considered obsolete.

In this work two types of on-chip inductors are used, namely bondwire and metal-track inductors. Figure 6.1(a) shows a schematic perspective view of a hollow-spiral rectangular bondwire inductor [Wen07], with a patterned capacitor underneath. A hollow-spiral octagonal metal-track inductor, above a silicon chip substrate, is shown in Fig. 6.1(b). The inductors used in the practical converter implementations

6.1 Monolithic Converter Components

Fig. 6.1 A schematic perspective view of (a) a hollow-spiral rectangular bondwire inductor with a patterned capacitor underneath and (b) an integrated hollow-spiral octagonal metal-track inductor, in top metal above the silicon substrate

are designed and optimized using the 2-D field-solver FastHenry [Kam94], the illustrations of Fig. 6.1 are generated by means of the input data of FastHenry.

Bondwire inductors [Cra97] have a number of advantages over metal-track inductors. First, they can achieve a higher inductance L, for the same parasitic series resistance, compared to metal-track inductors. In other words, bondwire inductors are able to achieve a higher Q-factor. This is due to the increased conductor thickness of the bondwire,[1] compared to on-chip metal layers. A second advantage is the increased distance of the windings of the bondwire inductor to the chip substrate, compared to metal-track inductors. This yields two benefits: 1) the parasitic substrate capacitance is mainly determined by the limited total area of the bonding-pads, being less than the area of a metal-track inductor, 2) the negative mutual inductance, due to the conductive substrate, and the associated eddy-currents (see Sect. 4.2.1), are significantly reduced. Last but not least, the space underneath the bondwire inductor can be used for circuits and/or the output/decouple capacitor of the converter [Zha06].

In the converter implementations discussed in the respective Sects. 6.3.1 and 6.4.1, the output capacitor is placed underneath the bondwire inductor. However, despite of the total area reduction of the converter, this results in a reduced total inductance. This problem is avoided by adding slots, underneath the bondwires and perpendicular to them, into the output capacitor. This concept is illustrated in Fig. 6.1(a). By adding these slots with an intermediate distance of 100 μm, an increase of approximately 40% of the total inductance and a decrease of approximately 20% of the parasitic series resistance is observed. This depends on the average height of the bondwires above the substrate, which is in the order of 100 μm. A patterned output capacitor will result in a slight increase of its parasitic series resistance, which is estimated in the order of approximately 20%. Nevertheless, this will result in an overall benefit for the converter's performance. A patterned output

---

[1] The standard bondwire diameter used in the practical implementations in this work is 25 μm.

capacitor is used in the practical implementation described in Sect. 6.4.5. Note that the influence of the chip substrate is negligible when an output capacitor is placed underneath the bondwire inductor.

Although hollow-spiral rectangular bondwire inductors have many benefits over metal-track inductors, they are not practical for footprints smaller than about 1 mm × 1 mm. This implies that they are not suited for realizing small inductance values, in the order of 10 nH and less. Alternatively, linear bondwire inductors may be used. However, the benefit of the positive mutual inductance between the windings is lost in that case, resulting in a lower Q-factor. For these reasons metal-track inductors are preferred for realizing inductance values lower than 10 nH. Although this type of inductor is not suited in combination with a capacitor underneath, it does prove to be beneficial in multi-phase converters (see Sects. 6.4.3 and 6.4.4). The design of on-chip metal-track inductors incorporates four fundamental degrees of freedom: the number of turns, the number of metal layers, the track width and the geometry. The first three degrees of freedom are dependent on the metal specifications of the used CMOS IC technology, of which the most important are: the number of available metals, the square resistance of the metals and the thickness of the oxide between the metals. For this reason the designs are performed through exhaustive optimization, using FastHenry, of which a qualitative discussion is omitted in this work.

6.1.2 Capacitor

Analogue to on-chip inductors, on-chip capacitors can essentially be realized by using standard CMOS technology features, or by performing additional processing steps. The best known example of micro-machined on-chip capacitors are deep-trench capacitors, which are nowadays able to achieve capacitance densities in the order of 100 nF/mm^2 [Joh09]. Another non-native CMOS, well known on-chip capacitor type is the MIM capacitor, which is quasi standardly available in modern deep-submicron CMOS IC technologies. A schematic perspective view of a MIM capacitor is shown in Fig. 6.2(a). The MIM capacitor is formed between metal layer ME X and an additional dedicated metal layer ME MIM, of which the distance to metal layer ME X is much smaller than the standard inter-metal distance. Therefore, the capacitance density of a MIM capacitor can be in the order of 1 nF/mm^2 to 2 nF/mm^2, depending on the CMOS process. The benefits of MIM capacitors are their low parasitic series resistance and their compatibility with placing circuits underneath them. As a drawback, their maximum operating voltage is usually in the order of the nominal technology supply voltage.

Apart from the on-chip capacitors, which require additional processing steps, native CMOS on-chip capacitors are also an option. The two most practical and most used types are the MOM capacitor and the MOS capacitor. For the MOM capacitor many implementation varieties exist [Apa02, Sam98], of which the interleaved wire variant is widely used. Figure 6.2(b) shows the schematic perspective view of the interleaved wire configuration, for two metal layers. It is clear

6.1 Monolithic Converter Components

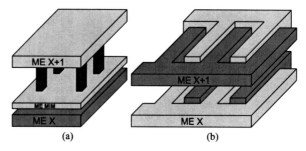

Fig. 6.2 (a) A schematic perspective view of a MIM capacitor and (b) a MOM capacitor in a parallel, interleaved wire configuration

that the capacitance density will strongly depend on the number of metals used, thus depending on the number of available, metal layers and their minimal allowed spacings and widths. In general, this type of capacitor will be beneficial for deep-submicron CMOS technologies, as the resolution of the masks and the number of metal layers tend to increase (see Sect. 1.3.1). Typical capacitance densities in deep-submicron technologies ranges from 100 pF/mm^2 to 1.5 nF/mm^2. The benefits of MOM capacitors are their ability to withstand higher voltages than the nominal technology supply voltage, their potentially low parasitic series resistance and the possibility to place circuits underneath them. As a drawback they have a low capacitance density. The second most used native on-chip capacitor type is the MOS capacitor, which is physically formed between the gate and the induced channel (see Fig. 1.17(b)). The capacitance of the MOS capacitor is dependent on the voltage over it, being quasi constant when this voltage exceeds V_t. The capacitance density depends on the gate-oxide thickness t_{ox} and its permittivity ϵ_{ox}, which in turn depend on the CMOS technology node. For current CMOS technologies this can vary anywhere between approximately 3 nF/mm^2 and 20 nF/mm^2. Obviously, the main advantage of MOS capacitors is their high capacitance density. However, this comes at a few caveats, such as: a potentially high parasitic series resistance, the incompatibility with circuits underneath them and the high gate-leakage in deep-submicron CMOS technologies,[2] which can be in the order of a few mA/mm^2.

In Sects. 4.2.2 and 4.2.5 it is explained that the parasitic series resistance R_{Cs} of both the output and input decouple capacitors is pernicious for the DC-DC converter's performance. For the MIM capacitor, having a structure similar as shown in Fig. 6.2(a), R_{Cs} is dependent on the two metal plates ME X and ME X + 1. More specifically, R_{Cs} is determined by the dimensions of these metal plates, their geometry and their square resistance. Knowing that a MOM capacitor will also be connected by two such plates, it can be intuitively understood that in this case these plates will also be dominant for the value of R_{Cs}. For large capacitors (>1 mm^2), the value of R_{Cs} for MIM and MOM capacitors can be in the order of 200 mΩ. For MOS capacitors the determination of R_{Cs} is less straightforward, because of

---

[2]The gate-leakage depends on the manufacturer of the CMOS process and the minimum feature size. Generally, this becomes abruptly significant from 90 nm CMOS technologies onwards.

218   6 Implementations

⊛CALCULATING THE ESR OF PLATE CAPACITORS

For plate capacitors with large aspect ratios the ESR can be reduced by a factor three, when connecting the plates at both sides. This is illustrated in Fig. 6.3. The series resistances R_{left} and R_{right}, seen from the left and the right, as a function of the length L of the plate, are shown by the respective black curves. The gray curve denotes $R_{left}//R_{right}$.

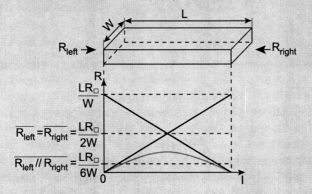

Fig. 6.3 The resistance of a conductive plate as a function of its length L, when the plate is connected from the left, the right and both sides

$$R_{left} = \frac{\ell R_\square}{W} \qquad (6.1)$$

$$R_{right} = \frac{L R_\square}{W} - \frac{\ell R_\square}{W} \qquad (6.2)$$

$$\overline{R_{left}} = \overline{R_{right}} = \int_0^\ell R_{left}\, d\ell = \int_0^\ell R_{right}\, d\ell = \frac{L R_\square}{2\cdot W} \qquad (6.3)$$

$$\overline{R_{left}//R_{right}} = \int_0^\ell R_{left}//R_{right}\, d\ell = \frac{L R_\square}{6\cdot W} \qquad (6.4)$$

R_{left} and R_{right} are calculated through (6.1) and (6.2). The mean value of R_{left} and R_{right} is equal to the ESR, caused by one plate, and is calculated through (6.3). When both sides of the plate are connected, the ESR is calculated through (6.4). The latter method yields the factor three.

the strong dependency on the aspect ratio of the gate and the total area. The optimal lay-out of a MOS capacitor is a finger-shaped structure, similar to that used for transistors (see Sect. 6.1.3). For this finger-structure the value of R_{Cs} for an n-MOS capacitor, without taking the connecting plates into account, is calculated through (6.5).

$$R_{Cs} = \frac{\frac{W}{6\cdot L}(R_{channel\square} + R_{poly\square}) + \frac{W_{drain}}{L}R_{n^+\square} + \frac{W_{source}}{L}R_{n^+\square} + \frac{R_{cont\_g}}{\#cont\_g} + \frac{R_{cont\_ds}}{\#cont\_ds}}{\#fingers}$$

(6.5)

Fig. 6.4 (a) The parasitic series resistance R_{Cs} and (b) the capacitance density C/A of a MOS capacitor, both as a function of the width W and the length L of the individual fingers

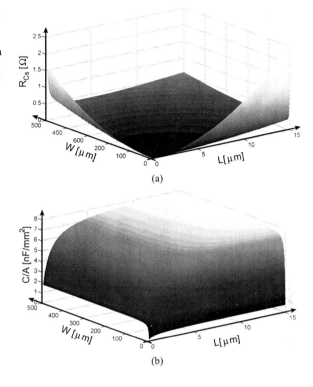

In (6.5) $R_{channel\square}$ denotes the square resistance of the induced channel,[3] $R_{poly\square}$ is the square resistance of the poly-silicon gate,[4] W_{drain} and W_{source} are the respective widths of the drain and the source, $R_{n+\square}$ is the square resistance of the n^+-doped regions,[5] R_{cont_g} denotes the resistance of a gate contact[6] and R_{cont_ds} is the resistance of a drain/source contact.[7] Note that $R_{channel\square}$ is the dominant factor for the value of R_{Cs} of a MOS capacitor.

Figures 6.4(a) and 6.4(b) show the graphical representations of R_{Cs} and the overall capacitance density C/A of a MOS capacitor, as a function of W and L of the individual fingers. The graphs are valid for a MOS capacitor having an area of 1 mm², in a 180 nm CMOS technology. In Fig. 6.4(b) the overhead of drain, source and bulk areas, as well as the gate contact areas, are taken into account. It can clearly be seen that a trade-off emerges between R_{Cs} and C/A. Indeed, a low value of R_{Cs} (wanted) will result in a low C/A (unwanted) and vice versa. The origin of this trade-off follows from the fact that $R_{channel\square}$ is the dominant factor in R_{Cs}. Hence, a smaller

---

[3] For deep-submicron technologies $R_{channel\square}$ is in the order of 5 k$\Omega/\square$, for $U_{gs} = U_{dd}$.
[4] For deep-submicron technologies $R_{poly\square}$ is in the order of 10 $\Omega/\square$.
[5] For deep-submicron technologies $R_{n+\square}$ is in the order of 10 $\Omega/\square$.
[6] For deep-submicron technologies R_{cont_g} is in the order of 15 $\Omega/\square$.
[7] For deep-submicron technologies R_{cont_ds} is in the order of 15 $\Omega/\square$.

value of L of the gate fingers will result in a lower R_{C_s}. However, a smaller L also results in more overhead due to drain and source areas, in turn resulting into a lower value of C/A. For the practical implementations in this work, of which the area of the output capacitors in the order of 1 mm^2, L is chosen approximately 10 μm and W is chosen approximately 100 μm. This results in a value of about 100 mΩ for the contribution of the MOS capacitor to R_{C_s}. For these values the associated reduction of C/A is in the range of 15% to 25%, depending on the CMOS technology. Similar to MIM and MOM capacitors, the total R_{C_s} is mostly determined by the overall metal connection plates of the MOS capacitor. In general this adds a resistance of 200 mΩ to 300 mΩ, in addition to the 100 mΩ of the MOS capacitor, giving a total R_{C_s} in the order of 300 mΩ to 400 mΩ. As expected, R_{C_s} for MOS capacitor is higher compared to a MIM or MOM capacitor.

Note that the estimation of the parasitic resistance of the connecting metal plates of an on-chip capacitor is performed by means of an approximation. This approximation is rather obvious for square-shaped plates, but becomes quite complex for other geometries. In those cases a mean value approximation is acquired.

6.1.3 Switches

Monolithic inductive DC-DC converters operate at high values of f_{SW}, due to their limited inductance and output capacitance. As explained in Sect. 4.2.3, this introduces stress on the switch devices, which need to be able to switch sufficiently fast to minimize the power losses. The combination of a high f_{SW} (>100 MHz), fast switching transients and relatively high currents (>100 mA), has led to previous implementations using specialized High Frequency (HF) Gallium Arsenide (GaAs) switches [Ajr01] and BiCMOS technologies [Gho04]. An additional problem is the fact that the DC-DC converters may have a higher in- or output voltage than the nominal supply voltage of the CMOS technology, which is the case for all the practical implementation in this work. One way to overcome this issue is to use optional thick-oxide devices that can cope with higher voltages.

The problem with specialized technologies and optional devices is their higher cost, compared to standard CMOS technologies. Therefore, the power switches of the DC-DC converters in this work are implemented using stacked MOSFETs [Kur05, Ser07]. The concept of stacked MOSFETs is shown in Fig. 6.6, for a dual-stack n-MOSFET example. The node voltages for the off- and on-state are shown in the respective Figs. 6.6(a) and 6.6(b). Note that M_2 is implemented using a triple-well n-MOSFET. It is observed that the gate of M_2 is constantly biased at U_{dd} and that the gate of M_1 is switched between GND and U_{dd}. By doing so, the voltage $2 \cdot U_{dd}$ is evenly distributed over both transistors in the off-state. The same concept is used for p-MOSFET switches. Also, higher voltages can be applied, by using additional stacked transistors.

The lay-out of MOSFETs for power applications can essentially be done in two ways: by means of a linear finger structure, or by using a waffle-shaped lay-out.

⚠ CMOS LATCH-UP

When dealing with CMOS circuits which need to handle large currents, transients and/or bulk conduction, latch-up becomes a potential issue, compromising the correct operation of the circuit. As these conditions are valid for monolithic DC-DC converters, care is to be taken to avoid latch-up.

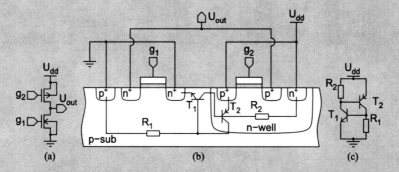

Fig. 6.5 (a) The considered CMOS circuit. (b) The physical cross-section of the CMOS circuit, with the parasitic thyristor structure. (c) The equivalent BJT circuit of a thyristor

Figure 6.5(a) shows the considered CMOS circuit, which is similar to a half-bridge driver of a buck converter. The physical cross-section of this circuit is shown in Fig. 6.5(b), together with the parasitic thyristor structure. When insufficient bulk contacts in the n-well and substrate contacts are present, the value of the parasitic resistances R_1 and R_2 potentially becomes sufficiently large such that a substrate or n-well current triggers the parasitic thyristor. The equivalent BJT circuit of this parasitic thyristor is shown in Fig. 6.5(c). Once triggered, a large current will flow from U_{dd} to GND, which keeps flowing due to positive feedback. Note that other latch-up mechanisms can also be identified, including capacitive triggered ones [Rec88].

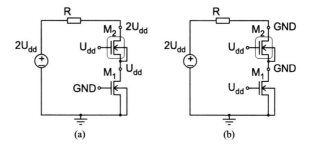

Fig. 6.6 The concept of stacked MOSFETs, for a dual-stack n-MOSFET example, (a) in the off-state and (b) in the on-state

Both lay-out concepts are shown in the respective Figs. 6.7(a) and 6.7(b). Waffle-shaped MOSFETs are known to have a reduced overall area [Mal00] and can potentially achieve a reduced parasitic drain capacitance [Lam01], compared to linear finger MOSFETs. Although the reduced drain capacitance is mainly an advantage in CMOS technologies which allow for minimal drain and source areas, the ability of

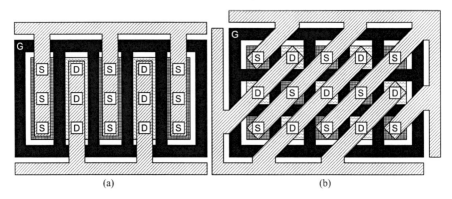

Fig. 6.7 The lay-out of a MOSFET using (**a**) a linear finger structure and (**b**) using a waffle-shaped structure

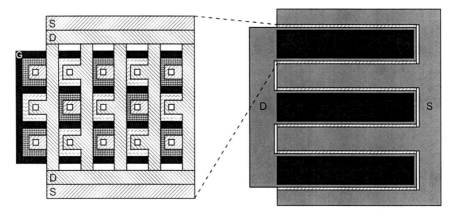

Fig. 6.8 The alternative lay-out of a waffle-shaped MOSFET, modified for large current handling. The figure on the *left* shows the detail of the waffle-shaped structure and the *right-hand* figure shows the entire transistor

lay-outing waffle-shaped MOSFETs in a compact fashion is considered their main general advantage. For this reason, all the power switches, except for those used in the high-voltage design discussed in Sect. 6.4.5, in this work are implemented using waffle-shaped MOSFETs.

The lay-out of the waffle-shaped MOSFET, shown in Fig. 6.7(b), is not optimal for high currents. This is because the angled drain and source connections are not symmetrical, leading to current unbalance. Moreover, the total width of the drain and source connections is too small for handling high peak currents. Therefore, the alternative lay-out of Fig. 6.8 is used. The left-hand figure shows the detail of the modified waffle-shaped structure. Instead of using 45° routing for the drain and source connections, 90° routing is used. The respective drain and source connections are lay-out in different metal-layers, connected on both sides. This waffle-structure is rectangular with a large aspect ratio, resulting in short gate, drain and source

6.1 Monolithic Converter Components

④LIDAR DRIVER: 10 A & 2.2 NS RISE-TIME

A Laser Imaging Detection And Ranging (LIDAR) driver chip, shown in Fig. 6.9(a), is an example of the high current capability (10 A) and short rise- and fall-times, of CMOS technologies [Wen09a]. Figure 6.9(b) shows both the driver chip and the laser-diode, mounted closely together on a PCB.

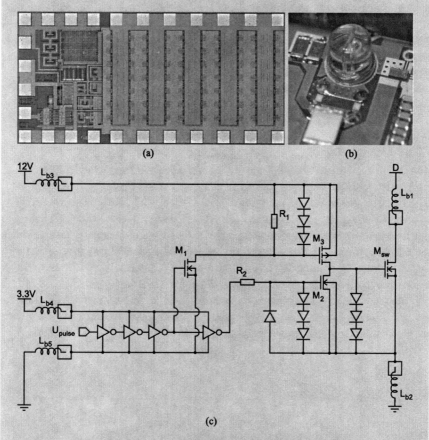

Fig. 6.9 (a) The micro-photograph of the driver chip. (b) The laser-diode together with the driver chip, mounted on a PCB. (c) The circuit of the driver chip

To cope withe the high surge-voltages and ground-bounce, the large n-MOSFET ($W = 38500$ μm) driver cannot be driven by a conventional digital tapered buffer circuit. Therefore, a custom pre-driver, of which the circuit is shown in Fig. 6.9(c) is designed to fulfill this task. Note that DC-DC converters in this work do not require such a pre-driver.

connections. The rectangular waffle-shaped MOSFET structure is duplicated, as illustrated by the right-hand drawing of Fig. 6.8. A large metal finger structure is used to provide the overall drain an source connections. The width of these metal fingers is also determined by the current that is to be coped with. Obviously, wider fingers also result in a lower parasitic drain and source series resistances, but requires more area and yields more parasitic drain and source capacitance. To minimize additional parasitic drain and source capacitance towards the substrate, the metal finger structure is lay-out in the upper metal(s). The space between the waffle areas is filled with substrate contacts, in addition to an overall guard ring, avoiding latch-up.

The typical values for W of the power switches, used in this work, are between 1000 µm and 6000 µm. The additional combined parasitic drain and source series resistance of the metal connections, achieved with the proposed structure, is in the order of 150 mΩ to 300 mΩ.

6.2 On Measuring DC-DC Converters

The measurement of DC-DC converters may seem rather straight-forward, as merely DC currents and voltages are easily measured. Basically, this perception is not wrong. However, care should be taken for DC-DC converter measurements, in order for them to be a correct and fair representation. Indeed, if certain rules are not followed, the obtained measurements results will either under- or overestimate the converter's performance. Therefore, a standardized measurement setup is introduced for measuring monolithic DC-DC converters. Obviously, this measurement setup is consequentially used throughout the presented work.

The main principles involving the measurements of monolithic DC-DC converters, are discussed in Sect. 6.2.1. A practical example of a measurement setup used in this work, in addition with some associated circuits, is provided in Sect. 6.2.2.

6.2.1 Main Principles

The circuit for measuring monolithic DC-DC converters is illustrated in Fig. 6.10. The resistors R_{in}, R_{out} and R_{GND} represent the parasitic resistances of the various metal PCB tracks, wires and connectors. U'_{in} is a voltage source with sense inputs, performing a voltage feedback function for the voltage source. By connecting the sense inputs directly to the power input terminals of the DC-DC converter chip, the U_{in} of the converter is maintained at a constant level, regardless of the voltage drop over the parasitic resistances R_{in} and R_{GND}. A high-ohmic voltage meter measures the mean value of voltage over the sense inputs, being an accurate representation of the converter's $\overline{U_{in}}$. An ampère meter to measure the mean $\overline{I_{in}}$ is placed in series with the voltage sources and the converter's positive input, followed by a

6.2 On Measuring DC-DC Converters

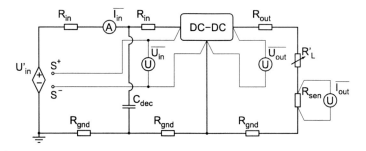

Fig. 6.10 The circuit for measuring monolithic DC-DC converters

decouple capacitor[8] C_{dec}. This C_{dec} makes sure that the correct value of $\overline{I_{in}}$ is measured, by effectively filtering ΔU_{in} away. The resulting static P_{in} is then calculated through (6.6).

$$P_{in} = \overline{U_{in}}\, \overline{I_{in}} \qquad (6.6)$$

The mean output voltage $\overline{U_{out}}$ is measured directly at the output of the converter, by means of a voltage meter. It is verified that the high frequency ΔU_{out} causes true RMS voltage meters to measure the mean value, rather than the RMS value of U_{out}. Nevertheless, for this measurement this behavior is desired. The load resistor at the output of the converter is formed by a variable (electronic) resistor R'_L in series with a fixed value sense resistor R_{sen}, which needs to have an accurately known resistance. The mean voltage U_{sen} over R_{sen} is directly measured over its physical terminals by a voltage meter, used to determine the mean output current $\overline{I_{out}}$, as stated by (6.7).

$$\overline{I_{out}} = \frac{\overline{U_{sen}}}{R_{sen}} \qquad (6.7)$$

Similar to the measurement of $\overline{U_{in}}$, a true RMS voltage meter will yield the mean value of U_{sen}, rather than its RMS value. This is due to the high frequency of ΔI_{out}, which is out of the range of the used meters. Also in this case this is a desired behavior. The resulting static P_{out} is calculated through (6.8).

$$P_{out} = \overline{U_{out}}\, \overline{I_{out}} \qquad (6.8)$$

Finally, the power conversion efficiency η_{SW} is obtained through (1.3).

For dynamic load an line regulation measurements, the respective voltage meters at the in- and the output of the converter are replaced by and oscilloscope. This allows the graphic determination of the load and line regulation. The circuits used for varying U_{in} and U_{out} of the converter are discussed in Sect. 6.2.2.

---

[8]In reality the decouple capacitor consists of multiple parallel capacitors of different values, ranging from 100 pF to a few hundred μF, for achieving an improved frequency response.

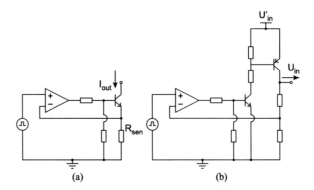

Fig. 6.11 The circuit of (a) an electronic controlled load and (b) an electronic controlled voltage source

6.2.2 Practical Example

One of the key measurements for characterizing DC-DC converters is η_{SW} as a function of P_{out}, at a constant U_{in} and U_{out}. This measurement requires a variable load, in order to sweep P_{out}. The circuit of such an electronic variable load is shown in Fig. 6.11(a). Basically, this circuit acts as a current I_{out} drain, which is controlled trough a voltage source or signal generator. This is achieved by feedback of the voltage over R_{sen}, which is proportional to I_{out}. This circuit is used for the measurements of both η_{SW} as a function of P_{out} and the load regulation. For the latter measurement a signal generator controls I_{out}, such that P_{out} is swept between P_{out_min} and P_{out_max} of the converter. This is preferably done by means of a square wave, having a frequency in the order of 10 kHz–100 kHz. Alternatively, it may be performed by using a sine wave, having a frequency in the order of a few MHz, depending on f_{SW} of the converter. The resulting load regulation is calculated through (5.1).

The line regulation requires a circuit which can vary the U_{in} of the converter, obtaining a square wave. The circuit to achieve this is shown in Fig. 6.11(b). This circuit acts as a linear series voltage converter, as explained in Sect. 2.1.1. An OPAMP compares a fraction of the output voltage with a reference voltage, in this case formed by a signal generator, and accordingly drives a PNP pass-transistor. Because U'_{in} can be larger than the supply voltage of the OPAMP, which is fed through a separate supply voltage, the PNP BJT is driven through a level-shifter. The resulting output voltage of the linear series voltage converter is used as U_{in} for the DC-DC converter. This U_{in} is then varied between U_{in_min} and U_{in_max}, at a frequency in the order of 1 kHz–10 kHz.

An example of a practical measurement setup is shown in Fig. 6.12. The measurement PCB contains the following building blocks:

- *Power & decoupling*: This block is responsible for providing power to both the converter and the measurement periphery on the PCB (OPAMPs). Sufficient decoupling capacitor is added to the power supply, in order to filter potential noise from the converter out before the measurement circuitry and to obtain an accurate measurement of $\overline{I_{in}}$ (see Sect. 6.2.1).

6.2 On Measuring DC-DC Converters

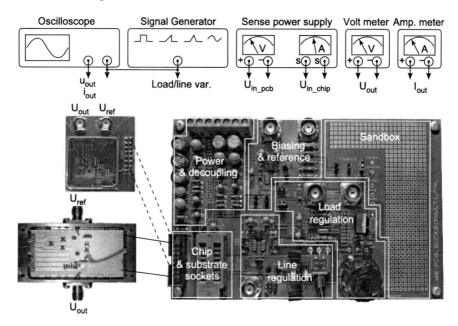

Fig. 6.12 The schematic representation of the measurement setup, containing: a DC-DC converter chip mounted on a substrate, a PCB with various measurement and biasing circuits, and laboratory measurement equipment

- *Chip & substrate sockets*: Various methods for mounting the chip are accommodated, including Al_2O_3 (aluminum-oxide ceramic) substrates, the PCB (FR4) substrates and Dual In Line (DIL) chip packages.
- *Biasing & reference*: This block provides the biasing currents for on-chip OPAMPs and comparators, enabling more degrees of freedom. Also, the static voltage reference(s) for controlling U_{out} are generated in this block.
- *Load regulation*: The circuit for varying the load of the converter, as shown in Fig. 6.11(a).
- *Line regulation*: The circuit for varying U_{in} of the converter, as shown in Fig. 6.11(b).
- *Sandbox*: Additional space for additional on-the-fly test circuits.

Apart from the measurement PCB and the chip substrate or package, some additional laboratory equipment is required. This is illustrated in Fig. 6.12 and comprises the following apparatus:

- *Oscilloscope*: Used for measuring ΔU_{out} and determining the load and line regulation.
- *Signal generator*: Used for driving the load and line regulation measurements circuits on the PCB.
- *Sense power supply*: A power supply with sense inputs, for measuring η_{SW} as a function of P_{out}, at a constant value of U_{in}.

- *Volt meter*: For measuring $\overline{U_{in}}, \overline{U_{out}}$ and $\overline{I_{out}}$, as explained in Sect. 6.2.1.
- *Ampere meter*: For measuring $\overline{I_{in}}$, as explained in Sect. 6.2.1.

6.3 Boost Converters

Two monolithic inductive boost DC-DC converters are realized in the presented work. In the following sections the circuit and the design parameters of these converters are discussed. It is noted that these converters are designed and optimized by means of the mathematical steady-state design model, which is explained in Chap. 4. The designs were optimized for maximal η_{SW} and P_{out}/A, starting from a given chip area A. Also, ΔU_{out} is kept lower than 10% of U_{in}.

The first implementation is a single-phase, single-output boost converter, using a bondwire inductor, which is discussed in Sect. 6.3.1. The second implementation is a single-phase, two-output SIMO converter, using a metal-track inductor and is discussed in Sect. 6.3.2. Both converters operate without using off-chip components.

6.3.1 Bondwire, Single-Phase, Single-Output

The first boost converter implementation is realized in a 180 nm, 1.8 V CMOS technology [Wen07]. In this technology six metal layers, including a thick-top (2 μm) copper layer, are available, in addition with a MIM capacitor (1 nF/mm²). The chip die dimensions are 1.5 mm × 1.5 mm. In the following sections the converter circuit, the input parameters and the measurements are discussed.

Circuit & Input Parameters

The circuit of the converter, together with the PWM feedback loop, is shown in Fig. 6.13. Both switches are implemented as two stacked transistors, in order to cope with voltages of $2 \cdot U_{dd}$. The gates of M_{1b} and M_{2b} are biased with U_{in}. The gate of M_{1a} is switched between U_{in} and GND, turning it on and off. The gate of M_{2a} is switched between U_{in} and U_{out}, turning it on and off. The bulk terminals of both M_{2a} and M_{2b} are connected to the output of the converter, allowing the body diodes to conduct i_L during the transient from the charge to the discharge phase. C_{out} is implemented as a sole MIM capacitor. A hollow-spiral bondwire inductor L, in addition with the connection bondwire between the package and the chip, yields the total inductance of 18 nH + 3 nH = 21 nH. This bondwire inductor measures 1.4 mm × 1.4 mm and it consists of four windings of 25 μm thick golden bonding wire. The center pitch of the bonding wires is 100 μm. An on-chip decouple capacitor is not implemented, enabling the inductance from the additional connection bondwire to be used. An overview of the most important circuit parameters is provided in Table 6.1.

6.3 Boost Converters

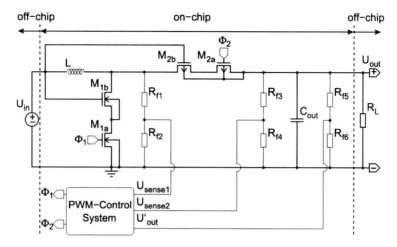

Fig. 6.13 The circuit of the implementation of the bondwire, single-phase, single-output boost DC-DC converter, with a PWM control system

Table 6.1 The circuit parameters of the bondwire, single-phase, single-output boost DC-DC converter implementation

Circuit parameter	Value
CMOS technology	180 nm
Width SW_1 W_{SW_1}	1800 μm
Width SW_2 W_{SW_2}	1800 μm
Switching frequency f_{SW}	100 MHz
Output capacitance C_{out_MIM}	1.3 nF
Inductance L	18 nH + 3 nH = 21 nH
Parasitic inductor series resistance R_{Ls} @ 100 MHz	1 Ω
Chip die area A	2.25 mm^2

The feedback resistors R_{f1}, R_{f2}, R_{f3} and R_{f4} are used to measure the voltage over M_{2a} and M_{2b}, providing current-sensing information to the PWM control system. The feedback resistors R_{f5} and R_{f6} are used to provide the feedback of U_{out}. The PWM control system implementation of this converter is discussed in Sect. 5.1.1. Note that this converter is self-starting, through the bulk conduction of M_{2a} and M_{2b}, and that consequentially no startup-circuit is required.

Measurements Results

Figure 6.14 shows the naked chip die micro-photograph of the bondwire, single-phase, single-output DC-DC boost converter. The building blocks: the switches, the

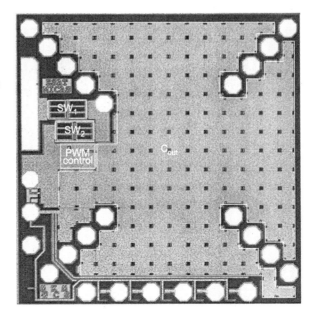

Fig. 6.14 The micro-photograph of the naked chip die of the bondwire, single-phase, single-output boost DC-DC converter, with the indication of the building blocks

output capacitor and the PWM control system are indicated. The bonding pads for the bondwire inductor are shaped octagonal, allowing a closer pitch of the bondwires. Multiple bondwires are used for the power supply connections, minimizing the parasitic input inductance and resistance.

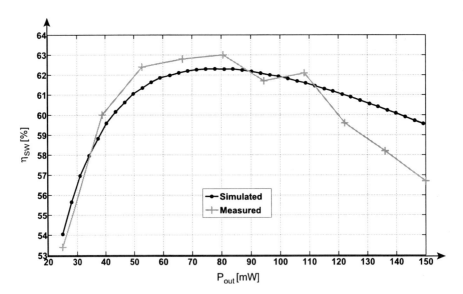

Fig. 6.15 The power conversion efficiency η_{SW} as a function of the output power P_{out}, of the bondwire, single-phase, single-output boost DC-DC converter implementation

6.3 Boost Converters

Fig. 6.16 The load regulation, measured for P_{out} varying between 25 mW and 150 mW, at a frequency of 1 kHz

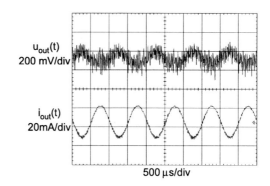

The measurement of η_{SW} as a function of P_{out} is shown by the gray curve in the graph of Fig. 6.15. For this measurement U_{in} and U_{out} where kept constant at 1.8 V and 3.3 V, respectively. For these nominal values, a maximal η_{SW} of 63% is achieved. As predicted in Sect. 5.1.3, η_{SW} tends to drop towards low values of P_{out}. At higher values of P_{out}, η_{SW} also drops due to temperature effects, as explained in Sect. 4.3. It is noted that the overall η_{SW} increases upon increasing values of U_{in} and vice versa. This is due to the fact that the inductor needs to deliver less energy E_L during the discharge phase, when U_{in} is higher ($U_{out} = U_{in} + U_L$). The black curve in the graph of Fig. 6.15 denotes the simulation, performed through the mathematical steady-state design model of Chap. 4. A maximal deviation between the measured and simulated η_{SW} curves of less than 3% is observed.

A measurement of the load regulation is shown in Fig. 6.16, P_{out} is varied between 25 mW and 150 mW, at a frequency of 1 kHz. The upper curve shows $u_{out}(t)$ and the lower curve shows $i_{out}(t)$. Apart from ΔU_{out}, an additional ripple due to the load regulation of 80 mV is observed. The converter can cope with sinusoidal load variations, having a frequency up to 9 MHz. The main measured parameters of the converter are summarized in Table 6.2.

Finally, the micro-photograph of the chip die of the bondwire, single-phase, single-output DC-DC boost converter implementation, with the bondwire inductor

Table 6.2 The main measured parameters of the bondwire, single-phase, single-output DC-DC boost converter implementation

Measured parameter	Value
Input voltage range U_{in}	1.6 V–2.2 V
Output voltage range U_{out}	2.5 V–4 V
Maximal power conversion efficiency η_{SW_max} @ $U_{in} = 1.8$ V; $U_{out} = 3.3$ V; $P_{out} = 80$ mW	63%
Output power range P_{out}	25 mW–150 mW
Power density P_{out_max}/A	67 mW/mm^2
Maximal output voltage ripple ΔU_{out_max}	200 mV
Load regulation	4‰/mA

Fig. 6.17 The micro-photograph of the chip die of the bondwire, single-phase, single-output DC-DC boost converter, with the bondwire inductor added

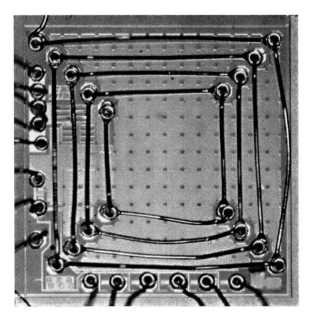

added, is shown in Fig. 6.17. Note that only minor variations ($\simeq 2\%$) of η_{SW} where observed between different samples, indicating that the variation of the inductance of the different bondwires is well under control and insignificant.

6.3.2 Metal-Track, Single-Phase, Two-Output SIMO

The second boost converter implementation is realized in a 130 nm, 1.2 V CMOS technology. In this technology nine metal layers, including a thick (2 µm) copper layer and a thick-top aluminum (1.2 µm), are available, in addition with a MIM capacitor (1.5 nF/mm^2). The chip die dimensions are 1.5 mm × 1.5 mm. The measurements of this chip revealed that it does not performs as expected, which is most likely due to technology issues. Nevertheless, a brief overview of the converter's circuit and the main (simulation) parameters is provided, giving the reader an idea of the possibilities of similar designs.

The circuit of the converter, together with the COOT feedback loop, is shown in Fig. 6.18. The power switches SW_1 and SW_3 are implemented as three stacked transistors, in order to cope with voltages of $3 \cdot U_{dd}$. The power switch SW_2 only needs to handle a voltage of $2 \cdot U_{dd}$, which is why it is implemented using only two stacked transistors. The gates of M_{1b} and M_{2b} are biased with U_{in} and the gate of M_{3b} is biased with U_{out1}. SW_1 is switched on by applying U_{in} to the gates of M_{1a} and M_{1c} and is switched off by switching these gates to GND and U_{out1}, respectively. SW_2 is switched on by applying U_{out1} to the gates of M_{3a} and M_{3c} and is switched off by switching these gates to U_{out2} and U_{in}, respectively. The gates of the transistors of SW_1 and SW_3 are driven by a specialized half-bridge stacked transistor driver [Ser05]. For further information on the design of this driver, which is

6.3 Boost Converters

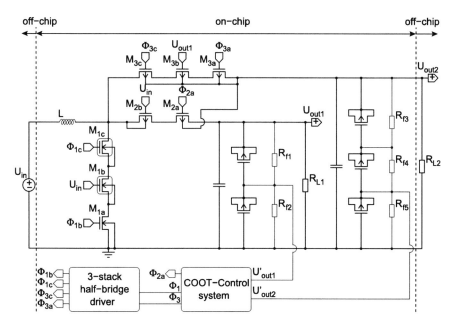

Fig. 6.18 The circuit of the implementation of the metal-track, single-phase, two-output SIMO boost DC-DC converter, with a COOT control system

beyond the scope of this dissertation, the reader is referred to [Ser07]. SW_2 is turned on and off by switching the gate of M_{2a} between U_{in} and U_{out2}. The bulk terminals of M_{3a}, M_{3b} and M_{3c} are connected to the second output of the converter, allowing the body diodes to conduct i_L during the transient from the charge to the discharge phases of either one of the outputs. C_{out1} is implemented as a MIM capacitor, parallel to the series circuit of two MOS capacitors. C_{out2} also is implemented as a MIM capacitor, parallel to the series circuit of three MOS capacitors. In both the output capacitors the MIM and MOS capacitors are physically stacked onto each other, reducing their total required area. A hollow-spiral octagonal metal-track inductor L, in addition with the connection bondwire between the package and the chip, yields the total inductance of 8 nH + 3 nH = 11 nH. This metal-track inductor measures 800 μm × 800 μm and it consists of four windings of 70 μm wide tracks. These tracks are formed by the thick copper layer and the thick-top aluminum layer. An on-chip input decouple capacitor is not implemented, enabling the inductance from the additional connection bondwire to be used. An overview of the most important circuit parameters is provided in Table 6.3.

The feedback resistors R_{f1} and R_{f2} are used to provide the feedback of U_{out1}, whereas the feedback resistors R_{f3} and R_{f3} are used to provide the feedback of U_{out2}. The COOT control system implementation of this converter is discussed in Sect. 5.2.3. Note that this converter is not self-starting and that a start-up circuit is not added to the design, requiring the converter to be started with the aid of external voltage sources.

Table 6.3 The circuit parameters of the metal-track, single-phase, two-output boost DC-DC converter implementation

Circuit parameter	Value
CMOS technology	130 nm
Width SW_1 W_{SW_1}	3700 µm
Width SW_2 W_{SW_2}	2800 µm
Width SW_3 W_{SW_3}	4200 µm
Maximal switching frequency f_{SW_max}	220 MHz
On-time t_{on}	3 ns
Real off-time output1 t_{off_real1}	1.2 ns
Real off-time output2 t_{off_real2}	1 ns
Output MIM capacitance1 C_{out1_MIM}	0.56 nF
Output MOS capacitance1 C_{out1_MOS}	1.07 nF
Output MIM capacitance2 C_{out2_MIM}	1 nF
Output MOS capacitance2 C_{out2_MOS}	0.84 nF
Inductance L	8 nH + 3 nH = 11 nH
Parasitic inductor series resistance R_{Ls} @ 1 GHz	1.6 Ω
Chip die area A	2.25 mm^2

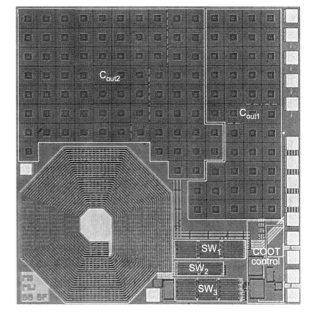

Fig. 6.19 The micro-photograph of the naked chip die of the metal-track, single-phase, two-output SIMO boost DC-DC converter, with the indication of the building blocks

Table 6.4 The main expected parameters of the metal-track, single-phase, two-output SIMO boost DC-DC converter implementation

Simulated parameter	Value
Input voltage U_{in}	1.2 V
Output1 voltage U_{out1}	2.4 V
Output2 voltage U_{out2}	3.3 V
Maximal power conversion efficiency η_{SW_max}	45%
Output1 power range P_{out1}	0 mW–26 mW
Output2 power range P_{out2}	0 mW–34 mW
Power density P_{out_max}/A	27 mW/mm^2
Maximal output1 voltage ripple ΔU_{out1_max}	240 mV
Maximal output2 voltage ripple ΔU_{out2_max}	330 mV

Figure 6.19 shows the naked chip die micro-photograph of the metal-track, single-phase, two-output SIMO DC-DC boost converter. The switches, the output capacitors and the COOT control system are indicated. Multiple bonding wires are used for the *GND* connection, minimizing the parasitic input inductance and resistance.

The main simulated output parameters, resulting from simulations with the mathematical steady-state design model (see Chap. 4), of the converter are summarized in Table 6.4. The rather low value of η_{SW_max} is due to the increased losses of the metal-track inductors, compared to a bondwire inductor. It is also inherent to the two-output SIMO topology, which requires an increased f_{SW}.

6.4 Buck Converters

Five monolithic inductive buck DC-DC converters are realized in the presented work. In the following sections, the circuit and the design parameters of these converters are discussed. It is noted that these converters are designed and optimized by means of the mathematical steady-state design model, which is explained in Chap. 4. The designs where optimized for maximal η_{SW} and P_{out}/A, starting from a given chip area A. Also, ΔU_{out} is kept lower than 10% of U_{in}.

The first implementation comprises a single-phase, single-output buck converter, using a bondwire inductor and it is discussed in Sect. 6.4.1. The second implementation is a single-phase, single-output converter, using a metal-track inductor and it is discussed in Sect. 6.4.2. The third implementation comprises a four-phase, single-output buck converter, using metal-track inductors, as discussed in Sect. 6.4.3. A four-phase, two-output SMOC buck converter, using metal-track inductors, is the fourth implementation and it is discussed in Sect. 6.4.4. Finally, the fifth implementation, comprising a high-voltage single-phase, two-output SMOC buck converter, using a bondwire inductor, is discussed in Sect. 6.4.5. All converters are designed to operate without the need for any off-chip components.

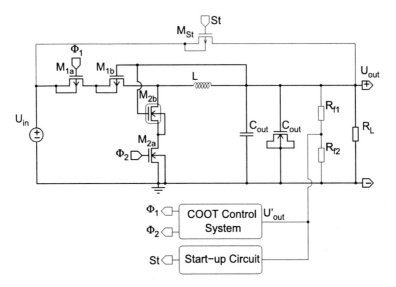

Fig. 6.20 The circuit of the implementation of the bondwire, single-phase, single-output buck DC-DC converter, with a COOT control system

6.4.1 Bondwire, Single-Phase, Single-Output

The first buck converter implementation is realized in a 180 nm, 1.8 V CMOS technology [Wen08b]. In this technology six metal layers, including a thick-top (2 μm) copper layer, are available, in addition with a MIM capacitor (1 nF/mm^2). The chip die dimensions are 1.5 mm × 1.5 mm. In the following sections the converter circuit, the input parameters and the measurements are discussed.

Circuit & Input Parameters

The circuit of the converter, together with the COOT feedback loop, is shown in Fig. 6.20. Both switches are implemented as two stacked transistors, in order to cope with voltages of $2 \cdot U_{dd}$. The gates of M_{1b} and M_{2b} are biased with U_{out}. The gate of M_{1a} is switched between U_{out} and U_{in}, turning it on and off. The gate of M_{2a} is switched between U_{out} and GND, turning it on and off. The bulk terminals of both M_{1a} and M_{1b} are connected to their respective source terminals, avoiding U_{gs} form exceeding $2 \cdot U_{dd}$. This approach still allows for the series connected body diodes to conduct i_L during the transient from the charge to the discharge phase. C_{out} is implemented as a MIM capacitor, parallel to a MOS capacitor. Both the MIM and MOS capacitor are physically stacked on each other, reducing their required area. A hollow-spiral bondwire inductor L yields the total inductance of 18 nH. This bondwire inductor measures 1.4 mm × 1.4 mm and it consists of four windings of 25 μm thick golden bondwire. The center pitch of the bondwire is 100 μm. An on-chip input decouple capacitor is not implemented, as the parasitic inductances of the

6.4 Buck Converters

Table 6.5 The circuit parameters of the bondwire, single-phase, single-output buck DC-DC converter implementation

Circuit parameter	Value
CMOS technology	180 nm
Width SW_1 W_{SW_1}	3800 μm
Width SW_2 W_{SW_2}	2000 μm
On-time t_{on}	4.4 ns
Real off-time t_{off_real}	2.7 ns
Output capacitance C_{out_MIM}	1.3 nF
Output capacitance C_{out_MOS}	9 nF
Inductance L	18 nH
Parasitic inductor series resistance R_{Ls} @ 100 MHz	1 Ω
Chip die area A	2.25 mm^2

power supplying bondwires is negligible, compared to the inductance of the bondwire inductor. An overview of the most important circuit parameters is provided in Table 6.5.

The feedback resistors R_{f1} and R_{f2} are used to provide the feedback of U_{out}. The COOT control system implementation of this converter is discussed in Sect. 5.2.2. Note that this converter is not self-starting and that it requires a start-up circuit, as explained in Sect. 5.5. This start-up circuit switches M_{St} on, during the initialization of the converter, until U_{out} reaches the value of 1.4 V. Afterwards the COOT control system enables the converter to take over.

Measurements Results

Figure 6.21 shows the naked chip die micro-photograph of the bondwire, single-phase, single-output DC-DC buck converter. The switches, the output capacitor and the COOT control system are indicated. The bonding pads for the bondwire inductor are shaped octagonal, allowing a closer pitch of the bondwires. Multiple bondwires are used for the power supply connections, minimizing the parasitic input inductance and resistance.

The measurement of η_{SW} as a function of P_{out} is shown by the gray curve in the graph of Fig. 6.22. For this measurement U_{in} and U_{out} where kept constant at 3.6 V and 1.8 V, respectively. For these nominal values, a maximal η_{SW} of 65% is achieved. This yields an *EEF* value of 23%, compared to a linear series voltage converter having the same voltage conversion ration k_{lin} of 0.5. As predicted in Sect. 5.1.3, η_{SW} tends to be higher at low values of P_{out}, compared to a PWM control system. Nevertheless, a decrease of η_{SW} is observed at low values of P_{out}, due to the static power consumption of the COOT control system. At higher values of P_{out}, η_{SW} also drops due to temperature effects, as explained in Sect. 4.3. It is noted that the overall η_{SW} decreases when the value of U_{in} and/or U_{out} differs from

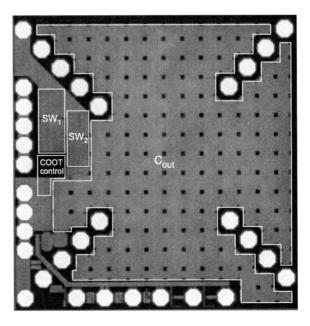

Fig. 6.21 The micro-photograph of the naked chip die of the bondwire, single-phase, single-output buck DC-DC converter, with the indication of the building blocks

the nominal value. This is due to the mismatch of t_{on} and t_{off_real}, as explained in Sect. 5.2.1. This decrease of η_{SW} is in the order of a few %, within the U_{in} and U_{out} operating range of the converter. The black curve in the graph of Fig. 6.22 denotes the simulation, performed through the mathematical steady-state design model of

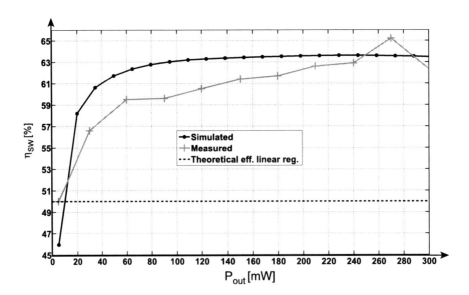

Fig. 6.22 The power conversion efficiency η_{SW} as a function of the output power P_{out}, of the bondwire, single-phase, single-output buck DC-DC converter implementation

6.4 Buck Converters

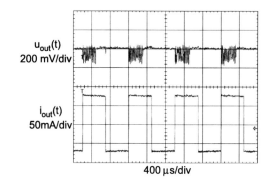

Fig. 6.23 The load regulation, measured for P_{out} varying between 30 mW and 300 mW, at a frequency of 1 kHz

Chap. 4. A maximal deviation between the measured and simulated η_{SW} curves of less than 4% is observed.

A measurement of the load regulation is shown in Fig. 6.23, P_{out} is varied between 30 mW and 300 mW, at a frequency of 1 kHz. The upper curve shows $u_{out}(t)$ and the lower curve shows $i_{out}(t)$. Apart from ΔU_{out}, an additional ripple due to the load regulation of 50 mV is observed. The converter can cope with steep transient load variations, without compromising its stability. The main measured parameters of the converter are summarized in Table 6.6.

Finally, the micro-photograph of the chip die of the bondwire, single-phase, single-output DC-DC buck converter implementation, with the bondwire inductor added, is shown in Fig. 6.24. Note that only minor variations ($\simeq 2\%$) where observed between different samples, indicating that the variation of the inductance of the different bondwires is well under control and insignificant.

Table 6.6 The main measured parameters of the bondwire, single-phase, single-output buck DC-DC converter implementation

Measured parameter	Value
Input voltage range U_{in}	3 V–4 V
Output voltage range U_{out}	1.5 V–2.1 V
Switching frequency range f_{SW}	20 Hz–140 MHz
Maximal power conversion efficiency η_{SW_max} @ $U_{in} = 3.6$ V; $U_{out} = 1.8$ V; $P_{out} = 270$ mW	65%
Efficiency Enhancement Factor EEF	+23%
Mean Efficiency Enhancement Factor $\overline{EEF}$	+17%
Output power range P_{out}	0 mW–300 mW
Power density P_{out_max}/A	133 mW/mm²
Maximal output voltage ripple ΔU_{out_max}	160 mV
Load regulation	0.18‰/mA
Line regulation	5.6‰/V

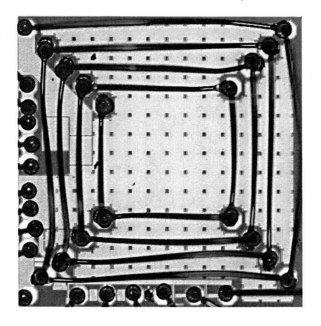

Fig. 6.24 The micro-photograph of the chip die of the bondwire, single-phase, single-output buck DC-DC converter, with the bondwire inductor added

6.4.2 Metal-Track, Single-Phase, Single-Output

The second buck converter implementation is realized in a 130 nm, 1.2 V CMOS technology [Wen08a]. In this technology nine metal layers, including a thick (2 μm) copper layer and a thick-top (1.2 μm) aluminum layer, are available, in addition with a MIM capacitor (1.5 nF/mm^2). The chip die dimensions are 1.5 mm × 2.25 mm. In the following sections the converter circuit, the input parameters and the measurement results are discussed.

Circuit & Input Parameters

The circuit of this converter, together with the COOT feedback loop, is similar to the circuit of the previous implementation, as shown in Fig. 6.20. The implementation of the switches and the output capacitor is also analogue to the description, provided in Sect. 6.4.1. The difference of this converter implementation, however, is the presence of an input decouple capacitor C_{dec}. This input decouple capacitor is implemented as a MIM capacitor, which can cope with higher voltages than the nominal technology supply voltage. It is physically stacked with a part (half) of the MOS capacitor part of C_{out}. Furthermore, a hollow-spiral metal-track inductor L yields the total inductance of 9.8 nH. This metal-track inductor measures 1 mm × 1.5 mm and it consists of three windings. The tracks of these windings have a width of 100 μm, consisting of both the thick copper layer and the thick-top aluminum layer. An overview of the most important circuit parameters is provided in Table 6.7.

Table 6.7 The circuit parameters of the metal-track, single-phase, single-output buck DC-DC converter implementation

Circuit parameter	Value
CMOS technology	130 nm
Width SW_1 W_{SW_1}	1800 μm
Width SW_2 W_{SW_2}	1100 μm
On-time t_{on}	2 ns
Real off-time t_{off_real}	0.75 ns
Output capacitance C_{out_MIM}	1.07 nF
Output capacitance C_{out_MOS}	14 nF
Decouple capacitance C_{dec}	1.07 nF
Inductance L	9.8 nH
Parasitic inductor series resistance R_{Ls} @ 1 GHz	1.6 Ω
Chip die area A	3.375 mm²

The COOT control system implementation of this converter is discussed in Sect. 5.2.2. Note that this converter is not self-starting and that it requires a start-up circuit, as explained in Sect. 5.5. This start-up circuit switches M_{St} on, during the initialization of the converter, until U_{out} reaches the value of 0.9 V. Afterwards the COOT control system enables the converter to take over.

Measurements Results

Figure 6.25 shows the naked chip die micro-photograph of the metal-track, single-phase, single-output buck DC-DC converter. The switches, the output capacitor, the input decouple capacitor and the COOT control system are indicated. Multiple bondwires are used for the power supply connections, minimizing the parasitic input inductance and resistance.

The measurement of η_{SW} as a function of P_{out} is shown by the gray curve in the graph of Fig. 6.26. For this measurement U_{in} and U_{out} where kept constant at 2.6 V and 1.2 V, respectively. For these nominal values, a maximal η_{SW} of 52% is achieved. This yields an *EEF* value of 12%, compared to a linear series voltage converter having the same voltage conversion ration k_{lin} of 0.46. As predicted in Sect. 5.1.3, η_{SW} tends to be maintained longer at low values of P_{out}, compared to a PWM control system. Nevertheless, a decrease of η_{SW} is observed at low values of P_{out}, due to the static power consumption of the COOT control system. The expected drop of η_{SW} at high values of P_{out} is not observed. This is due to the fact that the COOT control system limits P_{out} to a value at which this effect does not yet occurs. It is noted that the overall η_{SW} decreases when the value of U_{in} and/or U_{out} differ from the nominal values. This is due to the mismatch of t_{on} and t_{off_real}, as explained in Sect. 5.2.1. This decrease of η_{SW} is in the order of a few %, within

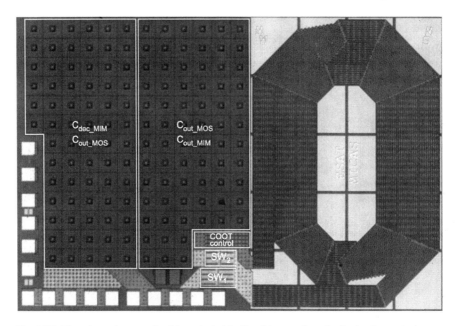

Fig. 6.25 The micro-photograph of the naked chip die of the metal-track, single-phase, single-output buck DC-DC converter, with the indication of the building blocks

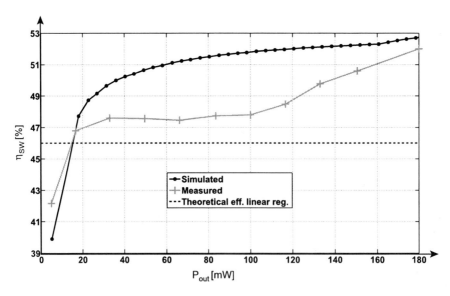

Fig. 6.26 The power conversion efficiency η_{SW} as a function of the output power P_{out}, of the metal-track, single-phase, single-output buck DC-DC converter implementation

6.4 Buck Converters

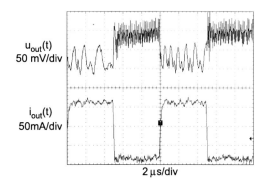

Fig. 6.27 The load regulation, measured for P_{out} varying between 5 mW and 180 mW, at a frequency of 100 kHz

the U_{in} and U_{out} operating range of the converter. The black curve in the graph of Fig. 6.26 denotes the simulation, performed through the mathematical steady-state design model of Chap. 4. A maximal deviation between the measured and simulated η_{SW} curves is less than 4% is observed.

A measurement of the load regulation is shown in Fig. 6.27, P_{out} is varied between 5 mW and 180 mW, at a frequency of 100 kHz. The upper curve shows $u_{out}(t)$ and the lower curve shows $i_{out}(t)$. Apart from ΔU_{out}, an additional ripple due to the load regulation of 75 mV is observed. The converter can cope with steep transient load variations, without compromising its stability. Finally, the main measured parameters of the converter are summarized in Table 6.8.

Table 6.8 The main measured parameters of the metal-track, single-phase, single-output buck DC-DC converter implementation

Measured parameter	Value
Input voltage range U_{in}	2 V–2.6 V
Output voltage range U_{out}	1.1 V–1.5 V
Switching frequency range f_{SW}	30 Hz–300 MHz
Maximal power conversion efficiency η_{SW_max} @ $U_{in} = 2.6$ V; $U_{out} = 1.2$ V; $P_{out} = 220$ mW	52%
Efficiency Enhancement Factor *EEF*	+12%
Mean Efficiency Enhancement Factor $\overline{EEF}$	+4%
Output power range P_{out}	0 mW–180 mW
Power density P_{out_max}/A	53 mW/mm^2
Maximal output voltage ripple ΔU_{out_max}	110 mV
Load regulation	0.18‰/mA
Line regulation	3.19%/V

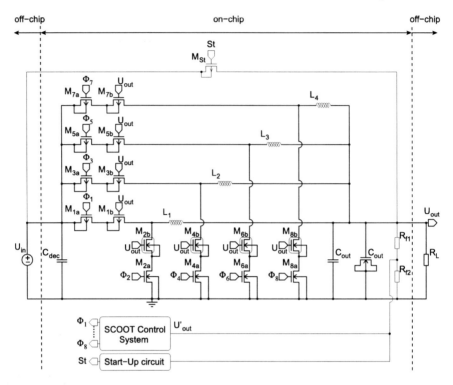

Fig. 6.28 The circuit of the implementation of the metal-track, four-phase, single-output buck DC-DC converter, with a SCOOT control system

6.4.3 Metal-Track, Four-Phase, Single Output

The third buck converter implementation is also realized in a 130 nm, 1.2 V CMOS technology [Wen09b]. In this technology nine metal layers, including a thick (2 μm) copper layer and a thick-top (1.2 μm) aluminum layer, are available, in addition with a MIM capacitor (1.5 nF/mm^2). The chip die dimensions are 1.6 mm × 2.35 mm. In the following sections the converter circuit, the input parameters and the measurements are discussed.

Circuit & Input Parameters

The circuit of the converter, together with the SCOOT feedback loop, is shown in Fig. 6.28. Both the high- and low-side switches are implemented as two stacked transistors, in order to cope with voltages of $2 \cdot U_{dd}$. The gates of M_{1b}–M_{8b} are biased with U_{out}. The gates of M_{1a}, M_{3a}, M_{5a} and M_{7a} are switched between U_{out} and U_{in}, turning them on and off. The gates of M_{2a}, M_{4a}, M_{6a} and M_{8a} are switched between U_{out} and GND, turning them on and off. All the bulk terminals of the MOSFETs are

6.4 Buck Converters

Table 6.9 The circuit parameters of the metal-track, four-phase, single-output buck DC-DC converter implementation

Circuit parameter	Value
CMOS technology	130 nm
Width SW_1–SW_7 W_{SW_1}–W_{SW_7}	3500 μm
Width SW_2–SW_8 W_{SW_2}–W_{SW_8}	1500 μm
On-time1 t_{on1}	0.8 ns
On-time2 t_{on2}	1.6 ns
Real off-time1 t_{off_real1}	0.3 ns
Real off-time2 t_{off_real2}	0.6 ns
Offset-time1 $t_{offset1}$	1.5 ns
Offset-time2 $t_{offste2}$	0.75 ns
Output capacitance C_{out_MIM}	1.13 nF
Output capacitance C_{out_MOS}	11.04 nF
Decouple capacitance C_{dec_MIM}	0.3 nF
Inductances L	3×3.9 nH
Parasitic inductor series resistance R_{Ls} @ 1 GHz	1 Ω
Chip die area A	3.76 mm^2

connected to their respective source terminals, avoiding U_{gs} form exceeding $2 \cdot U_{dd}$. For the low-side switches, this approach still allows for the series connected body diodes to conduct i_L during the transient from the charge to the discharge phase. C_{out} is implemented as a MIM capacitor, parallel to a MOS capacitor. Both the MIM and MOS capacitor are physically stacked on each other, reducing their required area. The input decoupling capacitor C_{dec} is implemented as a sole MIM capacitor, as it needs to cope with a voltage of $2 \cdot U_{dd}$. The hollow-spiral octagonal metal-track inductors L each yield a total inductance of 3.9 nH. These metal-track inductors consists of 2.875 windings, having a track width of 80 μm. They are implemented by using the thick copper layer and the thick-top aluminum layer. An overview of the most important circuit parameters is provided in Table 6.9.

The feedback resistors R_{f1} and R_{f2} are used to provide the feedback of U_{out}. The SCOOT control system implementation of this converter is discussed in Sect. 5.3.2. Note that this converter is not self-starting and that it requires a start-up circuit, as explained in Sect. 5.5. This start-up circuit switches M_{St} on, during the initialization of the converter, until U_{out} reaches the value of 0.8 V. Afterwards the SCOOT control system enables the converter to take over.

Measurements Results

Figure 6.29 shows the naked chip die micro-photograph of the metal-track, four-phase, single-output buck DC-DC converter. The switches, the output capacitor, the input decouple capacitor and the SCOOT control system are indicated. Multiple

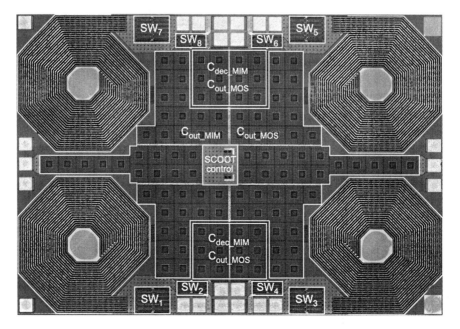

Fig. 6.29 The micro-photograph of the naked chip die of the metal-track, four-phase, single-output buck DC-DC converter, with the indication of the building blocks

bondwires are used for the power supply and output connections, minimizing the parasitic in- and output inductance and resistance.

The measurement of η_{SW} as a function of P_{out} is shown by the gray curve in the graph of Fig. 6.30. For this measurement U_{in} and U_{out} where kept constant at 2.6 V and 1.2 V, respectively. For these nominal values, a maximal η_{SW} of 58% is achieved. This yields an *EEF* value of 21%, compared to a linear series voltage converter having the same voltage conversion ration k_{lin} of 0.46. The region where the SCOOT control system applies t_{on1}, t_{off_real1} and $t_{offset1}$ extends until a value of P_{out} of 480 mW, for higher values of P_{out} the SCOOT control system applies t_{on2}, t_{off_real2} and $t_{offset2}$. For lower values of U_{in} this cross-over point occurs at lower values of P_{out}, which is due to the increasing f_{SW}. At higher values of P_{out} η_{SW} drops significantly, which is partly due to temperature effects, as explained in Sect. 4.3. An additional cause of this drop is due to the parasitic resistance of the metal-tracks and bondwires at the output of the converter. As can be seen in Fig. 6.29 these-metal tracks at the output of the power converter stages, located between the upper and lower inductors (left and right), are rather long and slim. It is estimated that, in combination with the bondwires, an additional parasitic output resistance of 250 mΩ is introduced. At P_{out_max} of 800 mW, this causes an additional power loss of 110 mW. An optimized lay-out can reduce this power loss by 60%, resulting in an increase of η_{SW} of more than 3%. It is noted that the overall η_{SW} decreases when the value of U_{in} and/or U_{out} differ from the nominal values. This is due to the mismatch of t_{on} and t_{off_real}, as explained in Sect. 5.2.1. This decrease of η_{SW} is in

6.4 Buck Converters

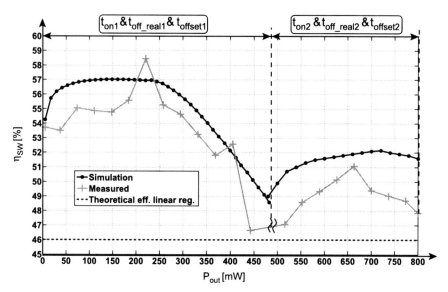

Fig. 6.30 The power conversion efficiency η_{SW} as a function of the output power P_{out}, of the metal-track, four-phase, single-output buck DC-DC converter implementation

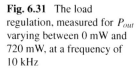

Fig. 6.31 The load regulation, measured for P_{out} varying between 0 mW and 720 mW, at a frequency of 10 kHz

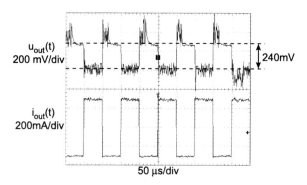

the order of a few %, within the U_{in} and U_{out} operating range of the converter. The black curve in the graph of Fig. 6.30 denotes the simulation, performed through the mathematical steady-state design model of Chap. 4. A maximal deviation between the measured and simulated η_{SW} curves is less than 4% is observed.

A measurement of the load regulation is shown in Fig. 6.31, where P_{out} is varied between 0 mW and 720 mW, at a frequency of 10 kHz. The upper curve shows $u_{out}(t)$ and the lower curve shows $i_{out}(t)$. Apart from ΔU_{out}, an additional ripple due to the load regulation of 240 mV is observed, which is significantly more compared to the COOT converters, described in the respective Sects. 6.4.1 and 6.4.2. The reason for this high additional load regulation is due to the high parasitic output resistance ($\simeq 250$ mΩ) of the converter, causing 150 mV of additional voltage drop in this measurement. Note that the converter can cope with steep transient load vari-

Table 6.10 The main measured parameters of the metal-track, four-phase, single-output buck DC-DC converter implementation

Measured parameter	Value
Input voltage range U_{in}	2 V–2.6 V
Output voltage range U_{out}	1.1 V–1.5 V
Switching frequency range f_{SW}	75 kHz–225 MHz
Maximal power conversion efficiency η_{SW_max} @ $U_{in} = 2.6$ V; $U_{out} = 1.2$ V; $P_{out} = 220$ mW	58%
Efficiency Enhancement Factor EEF	+21%
Mean Efficiency Enhancement Factor $\overline{EEF}$	+11%
Output power range P_{out}	0 mW–800 mW
Power density P_{out_max}/A	213 mW/mm^2
Maximal output voltage ripple ΔU_{out_max}	120 mV
Load regulation	0.33‰/mA
Line regulation	7.7%/V

ations, without compromising its stability. Finally, the main measured parameters of the converter are summarized in Table 6.10.

6.4.4 Metal-Track, Four-Phase, Two-Output SMOC

The fourth buck converter implementation is realized in a 90 nm, 1.2 V CMOS technology. In this technology ten metal layers are available, in addition with a MIM capacitor (2 nF/mm^2). The chip die dimensions are 1.8 mm × 1.8 mm. The measurements of this chip revealed that it does not performs as expected, which is due to a problem in the start-up circuit, causing the converter's timing to be disturbed. Nevertheless, a brief overview of the converter's circuit and the main (simulation) parameters is provided, giving the reader an idea of the possibilities of similar designs.

The circuit of the converter, together with the SCOOT feedback loop, is shown in Fig. 6.32. More information on the proposed SMOC topology is provided in Sect. 3.5.3. The power switches SW_1 and SW_2 are implemented as three stacked transistors, in order to cope with voltages of $3 \cdot U_{dd}$. The power switches SW_3 and SW_4 only need to handle a voltage of U_{dd}, requiring only one transistor. The gate of M_{1b} is biased with U_{out2} and the gate of M_{2b} is biased with U_{out1}. SW_1 is switched on by applying U_{out2} to the gates of M_{1a} and M_{1c} and is switched off by switching these gates to U_{in} and U_{out1}, respectively. SW_2 is switched on by applying U_{out1} to the gates of M_{2a} and M_{2c} and is switched off by switching these gates to GND and U_{out2}, respectively. SW_3 is turned on and off by switching the gate of M_3 between U_{out2} and U_{out1}. SW_4 is turned on and off by switching the gate of M_4 between U_{out1} and U_{out2}. The bulk terminals of all the MOSFET switches are connected

6.4 Buck Converters

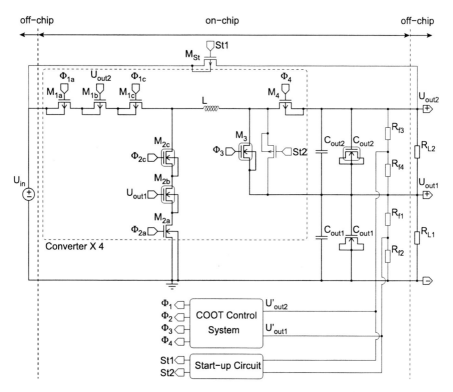

Fig. 6.32 The circuit of the implementation of the metal-track, four-phase, two-output SMOC buck DC-DC converter, with a SCOOT control system

to their respective source terminals, avoiding U_{gs} from exceeding U_{dd}. In this way the body diodes of M_{1a}, M_{1b} and M_{1c} are still allowed to conduct i_L during the transient from the charge to the discharge phases of either one of the outputs. C_{out1} and C_{out2} are implemented as a MIM capacitor, parallel to a MOS capacitor. In both the output capacitors the MIM and MOS capacitors are physically stacked onto each other, reducing their total required area. The input decouple capacitor C_{dec} is implemented as a sole MIM capacitor, since it needs to cope with voltages of $3 \cdot U_{dd}$. The inductors are implemented as rounded[9] hollow-spiral metal-track inductors, each having an inductance of 4.2 nH. These metal-track inductors measure 720 μm × 720 μm and they consist of three windings, having a track width of 80 μm. These tracks are formed by the second-last copper layer and the thick-top aluminum layer. An overview of the most important circuit parameters is provided in Table 6.11.

The feedback resistors R_{f1} and R_{f2} are used to provide the feedback of U_{out1}, whereas the feedback resistors R_{f3} and R_{f3} are used to provide the feedback of U_{out2}. The SCOOT control system implementation of this converter is discussed in Sect. 5.3.2.

---

[9]In reality the inductors are polygons, consisting of 32 corners.

Table 6.11 The circuit parameters of the metal-track, four-phase, two-output SMOC buck DC-DC converter implementation

Circuit parameter	Value
CMOS technology	90 nm
Width SW_1 W_{SW_1}	5000 µm
Width SW_2 W_{SW_2}	2000 µm
Width SW_3 W_{SW_3}	2000 µm
Width SW_4 W_{SW_4}	5000 µm
Maximal switching frequency f_{SW_max}	400 MHz
On-time 1a t_{on1a}	0.6 ns
On-time 1b t_{on1b}	0.8 ns
On-time 2a t_{on2a}	0.6 ns
On-time 2b t_{on2b}	1.4 ns
Real off-time 1a t_{off_real1a}	0.8 ns
Real off-time 1b t_{off_real1b}	1 ns
Real off-time 2a t_{off_real2a}	0.4 ns
Real off-time 2b t_{off_real2b}	0.6 ns
Output MIM capacitance1 C_{out1_MIM}	0.7 nF
Output MOS capacitance1 C_{out1_MOS}	5 nF
Output MIM capacitance2 C_{out2_MIM}	0.7 nF
Output MOS capacitance2 C_{out2_MOS}	5 nF
Decouple MIM capacitance C_{dec_MIM}	0.6 nF
Inductance L	4.2 nH
Parasitic inductor series resistance R_{Ls} @ 1 GHz	1.2 Ω
Chip die area A	3.24 mm^2

Figure 6.33 shows the naked chip die micro-photograph of the metal-track, four-phase, two-output SMOC DC-DC buck converter. The switches, the output capacitors, the input decouple capacitors and the SCOOT control system are indicated. Multiple bonding wires are used for the power supply and output connections, minimizing the parasitic input inductance and resistance. The main expected output parameters, resulting from simulations with the mathematical steady-state design model (see Chap. 4), of the converter are summarized in Table 6.12.

6.4.5 Bondwire, Single-Phase, Two-Output SMOC

The fifth buck converter implementation is realized in a 350 nm, 3.3 V CMOS technology, with vertical n-DMOSFETs and lateral p-DMOSFETs, capable of handling $U_{ds} = 80$ V. In this technology four metal layers are available, in addition with a MIM capacitor (1 nF/mm^2). The chip die dimensions are 2.2 mm × 3.2 mm. In

6.4 Buck Converters

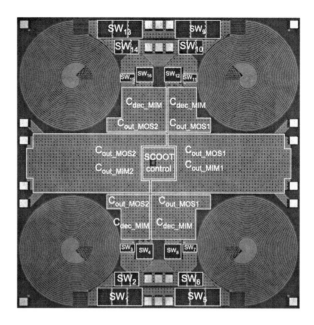

Fig. 6.33 The micro-photograph of the naked chip die of the metal-track, four-phase, two-output SMOC buck DC-DC converter, with the indication of the building blocks

the following sections the converter circuit, the input parameters and the simulation results are discussed.

Circuit & Input Parameters

The circuit of the converter, together with the F^2SCOOT feedback loop, is shown in Fig. 6.34. Both the high- and low-side switches are implemented as a p-DMOSFET and n-DMOSFET, respectively. The gate of M_1 is switched between $U_{in} - 3.3$ V and U_{in}, turning it on and off. The gate of M_2 is switched between U_{out1} and GND, turn-

Table 6.12 The main expected parameters of the metal-track, four-phase, two-output SMOC buck DC-DC converter implementation

Simulated parameter	Value
Input voltage U_{in}	3.6 V
Output1 voltage U_{out1}	1.2 V
Output2 voltage U_{out2}	2.4 V
Maximal power conversion efficiency η_{SW_max}	60%
Output1 power range P_{out1}	0 mW–400 mW
Output2 power range P_{out2}	0 mW–800 mW
Power density P_{out_max}/A	370 mW/mm^2
Maximal output1 voltage ripple ΔU_{out1_max}	120 mV
Maximal output2 voltage ripple ΔU_{out2_max}	240 mV

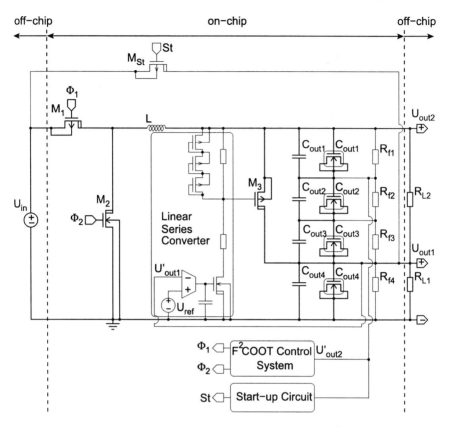

Fig. 6.34 The circuit of the implementation of the bondwire, single-phase, two-output SMOC DC-DC buck converter, with a F²SCOOT control system

ing it on and off. C_{out1}, C_{out2}, C_{out3} and C_{out4} are implemented as MIM capacitors parallel to MOS capacitors and all four output capacitors have an equal capacitance. Both the MIM and MOS capacitors are physically stacked on each other, reducing their required area. Also, slots underneath and perpendicular to the bondwires of the bondwire inductor are made into the output capacitor and its metal connecting plates, resulting in a patterned output capacitor. This reduces the eddy-currents losses and increases the inductance and Q-factor of the bondwire inductor, as explained in Sect. 6.1.1. The input decoupling capacitor C_{dec} is implemented as a sole MOM capacitor, as it needs to cope with a voltage up to 80 V. A hollow-spiral bondwire inductor L yields an inductance of 35 nH. This bondwire inductor measures 2 mm × 2 mm and it consists of four windings of 25 μm thick golden bondwire. The center pitch of the bondwires is 100 μm. An overview of the most important circuit parameters is provided in Table 6.13.

The feedback resistors R_{f1}, R_{f2}, R_{f3} and R_{f4} have two functions. The first function is to provide feedback of U_{out1} and U_{out2} to the F²SCOOT control system. The second function is to ensure the correct biasing of the intermediate nodes between

6.4 Buck Converters

Table 6.13 The circuit parameters of the bondwire, single-phase, two-output SMOC buck DC-DC converter implementation

Circuit parameter	Value
CMOS technology	350 nm
Width SW_1 W_{SW_1}	6000 μm
Width SW_2 W_{SW_2}	4000 μm
Width SW_1 W_{SW_1}	1000 μm
On-time range t_{on}	3.6 ns–0.65 ns
Real off-time range t_{off_real}	1 ns–1.55 ns
Output capacitance1–4 C_{out_MIM1-4}	1.1 nF
Output capacitance1–4 C_{out_MOS1-4}	3.2 nF
Decouple capacitance C_{dec_MOM}	75 pF
Inductance L	35 nH
Parasitic inductor series resistance R_{Ls} @ 100 MHz	2.7 Ω
Chip die area A	7.04 mm^2

C_{out1}, C_{out2} and C_{out1}, preventing them from operating out of their safe-operating region U_{dd}. The second output is fed by the buck converter itself, while the first output is fed by the second output by means of a linear series voltage converter. The reason for this approach is the lack of switching speed of this technology, which is needed for the correct timing of two switched outputs. As such, the U_{out1} is mainly used to power the F^{2}SCOOT control system and it will only provide limited additional output power. The F^{2}SCOOT control system implementation of this converter is discussed in Sect. 5.4.2. Note that this converter is not self-starting and that it requires a start-up circuit, as explained in Sect. 5.5. This start-up circuit switches M_{St} on, during the initialization of the converter, until U_{out1} reaches the value of 2.5 V. Afterwards the F^{2}SCOOT control system enables the converter to take over, powering up U_{out2} to 12 V and U_{out1} to 3.3 V.

Simulation Results

Figure 6.35 shows the naked chip die micro-photograph of the bondwire, single-phase, two-output SMOC DC-DC buck converter. The switches, the output capacitor, the input decouple capacitor and the F^{2}SCOOT control system are indicated. Multiple bondwires are used for the power supply connections, minimizing the parasitic input inductance and resistance. Octagonal bonding pads are used for the bondwire inductor, enabling a smaller pitch between the consecutive wires, which is beneficial for the total inductance.

Table 6.14 lists the main simulated output parameters of the bondwire, single-phase, two-output SMOC buck DC-DC converter implementation. The microphotograph of the assembled chip, with the bondwire inductor added, is shown in Fig. 6.36.

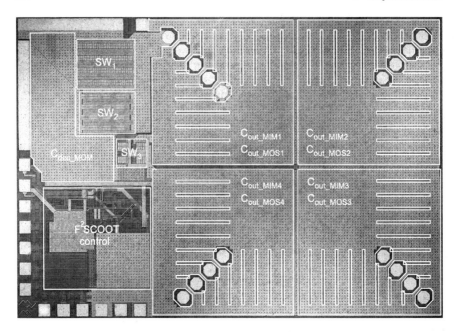

Fig. 6.35 The micro-photograph of the naked chip die of the bondwire, single-phase, two-output SMOC DC-DC buck converter, with the indication of the building blocks

Table 6.14 The main simulated parameters of the bondwire, single-phase, two-output SMOC buck DC-DC converter implementation

Simulated parameter	Value
Input voltage range U_{in}	20–70 V
Output1 voltage U_{out1}	3.3 V
Output2 voltage U_{out2}	12 V
Maximal power conversion efficiency η_{SW_max}	75%
Output1 power range P_{out1}	0 mW–200 mW
Output2 power range P_{out2}	0 mW–1800 mW
Power density P_{out_max}/A	284 mW/mm^2
Maximal output1 voltage ripple ΔU_{out1_max}	200 mV
Maximal output2 voltage ripple ΔU_{out2_max}	800 mV

6.5 Comparison to Other Work

In order to appreciate and situate the performance of the realized monolithic inductive DC-DC converters in the presented work, a comparison to other work is necessary. This comprises the side-by-side comparison of the most important measured parameters, including η_{SW}, P_{out_max}, P_{out}/A, EEF, $\overline{EEF}$, ... As such, the strengths and weaknesses of the designs, discussed in the previous sections, are unveiled.

6.5 Comparison to Other Work

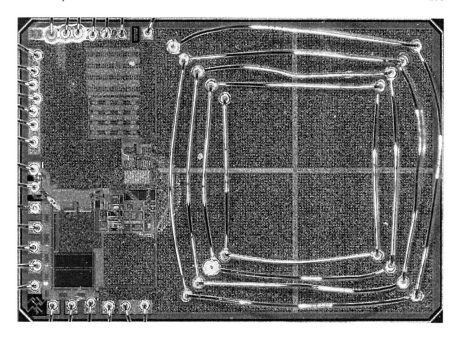

Fig. 6.36 The micro-photograph of the chip die of the bondwire, single-phase, two-output SMOC DC-DC buck converter, with the bondwire inductor added

The comparison of the most important measured parameters of the monolithic inductive DC-DC step-up and step-down converters, is provided in the respective Sects. 6.5.1 and 6.5.2.

6.5.1 Inductive Step-Up Converters

The most important measured parameters[10] of the monolithic DC-DC step-up converter [Wen07], presented in Sect. 6.3.1 in this work, to other known work [Sav03, Ric04], is listed in Table 6.15. All three converters are implemented in a similar 180 nm CMOS technology. It is observed that the presented converter implementation outperforms the other converters, in terms of η_{SW}, P_{out_max} and P_{out}/A. This is most probably due to two facts. First, the voltage conversion ratio k_{SW} of [Wen07] is lower. Because of this, the inductor needs to deliver proportionally less energy E_L to the output. The second reason is the fact that [Wen07] uses a bondwire inductor, while [Sav03] uses a metal-track transformer and [Ric04] a metal-track inductor. The latter two have an intrinsically lower Q-factor, resulting in a larger parasitic series resistance and in turn higher power losses. Note that [Sav03, Ric04] do not incorporate a control system, as does [Wen07].

---

[10]The capacitance C of the output capacitor is not provided in either [Sav03] nor [Ric04].

Table 6.15 The comparison of the most important measured parameters of the monolithic inductive DC-DC step-up converter, presented in this work, to the other known work

Ref.	[Sav03]	[Ric04]	[Wen07]
Year	2003	2004	2007
Process	CMOS	CMOS	CMOS
L_{min} (nm)	180	180	180
Topology	Fly-back	Boost	Boost
Control system	Open-loop	Open-loop	PWM
U_{in} range (V)	1.8	1.6–2.6	1.6–2
U_{out} range (V)	8–10	2–15	2.5–4
U_{in} (V)	1.8	1.6	1.8
U_{out} (V)	8	6	3.3
$k_{SW} = U_{out}/U_{in}$	4.44	3.75	1.83
ΔU_{out} (mV)	60	450	200
$P_{out,max}$ (mW)	6.4	3.6	150
P_{out} (mW)	6.4	3.6	80
η_{SW} (%) @ U_{in} & U_{out} & P_{out}	19	28	63
f_{SW} (Hz)	1.4e9	120e6	100e6
L (nH)	$L_{prim} = 4.8\ L_{sec} = 34.4$	20	$18 + 3 = 21$
C (nF)	–	–	1.3
A (mm^2)	0.127	1.69	2.25
P_{out}/A (mW/mm^2)	50	2.1	67
Level of integration	full	full	full bondwire L

From these observations it is concluded that a trade-off between k_{SW} and η_{SW} exists for monolithic boost converters. Moreover, the implementation of the inductor plays a key role in both the achievable η_{SW}, P_{out_max} and P_{out}/A. Finally, it is noted that for low values of P_{out_max} (< 10 mW) monolithic charge pump DC-DC step-up converters are a better choice, in terms of η_{SW} (> 80%) [Bre09a]. Whereas, monolithic inductive DC-DC step-up converters are better suited for relatively large values of P_{out} (> 100 mW).

6.5.2 Inductive Step-Down Converters

The most important measured parameters of the monolithic DC-DC step-up converters [Wen08a, Wen08b, Wen09b], presented in Sects. 6.4.1, 6.4.2 and 6.4.3 in this work, compared to other known work [Haz05, Mus05b, Ali07, Abe07, Oni07, Wib08, Ber08, Ali09, Jin09, Ber09], is listed in Table 6.16. All the compared converters use relatively low values of U_{in}, in the range of 1 V–5 V, and convert this into relatively low values of U_{out}, in the range of 0.3 V–3.5 V. The maximal reported value for P_{out}, of 800 mW, is achieved in this work [Wen09b]. This design also achieves the highest P_{out}/A, of 213 mW/mm^2, which is the same as achieved in [Haz05]. However, [Haz05] uses off-chip air-core inductors, which are not taken into account for the occupied converter area A. In other words, the achieved power density in [Wen09b] is in the same order of magnitude as converters that make use of external components. Moreover, the respective EEF and $\overline{EEF}$ values of [Wen09b] are higher than those of [Haz05]. The values of ΔU_{out} cannot be compared, as they are not provided in [Haz05].

In terms of the EEF (see Sect. 2.4), both the implementations of [Wen08a] and [Jin09] achieve a value of 23%. It is noted that, despite the bondwire inductor used in [Wen08b], [Jin09] uses thick-oxide switches instead of stacked transistors. This yields a small advantage in terms of on-resistance, but it is not considered standard CMOS. All the designs, presented in this work, in Table 6.16 show an EEF having a significant positive value. This indicates that these designs all have a better η_{SW}, compared to the η_{lin} of a linear series voltage converter, having the same voltage conversion ratio $k_{lin} = k_{SW}$. This is certainly not the case for all designs listed in the literature, such as [Ali07, Abe07, Oni07, Ali09]. Obviously, these switched DC-DC converters are obsolete and should be replaced by linear voltage converters, in a real application.

When comparing the overall mean Efficiency Enhancement Factor $\overline{EEF}$, it is noted that some designs show a significantly worse performance, compared to the EEF. Some designs [Mus05b, Wib08, Jin09] even show a negative value for $\overline{EEF}$, whilst the value of the EEF is positive. This indicates that when connecting a load to these converters of which the probability distribution function $\alpha(P_{out})$ over the P_{out} range of the converters is equal to 1, a linear voltage converter would achieve a higher overall η_{lin}. It is clear the designs in this work [Wen08a, Wen08b, Wen09b] shown a superior behavior in terms of the $\overline{EEF}$. Thus, implying that their overall η_{SW} is higher, compared to a linear voltage converter, with the same voltage conversion ratio. This may also be verified in the measurements of these converters.

The main reason for the superior performance of the presented converters is due to two reasons. First, the converters in this work are optimized by means of the mathematical steady-state design model, described in Chap. 4. This allows for a thorough exploration of the design space, yielding near-optimal designs, having a high relative value of η_{SW}. Second, the presented converters make use of the specialized control system techniques, COOT and SCOOT, which are optimized for monolithic inductive DC-DC converters. This causes the overall value of η_{SW}, over a large P_{out} range, to be optimized. This is not the case for many other designs, which make use of the PWM technique. Consequentially, these designs yield a poor EEF at the lower end of the P_{out} range.

In general, it is concluded that the bondwire inductor design of [Wen08b] achieves slightly higher values of the EEF and the $\overline{EEF}$, due to the higher associated Q-factor, compared to metal-track inductors. Also, multi-phase converters [Wib08, Wen09b] are capable of achieving the highest power densities of monolithic inductive DC-DC converters. Finally, the type of control system is of crucial importance for obtaining a high overall $\overline{EEF}$, in addition to a high value of the EEF.

Notice that monolithic charge-pump DC-DC step-down converters, making use of specialized (not standard CMOS) IC technologies, can reach power densities in

Table 6.16 The comparison of the most important measured parameters of the monolithic inductive DC-DC step-down converters, presented in this work, to the other known work

Ref.	[Haz05]	[Mus05a]	[Ali07]	[Abe07]	[Oni07]	[Wib08]	[Ber08]	[Wen08a]	[Wen08b]	[Ali09]	[Jin09]	[Ber09]	[Wen09b]
Year	2005	2005	2007	2007	2007	2008	2008	2008	2008	2009	2009	2009	2009
Process	CMOS	CMOS	CMOS	BiCMOS	CMOS	CMOS	CMOS	CMOS	CMOS	CMOS	CMOS	CMOS	CMOS
L_{min} (nm)	90	1500	90	180	350	130	180	130	180	180	130	65	130
Topology	4-phase Buck	1-phase Buck	1-phase Buck	2-phase Buck	1-phase Buck	2-phase Buck	1-phase Buck	1-phase Buck	1-phase Buck	1-phase Buck	1-phase Buck	1-phase Buck	4-phase Buck
Control system	PFM	–	PWM	PWM	PWM	PWM	PWM	COOT	COOT	–	PWM	PWM	SCOOT
U_{in} range (V)	1.2–1.4	5	1	2.8	3.3	1.2	1.8	2–2.6	3–4	2.2	3.3	1.2	2–2.6
U_{out} range (V)	0.9–1.1	1.5–3.5	0.5–0.7	1.5–2	2.3	0.3–0.9	1.1	1.1–1.5	1.5–2.1	0.75–1	1.2–2.5	0.7–1.2	1.1–1.5
U_{in} (V)	1.2	5	1	2.8	3.3	1.2	1.8	2.6	3.6	2.2	3.3	1.2	2.6
U_{out} (V)	0.9	2.5	0.52	1.8	2.3	0.9	1.1	1.2	1.8	1	1.8	0.95	1.2
ΔU_{out} (mV)	–	50	26	250	230	40	35	110	160	–	90	–	120
$P_{out,max}$ (mW)	270	75	53	360	161	315	165	180	300	55	720	180	800
P_{out} (mW)	225	70	34	360	138	180	110	180	270	55	288	86	220
η_{lin} (%) @ U_{in} & U_{out} & P_{out}	75	50	52	64	70	75	61	46	50	45	55	79	46
η_{sw} (%) @ U_{in} & U_{out} & P_{out}	83	53	48	64	62	78	65	52	65	31	70	87.5	58
EEF (%)	+9.6	+5.7	−8.3	0	−13	+3.7	+6.2	+12	+23	−45	+23	+9.7	+21
$\overline{EEF}$ (%)	+3.2	−42	−10	−32	−30	−60	–	+4	+17	−43	−0.4	+4	+11
f_{sw} (Hz)	100e6–600e6	10e6	3e9	45e6	200e6	170e6	80e6	30–300e6	20–140e6	660e6	180e6	100e6	75e3–225e6
L (nH)	4 × 6.8†	80†	0.32	2 × 11	22†	2 × 2 coupled	20†	9.8	†18	4.38	10.5	†10.1	4 × 3.9
C (nF)	2.5	3	0.35	6	1	5.2	75†	15	10.3	1.1	3.6	†30	12
A (mm²)	1.267†	16	0.81	6.75	4	1.5	3.24	3.375	2.25	2.5	4	23	3.76
P_{out}/A (mW/mm²)	213†	4.7	65	53	40	210	51	53	133	22	180	7.8	213
Level of integration	†off-chip SMT L	†thick film L	full	full	†stacked-chip L & C	full	†pass. die L & C	full	†bondwire L	full	full	†pass die L & C	full

the order of 550 mW/mm^2 [Le10]. Also, thick-film hybrid technologies can be used for compact DC-DC converters [Per04]. This type of technology allows for a higher η_{SW} ($\simeq 90\%$), at a lower power density (62.5 mW/mm^2). However, the cost-effectiveness of this kind of implementations may be questioned.

6.6 Conclusions

The on-chip implementation of the main inductive DC-DC converter's components: the inductor(s), the capacitor(s) and the switches, are discussed in Sect. 6.1. For the inductor two types are considered: the bondwire inductor and the metal-track inductor. The latter is truly fully-integrated, but yields a lower Q-factor, compared to the bondwire inductor. These inductors are optimized using the FastHenry software. The on-chip capacitor may be implemented as a MIM, MOM or MOS capacitor, of which only the last two are native in standard CMOS technologies. Nevertheless, the designs in this work mostly use the combination of MIM and MOS capacitors, because they can effectively be physically stacked on top of each other. This increases the overall capacitance density. The switches need to handle both high peak currents and high voltages, higher that the nominal technology supply voltage. For these reasons, stacked and waffle-shaped MOSFET switches are used.

In Sect. 6.2 the main principles for the accurate and correct measurement of the key parameters of monolithic DC-DC converters are provided. These measurements include the power conversion efficiency as a function of the output power, the line regulation and the load regulation. For sake of completeness, the reader is also provided with a discussion on a practical measurement setup.

The practical chip implementations of the monolithic inductive DC-DC converters, realized in this work, are discussed in the respective Sects. 6.3 and 6.4. These practical implementations incorporate the following designs:

1. A bondwire, single-phase, single-output boost converter, using a PWM control system, in a 180 nm 1.8 V CMOS technology.
2. A metal-track, single-phase, two-output SIMO boost converter, using a COOT control system, in a 130 nm 1.2 V CMOS technology.
3. A bondwire, single-phase, single-output buck converter, using a COOT control system, in a 180 nm 1.8 V CMOS technology.
4. A metal-track, single-phase, single-output buck converter, using a COOT control system, in a 130 nm 1.2 V CMOS technology.
5. A metal-track, four-phase, single-output buck converter, using a SCOOT control system, in a 130 nm 1.2 V CMOS technology.
6. A metal-track, four-phase, two-output SMOC buck converter, using a SCOOT control system, in a 90 nm 1.2 V CMOS technology.
7. A bondwire, single-phase, two-output SMOC buck converter, using an F^{2}SCOOT control system, in a 350 nm 3.3 V/80 V CMOS technology.

Compared to other work in the literature, provided in Sect. 6.5, the designs in this work achieve the highest power density and output power, in addition to the highest EEF and $\overline{EEF}$ for DC-DC step-down converters. This is achieved through both the optimized design procedure and the optimized novel control systems.

Chapter 7
General Conclusions

7.1 Conclusions

The reader is introduced into the domain of DC-DC converters in Chap. 1, with a strong affiliation towards monolithic inductive DC-DC converters. A brief historical overview of the use of DC-DC converters, from the 20th century onwards, provides a background on the subject. The main applications areas of DC-DC converters are discussed by distinguishing three plain types of DC-DC converters, namely the mains-operated step-down converter, the battery-operated step-down converter and the battery-operated step-up converter. This introduction is concluded with a discussion on the basics of standard CMOS IC technologies and the challenges, involved in the realization of monolithic inductive DC-DC converters.

In Chap. 2 a comprehensive overview is given on the different DC-DC converter types, including linear voltage converters, charge-pump voltage converters and inductive voltage converters. The individual analysis of these converter types reveals their potential benefits and drawbacks. In general, it can be stated that linear DC-DC converters are useful for step-down converters, having a high voltage conversion ratio and a low output power. Charge-pump DC-DC converters are useful for both step-up and down converters, having a fixed or predefined voltage conversion ratio. Inductive DC-DC converters may be used for both step-up and/or step-down converters, for virtually any voltage conversion ratio. In addition, the mathematical small-ripple approximation, for calculating ideal inductive DC-DC converters, is explained. This calculation method is acquired for a fundamental understanding of the important concepts of inductive DC-DC converters.

An overview of commonly used inductive DC-DC converter topologies is provided in Chap. 3, in addition to a comparison of non-galvanic separated step-down, step-up and step-up/down converters. This comparison is performed on the basis of the required total capacitance for achieving a certain output voltage ripple. As a result, it is concluded that the buck step-down, the boost step-up and the zeta step-up/down converter require the smallest capacitance and hence chip area. In addition to these comparisons a brief discussion is provided on galvanic separated and resonant DC-DC converters. Although galvanic separated DC-DC converters are proven feasible for the purpose of monolithic integration, they are beyond the scope of this work. Resonant DC-DC converters are not considered practical for monolithic integration, due to the high expected losses in their switch-network and rectifier. The primary types of buck, boost and buck/boost DC-DC converters may be extended by a number of modifications, yielding multi-phase and SIMO converters (and their combinations). These extensions will prove to be beneficial for monolithic integration. In addition to these known extensions, a novel modification is proposed: the SMOC converter. This is a type of SIMO converter, which is especially suited for monolithic integration, by reducing the required output capacitance and hence the required chip area.

An accurate mathematical steady-state model, for both monolithic boost and buck converters is discussed in Chap. 4. The flow of this model comprises four steps. First, a second-order model, based on the differential equations of the converter, is derived. This model takes all the significant resistive losses into account. Secondly, the dynamic power losses of the converter components are modeled. Thirdly, temperature effects due to self-heating are modeled. Finally, all the elements from the first three steps are joined into a final model flow. In addition to the description of this model, the effect of the parasitics of the converter components on the general performance of the converter is discussed extensively. Afterwards, this mathematical model is used for the deduction of some important design trade-offs, involved in monolithic inductive DC-DC converters.

One of the main tasks of a DC-DC converter is to obtain a constant output voltage, at the desired level, under varying load and line conditions. To achieve this a control system is required, providing regulation through feedback and/or feedforward. Both known and novel control techniques, in addition with some practical implementations, are discussed in Chap. 5. First, a discussion is provided on the known PWM and PFM control techniques, in addition with a comparison of both techniques. The PFM control system is found to be able to achieve higher overall power conversion efficiencies, compared to the fixed-frequency PWM technique, due to the adaptive switching frequency. However, the use of the PWM technique results in a lower overall output voltage ripple, for the same output capacitance. Nevertheless, apart from one design, all the designs in this work will acquire variants of the PFM control technique. These novel variants are the COOT, the SCOOT and the F^2SCOOT timing schemes, which are especially designed and optimized for controlling monolithic DC-DC converters. Finally, the concept of start-up circuits is explained, together with a practical example.

The practical chip implementations of monolithic inductive DC-DC converters are discussed in Chap. 6. First, some practical considerations are provided on the lay-out of the main converter components, namely the inductor, the capacitor and the power switches. Secondly, the main principles of measuring the key parameters of monolithic DC-DC converters are discussed. Also, a practical measurement setup is discussed, which is used throughout the measurements. Furthermore, the discussion on the practical chip implementations, including the measurements results, is provided. These implementations incorporate buck, boost, SIMO boost, multi-phase buck, multi-phase SMOC buck and SMOC buck converters. Finally, the obtained measurements are compared to other work in the literature. The tedious design in combination with the novel control systems leads to the observation that the converters in this work outperform many other implementations, in terms of: output power, power density, EEF and $\overline{EEF}$. Note, that the maximization of these parameters was the main goal of this work.

7.2 Remaining Challenges

The practical implementations, provided in Chap. 6, lead to the conclusion that there is room for improvement on certain parameters. These parameters are mainly: the EEF, the $\overline{EEF}$, the output voltage ripple and the line- and load regulation.

Both the EEF and the $\overline{EEF}$ can be improved by reducing the overall power losses in the converter's components. These power losses are mainly located in the switches, the inductor and the metal interconnect. The trend towards deep-sub-nanometer CMOS technologies may reduce the switching losses, due to the increased switch speed. Also, the increasing number of available metal layers will reduce the interconnect losses. Remain the on-chip inductors, which may be improved by the increasing number of metals. Nevertheless, additional research on improving the inductors, through micro-machining for instance, may prove to be worth the effort.

The output voltage ripple can be decreased in deep-sub-nanometer CMOS technologies, through the higher capacitance densities of MOS capacitors. Also, the increasing achievable switching speeds of these technologies could solve this problem. Obviously, the trade-off with the power conversion efficiency should be kept in mind. An additional possible solution is the use of a post low-dropout linear regulator (LDO), to filter out the excess output voltage ripple of the converter [EN10]. Again, this will be at the cost of a lower power conversion efficiency.

The line- and load regulation can be strongly improved by improving the lay-out and minimizing the parasitic connection resistances. However, a further improvement is to be performed on the level of the control systems. For this purpose, additional dynamic modeling of both the converter and the control systems is required. In this way the potential bottle-necks can be found for solving these issues.

Finally, it is noted that a number of EMI related questions remain to be answered. This additional research may be conducted on two levels. The first level is the noise generated by the voltage and current ripple at the in- and outputs of the power converter. In case this poses EMI problems, the solution may be adding additional in- and/or output filtering. Also, the spreading or de-spreading of the energy could be achieved, by adapting the switching scheme (e.g. adding random sequences). The second level is the radiated EMI of the inductor(s). In case this poses a problem, the solution might be provided through magnetic shielding (e.g. by means of the chip package).

References

[Abe07] S. Abedinpour, B. Bakkaloglu, S. Kiaei, A Multistage Interleaved Synchronous Buck Converter with Integrated Output Filter in 0.18 μm SiGe Process. IEEE Trans. Power Electron. **22**(6), 2164–2175 (2007). doi:10.1109/TPEL.2007.909288

[AH09] W. Al-Hoor, J.A. Abu-Qahouq, L. Huang, W.B. Mikhael, I. Batarseh, Adaptive Digital Controller and Design Considerations for a Variable Switching Frequency Voltage Regulator. IEEE Trans. Power Electron. **24**(11), 2589–2602 (2009). doi:10.1109/TPEL.2009.2031439

[Ahn96] C.H. Ahn, M.G. Allen, A Comparison of Two Micromachined Inductors (Bar- and Meander-Type) for Fully Integrated Boost DC/DC Power Converters. IEEE Trans. Power Electron. **11**(2), 239–245 (1996). doi:10.1109/63.486171

[Ajr01] S. Ajram, G. Salmer, Ultrahigh Frequency DC-to-DC Converters Using GaAs Power Switches. IEEE Trans. Power Electron. **16**(5), 594–602 (2001). doi:10.1109/63.949492

[Ali07] M. Alimadadi, S. Sheikhaei, G. Lemieux, S. Mirabbasi, P. Palmer, A 3 GHz Switching DC-DC Converter Using Clock-Tree Charge-Recycling in 90 nm CMOS with Integrated Output Filter, in *IEEE International Solid-State Circuits Conference ISSCC: Digest of Technical Papers*, 2007, pp. 532–620. doi:10.1109/ISSCC.2007.373529

[Ali09] M. Alimadadi, S. Sheikhaei, G. Lemieux, S. Mirabbasi, W.G. Dunford, P.R. Palmer, A Fully Integrated 660 MHz Low-Swing Energy-Recycling DC-DC Converter. IEEE Trans. Power Electron. **24**(6), 1475–1485 (2009). doi:10.1109/TPEL.2009.2013624

[Apa02] R. Aparicio, A. Hajimiri, Capacity Limits and Matching Properties of Integrated Capacitors. IEEE J. Solid-State Circuits **37**(3), 384–393 (2002). doi:10.1109/4.987091

[Ber08] H.J. Bergveld, R. Karadi, K. Nowak, An Inductive Down Converter System-in-package for Integrated Power Management in Battery-Powered Applications, in *IEEE Power Electronics Specialists Conference PESC*, 2008, pp. 3335–3341. doi:10.1109/PESC.2008.4592470

[Ber09] H.J. Bergveld, K. Nowak, R. Karadi, S. Iochem, J. Ferreira, S. Ledain, E. Pieraerts, M. Pommier, A 65-nm-CMOS 100-MHz 87-Efficient DC-DC Down Converter Based on Dual-Die System-in-Package Integration, in *IEEE Energy Conversion Congress and Exposition ECCE*, 2009, pp. 3698–3705. doi:10.1109/ECCE.2009.5316334

[Bio06] T. Biondi, A. Scuderi, E. Ragonese, G. Palmisano, Analysis and Modeling of Layout Scaling in Silicon Integrated Stacked Transformers. IEEE Trans. Microw. Theory Tech. **54**(5), 2203–2210 (2006). doi:10.1109/TMTT.2006.872788

[Boh09] M. Bohr, The New Era of Scaling in an SoC World, in *IEEE International Solid-State Circuits Conference Digest of Technical Papers*, 2009, pp. 23–28

[Boh10] M. Bohr, The New Era of Scaling in an SoC World (2010), http://download.intel.com/technology/architecture-silicon/ISSCC_09_plenary_bohr_presentation.pdf. Presentation ISSCC 2009 from eponymous paper

[Bre08] T.V. Breussegem, M. Wens, E. Geukens, D. Geys, M. Steyaert, Area-Driven Optimization of Switched-Capacitor DC/DC Converters. IEEE Electron. Lett. **44**(25), 1488–1490 (2008)

[Bre09a] T.V. Breussegem, M. Steyaert, A 82% Efficiency 0.5% Ripple 16-Phase Fully Integrated Capacitive Voltage Doubler, in *IEEE Proceedings of the 2009 Symposium on VLSI Circuits*, vol. 1, 2009, pp. 198–199

[Bre09b] T.V. Breussegem, M. Wens, E. Geukend, D. Geys, M. Steyaert, A DMOS Integrated 320 mW Capacitive 12 V to 70 V DC/DC-Converter for Lidar Applications, in *IEEE Energy Conversion Congress and Exposition*, vol. 1, 2009, pp. 3865–3869

[Bur98] J.N. Burghartz, D.C. Edelstein, M. Soyuer, H.A. Ainspan, K.A. Jenkins, RF Circuit Design Aspects of Spiral Inductors on Silicon. IEEE J. Solid-State Circuits **33**(12), 2028–2034 (1998). doi:10.1109/4.735544

[Cao03] Y. Cao, R.A. Groves, X. Huang, N.D. Zamdmer, J.O. Plouchart, R.A. Wachnik, T.J. King, H. Chenming, Frequency-Independent Equivalent-Circuit Model for On-Chip Spiral Inductors. IEEE J. Solid-State Circuits **38**(3), 419–426 (2003). doi:10.1109/JSSC.2002.808285

[Cha10] L. Chang, R.K. Montoyeand, B.L. Ji, A.J. Weger, K.G. Stawiasz, R.H. Dennard, A Fully-Integrated Switched-Capacitor 2:1 Voltage Converter with Regulation Capability and 90% Efficiency at 2.3 A/mm^2, 2010, pp. 55–56. doi:10.1109/VLSIC.2010.5560267

[Cra97] J. Craninckx, Low-Phase-Noise Fully Integrated CMOS Frequency Synthesizers, PhD thesis, ESAT-MICAS, K.U. Leuven, Belgium, 1997

[Cro96] J. Crols, P. Kinget, J. Craninckx, M. Steyaert, An Analytical Model of Planar Inductors on Lowly Doped Silicon Substrates for High Frequency Analog Design up to 3 GHz, in *Symposium on VLSI Circuits, Digest of Technical Papers*, 1996, pp. 28–29. doi:10.1109/VLSIC.1996.507703

[Dan99] L. Daniel, C.R. Sullivan, S.R. Sanders, Design of Microfabricated Inductors. IEEE Trans. Power Electron. **14**(4), 709–723 (1999). doi:10.1109/63.774209

[Dav07] A. Davoudi, J. Jatskevich, Parasitics Realization in State-Space Average-Value Modeling of PWM DC-DC Converters Using an Equal Area Method. IEEE Trans. Circuits Syst. I, Regul. Pap. **54**(9), 1960–1967 (2007). doi:10.1109/TCSI.2007.904686

[Dav09] A. Davoudi, J. Jatskevich, P.L. Chapman, Numerical Dynamic Characterization of Peak Current-Mode-Controlled DC-DC Converters. IEEE Trans. Circuits Syst. II, Express Briefs **56**(12), 906–910 (2009). doi:10.1109/TCSII.2009.2035272

[Dei78] C. Deisch, Simple Switching Control Method Changes Power Converter into a Current Source, in *IEEE Power Electronics Specialists Conference*, 1978, pp. 300–306

[Del09] O. Deleage, J.C. Crebier, M. Brunet, Y. Lembeye, H.T. Manh, Design and Realization of Highly Integrated Isolated DC/DC Micro-Converter, in *IEEE Energy Conversion Congress and Exposition*, vol. 1, 2009, pp. 3690–3697

[Den74] R.H. Dennard, F.H. Gaensslen, V.L. Rideout, E. Bassous, A.R. LeBlanc, Design of Ion-Implanted MOSFET's with Very Small Physical Dimensions. IEEE J. Solid-State Circuits **9**(5), 256–268 (1974)

[Ebe09] W. Eberle, Z. Zhang, Y.F. Liu, P.C. Sen, A Practical Switching Loss Model for Buck Voltage Regulators. IEEE Trans. Power Electron. **24**(3), 700–713 (2009). doi:10.1109/TPEL.2008.2007845

[Eir08] G. Eirea, S.R. Sanders, Phase Current Unbalance Estimation in Multiphase Buck Converters. IEEE Trans. Power Electron. **23**(1), 137–143 (2008). doi:10.1109/TPEL.2007.911840

[EN10]	M. El-Nozahi, A. Amer, J. Torres, K. Entesari, E. Sanchez-Sinencio, High PSR Low Drop-Out Regulator with Feed-Forward Ripple Cancellation Technique. IEEE J. Solid-State Circuits **45**(3), 565–577 (2010). doi:10.1109/JSSC.2009.2039685
[Eri04]	R.W. Erickson, D. Maksimović, *Fundamentals of Power Electronics*, 2nd edn. (Kluwer Academic, New York, 2004). First edition: 2001
[Gar07]	D.S. Gardner, G. Schrom, P. Hazucha, F. Paillet, T. Karnik, S. Borkar, Integrated On-Chip Inductors with Magnetic Films. IEEE Trans. Magn. **43**(6), 2615–2617 (2007). doi:10.1109/TMAG.2007.893794
[Gho04]	M. Ghovanloo, K. Najafi, Fully Integrated Wideband High-Current Rectifiers for Inductively Powered Devices. IEEE J. Solid-State Circuits **39**(11), 1976–1984 (2004). doi:10.1109/JSSC.2004.835822
[Gra29]	Graham Amplion Limited, Improvements in and Relating to Electrical Supply Means for Use with Wireless Apparatus, UK Patent No. 329.113, 1929. DC-DC step-up converter, using an electromotor and commutators, for the purpose of battery-operated vacuum-tube transceivers
[Gre74]	H.M. Greenhouse, Design of Planar Rectangular Microelectronic Inductors. IEEE Trans. Parts Hybrids Packag. **10**(2), 101–109 (1974)
[Gro62]	F.W. Grover, *Inductance Calculations: Working Formulas and Tables*, 2nd edn. (Dover, New York, 1962). First edition: 1946
[Gue10]	R. Guevremont, FAIMS: High-Field Asymmetric Waveform Ion Mobility Spectrometry (2010), http://www.faims.com/index.htm. Separation of ions at atmospheric pressure and room temperature
[Ham88]	D.C. Hamill, D.J. Jeffries, Subharmonics and Chaos in a Controlled Switched-Mode Power Converter. IEEE Trans. Circuits Syst. **35**(8), 1059–1061 (1988). doi:10.1109/31.1858
[Haz40]	Hazeltine Corporation, Improvements in High-Voltage Power-Supply Systems, US Patent No. 549.865, 1940. Single triode fly-back converter for DC high-voltage generation out of a low DC supply voltage, for the purpose of supplying vacuum-tube circuits
[Haz05]	P. Hazucha, G. Schrom, J. Hahn, B.A. Bloechel, P. Hack, G.E. Dermer, S. Narendra, D. Gardner, T. Karnik, V. De, S. Borkar, A 233-MHz 80-87 Efficient Four-Phase DC-DC Converter Utilizing Air-Core Inductors on Package. IEEE J. Solid-State Circuits **40**(4), 838–845 (2005). doi:10.1109/JSSC.2004.842837
[Haz07]	P. Hazucha, S.T. Moon, G. Schrom, F. Paillet, D. Gardner, S. Rajapandian, T. Karnik, High Voltage Tolerant Linear Regulator with Fast Digital Control for Biasing of Integrated DC-DC Converters. IEEE J. Solid-State Circuits **42**(1), 66–73 (2007)
[Hon00]	S.S. Hong, B. Choi, Technique for Developing Averaged Duty Ratio Model for DC-DC Converters Employing Constant On-Time Control. IEEE Electron. Lett. **36**(5), 397–399 (2000). doi:10.1049/el:20000331
[Hua06]	F. Huang, J. Lu, N. Jiang, X. Zhang, W. Wu, Y. Wang, Frequency-Independent Asymmetric Double-Pi Equivalent Circuit for On-Chip Spiral Inductors: Physics-Based Modeling and Parameter Extraction. IEEE J. Solid-State Circuits **41**(10), 2272–2283 (2006). doi:10.1109/JSSC.2006.881574
[Ing97]	M. Ingels, M.S.J. Steyaert, Design Strategies and Decoupling Techniques for Reducing the Effects of Electrical Interference in Mixed-Mode IC's. IEEE J. Solid-State Circuits **32**(7), 1136–1141 (1997). doi:10.1109/4.597306
[Ins10]	Institution of Engineering and Technology, Archives Biographies: Michael Faraday (2010), http://www.theiet.org/about/libarc/archives/biographies/faraday.cfm
[Jen02]	S. Jenei, B.K.J.C. Nauwelaers, S. Decoutere, Physics-Based Closed-Form Inductance Expression for Compact Modeling of Integrated Spiral Inductors. IEEE J. Solid-State Circuits **37**(1), 77–80 (2002). doi:10.1109/4.974547
[Jin09]	N. Jinhua, Z. Hong, B.Y. Liu, Improved On-Chip Components for Integrated DC-DC Converters in 0.13 m CMOS, in *Proceedings of the European Solid-State Circuits Conference ESSCIRC*, 2009, pp. 448–451. doi:10.1109/ESSCIRC.2009.5325987

[Joh09]   H. Johari, F. Ayazi, High-Density Embedded Deep Trench Capacitors in Silicon with Enhanced Breakdown Voltage. IEEE Trans. Compon. Packag. Technol. **32**(4), 808–815 (2009). doi:10.1109/TCAPT.2009.2024210

[Kam94]   M. Kamon, M.J. Tsuk, J.K. White, FASTHENRY: A Multipole-Accelerated 3-D Inductance Extraction Program. IEEE Trans. Microw. Theory Tech. **42**(9), 1750–1758 (1994). doi:10.1109/22.310584

[Kar06]   T. Karnik, P. Hazucha, G. Schrom, F. Paillet, D. Gardner, High-Frequency DC-DC Conversion: Fact or Fiction, in *2006 IEEE International Symposium on Circuits and Systems, ISCAS 2006, Proceedings*, 2006, pp. 245–248. doi:10.1109/ISCAS.2006.1692568

[Kaz99]   M.K. Kazimierczuk, L.A. Starman, Dynamic Performance of PWM DC-DC Boost Converter with Input Voltage Feedforward Control. IEEE Trans. Circuits Syst. I, Fundam. Theory Appl. **46**(12), 1473–1481 (1999). doi:10.1109/81.809549

[Ki98]   W.H. Ki, Analysis of Subharmonic Oscillation of Fixed-Frequency Current-Programming Switch Mode Power Converters. IEEE Trans. Circuits Syst. I, Fundam. Theory Appl. **45**(1), 104–108 (1998). doi:10.1109/81.660771

[Kos68]   O. Kossov, Comparative Analysis of Chopper Voltage Regulators with LC Filter. IEEE Trans. Magn. **4**(4), 712–715 (1968)

[Kre57]   O. Kreutzer, Plug-In Transistorised Current Chopper or Inverter Unit, UK Patent No. 836.765, 1957. Transistorized vibrator to drive a transformer with an alternating voltage, for the purpose of generating high voltages out of a battery

[Kur02]   V. Kursun, S.G. Narendra, V.K. De, E.G. Friedman, Efficiency Analysis of a High Frequency Buck Converter for On-Chip Integration with a Dual Vdd Microprocessor, in *Proceedings of the 28th European Solid-State Circuits Conference ESSCIRC*, 2002, pp. 743–746

[Kur05]   V. Kursun, G. Schrom, V.K. De, E.G. Friedman, S.G. Narendra, Cascode Buffer for Monolithic Voltage Conversion Operating at High Input Supply Voltages, in *IEEE International Symposium on Circuits and Systems ISCAS*, vol. 1, 2005, pp. 464–467. doi:10.1109/ISCAS.2005.1464625

[Kwo09]   D. Kwon, G.A. Rincón-Mora, Single-Inductor-Multiple-Output Switching DC-DC Converters. IEEE Trans. Circuits Syst. II **59**(8), 614–618 (2009)

[Lam01]   S. Lam, W.H. Ki, K.C. Kwok, M. Chan, Realization of Compact MOSFET Structure by Waffle-Layout, in *Proceeding of the 31st European Solid-State Device Research Conference*, 2001, pp. 119–122

[Le07]   H.P. Le, C.S. Chae, K.C. Lee, S.W. Wang, G.H. Cho, G.H. Cho, A Single-Inductor Switching DC-DC Converter with Five Outputs and Ordered Power-Distributive Control. IEEE J. Solid-State Circuits **42**(12), 2706–2714 (2007). doi:10.1109/JSSC.2007.908767

[Le10]   H.P. Le, M. Seeman, S.R. Sanders, V. Sathe, S. Naffziger, E. Alon, A 32 nm Fully Integrated Reconfigurable Switched-Capacitor DC-DC Converter Delivering 0.55 W/mm^2 at 81 Efficiency, in *IEEE International Solid-State Circuits Conference ISSCC: Digest of Technical Papers*, 2010, pp. 210–211. doi:10.1109/ISSCC.2010.5433981

[Lee00]   P.W. Lee, Y.S. Lee, D.K.W. Cheng, X.C. Liu, Steady-State Analysis of an Interleaved Boost Converter with Coupled Inductors. IEEE Trans. Ind. Electron. **47**(4), 787–795 (2000). doi:10.1109/41.857959

[Lee10a]   T.H. Lee, The (Pre-) History of the Integrated Circuit: A Random Walk (2010), http://www.ieee.org/portal/site/sscs/menuitem.f07ee9e3b2a01d06bb9305765bac26c8/index.jsp?&pName=sscs_level1_article&TheCat=2171&path=sscs/07Spring&file=Lee.xml

[Lee10b]   T.H. Lee, The (Pre-) History of the Integrated Circuit: A Random Walk (2010), http://www.ieee.org/portal/cms_docs_sscs/sscs/07Spring/Lee5.jpg

[Lee10c]   T.H. Lee, The (Pre-) History of the Integrated Circuit: A Random Walk (2010), http://www.ieee.org/portal/cms_docs_sscs/sscs/07Spring/Lee6.jpg

[Leu05]	C.Y. Leung, P.K.T. Mok, K.N. Leung, A 1-V Integrated Current-Mode Boost Converter in Standard 3.3/5-V CMOS Technologies. IEEE J. Solid-State Circuits **40**(11), 2265–2274 (2005). doi:10.1109/JSSC.2005.857374
[Li07]	H. Li, Z. Li, W.A. Halang, B. Zhang, G. Chen, Analyzing Chaotic Spectra of DC-DC Converters Using the Prony Method. IEEE Trans. Circuits Syst. II, Express Briefs **54**(1), 61–65 (2007). doi:10.1109/TCSII.2006.883100
[Lu10]	J. Lu, H. Jia, X. Wang, K. Padmanabhan, W.G. Hurley, Z.J. Shen, Modeling, Design, and Characterization of Multiturn Bondwire Inductors with Ferrite Epoxy Glob Cores for Power Supply System-on-Chip or System-in-Package Applications. IEEE Trans. Power Electron. **25**(8), 2010–2017 (2010). doi:10.1109/TPEL.2010.2045514
[Ma03]	D. Ma, W.H. Ki, C.Y. Tsui, P.K.T. Mok, Single-Inductor Multiple-Output Switching Converters with Time-Multiplexing Control in Discontinuous Conduction Mode. IEEE J. Solid-State Circuits **38**(1), 89–100 (2003)
[Mak91]	D. Maksimovic, S. Cuk, Switching Converters with Wide DC Conversion Range. IEEE Trans. Power Electron. **6**(1), 151–157 (1991)
[Mak95]	M.S. Makowski, D. Maksimovic, Performance Limits of Switched-Capacitor DC-DC Converters, in *Proceedings of IEEE Power Electronics Specialists Conference*, 1995, pp. 1215–1221
[Mal00]	S.Q. Malik, R.L. Geiger, Minimization of Area in Low-Resistance MOS Switches, in *Proceedings of the IEEE Midwest Symposium on Circuits and Systems*, vol. 3, 2000, pp. 1392–1395. doi:10.1109/MWSCAS.2000.951473
[Mas77]	R.P. Massey, E.C. Snyder, High-Voltage Single-Ended DC-DC Converter, in *IEEE Power Electronics Specialists Conference*, 1977, pp. 156–159
[Max54]	J.C. Maxwell, *A Treatise on Electricity and Magnetism, Parts III and IV* (Dover, New York, 1954). Reprint, first edition: 1873
[Mey92]	T.A. Meynard, H. Foch, Multi-Level Conversion: High Voltage Choppers and Voltage-Source Inverters, in *IEEE Power Electronics Specialists Conference*, vol. 1, 1992, pp. 397–403
[Mid76]	R.D. Middlebrook, S. Cúk, A General Unified Approach to Modelling Switching-Converter Power Stages, in *IEEE Power Electronics Specialists Conference*, 1976, pp. 73–86
[Moe08]	R. Moers, *Fundamentele Versterkertechniek met Elektrononbuizen*, 1st edn. (Elektor International Media BV, Susteren, 2008)
[Moh99]	S.S. Mohan, M. del Mar Hershenson, S.P. Boyd, T.H. Lee, Simple Accurate Expressions for Planar Spiral Inductances. IEEE J. Solid-State Circuits **34**(10), 1419–1424 (1999). doi:10.1109/4.792620
[Moo65]	G.E. Moore, Cramming More Components onto Integrated Circuits. Electronics **38**(8), 114–117 (1965)
[Moo75]	G.E. Moore, Progress in Digital Integrated Electronics, in *International Electron Devices Meeting*, vol. 21, 1975, pp. 11–13
[Mus05a]	S. Musunuri, P.L. Chapman, Design of Low Power Monolithic DC-DC Buck Converter with Integrated Inductor, in *IEEE Power Electronics Specialists Conference PESC*, 2005, pp. 1773–1779. doi:10.1109/PESC.2005.1581871
[Mus05b]	S. Musunuri, P.L. Chapman, J. Zou, L. Chang, Design Issues for Monolithic DC-DC Converters. IEEE Trans. Power Electron. **20**(3), 639–649 (2005). doi:10.1109/TPEL.2005.846527
[Nam09]	A. Nami, A.A. Boora, F. Zare, A. Gosh, F. Blaabjerg, A Novel Configuration for Voltage Sharing in DC-DC Converters, in *International Conference on Renewable Energies and Power Quality*, 2009
[Nob00]	G.T. Nobauera, H. Moser, Analytical Approach to Temperature Evaluation in Bonding Wires and Calculation of Allowable Current. IEEE Trans. Adv. Packaging **23**(3), 426–435 (2000). doi:10.1109/6040.861557
[Oni07]	K. Onizuka, K. Inagaki, H. Kawaguchi, M. Takamiya, T. Sakurai, Stacked-Chip Implementation of On-Chip Buck Converter for Distributed Power Sup-

	ply System in SiPs. IEEE J. Solid-State Circuits **42**(11), 2404–2410 (2007). doi:10.1109/JSSC.2007.906204
[Pap04]	G.A. Papafotiou, N.I. Margaris, Calculation and Stability Investigation of Periodic Steady States of the Voltage Controlled Buck DC-DC Converter. IEEE Trans. Power Electron. **19**(4), 959–970 (2004). doi:10.1109/TPEL.2004.830040
[Per04]	A.M. Pernia, M.J. Prieto, J.M. Lopera, J. Reilly, S.S. Linton, C. Quinones, Thick-Film Hybrid Technology for Low-Output-Voltage DC/DC Converter. IEEE Trans. Ind. Appl. **40**(1), 86–93 (2004). doi:10.1109/TIA.2003.821814
[Phi53]	Philips Electrical Industries Limited, Improvements in or Relating to Transistor Circuits, UK Patent No. 769.445, 1953. Transistorized fly-back converter, for the purpose of supplying a vacuum-tube amplifier out of a battery
[Qi00]	X. Qi, P. Yue, T. Arnborg, H.T. Soh, H. Sakai, Y. Zhiping, R.W. Dutton, A Fast 3-D Modeling Approach to Electrical Parameters Extraction of Bonding Wires for RF Circuits. IEEE Trans. Adv. Packaging **23**(3), 480–488 (2000). doi:10.1109/6040.861564
[Qiu06]	Y. Qiu, M. Xu, K. Yao, J. Sun, F.C. Lee, Multifrequency Small-Signal Model for Buck and Multiphase Buck Converters. IEEE Trans. Power Electron. **21**(5), 1185–1192 (2006). doi:10.1109/TPEL.2006.880354
[Rab03]	J.M. Rabaey, A. Chandrakasan, B. Nikolić, *Digital Integrated Circuits: A Design Perspective*, 2nd edn. (Pearson Education, Upper Saddle River, 2003). First edition: 1996
[Ram10]	Y. Ramadass, A. Fayed, B. Haroun, A. Chandrakasan, A 0.16 mm^2 Completely On-Chip Switched-Capacitor DC-DC Converter Using Digital Capacitance Modulation for LDO Replacement in 45 nm CMOS, in *IEEE International Solid-State Circuits Conference Digest of Technical Papers*, 2010, pp. 208–209
[Ran34]	A.R. Rangabe, S.J. Eintracht, Apparatus for the Supply of Electric Current to Neon and Like Luminous Discharge Tubes, UK Patent No. 432.726, 1934. DC-AC inverter, which uses an electromotor, commutator and transformer to increase a battery voltage for the purpose of gas-discharge tubes
[Rec88]	W. Reczek, W. Pribyl, Guidelines for Latch-Up Characterization Techniques, in *Proceedings of the 1988 IEEE International Conference on Microelectronic Test Structures ICMTS*, 1988, pp. 120–125
[Red09]	R. Redl, S. Jian, Ripple-Based Control of Switching Regulators, An Overview. IEEE Trans. Power Electron. **24**(12), 2669–2680 (2009). doi:10.1109/TPEL.2009.2032657
[Ric04]	A. Richelli, L. Collalongo, M. Quarantelli, M. Caramina, Z.M. Kovács-Vanja, A Fully-Integrated Inductor-Based 1.8-6-V Step-Up Converter. IEEE J. Solid-State Circuits **39**(1), 242–245 (2004)
[Rin98]	G.A. Rincon-Mora, P.E. Allen, A Low-Voltage, Low Quiescent Current, Low Drop-Out Regulator. IEEE J. Solid-State Circuits **33**(1), 36–44 (1998)
[Rio10a]	M. Riordan, L. Hoddeson, Crystal Fire: The Invention, Development and Impact of the Transistor (2010), http://www.ieee.org/portal/site/sscs/menuitem.f07ee9e3b2a01d06bb9305765bac26c8/index.jsp?&pName=sscs_level1_article&TheCat=6010&path=sscs/07Spring&file=Rior-Hodd.xml
[Rio10b]	M. Riordan, L. Hoddeson, Crystal Fire: The Invention, Development and Impact of the Transistor (2010), http://www.ieee.org/portal/cms_docs_sscs/sscs/07Spring/HR-1stTransistor.jpg
[Roy10]	Royal Institution of Great Britain, Faraday's Induction Ring (2010), http://www.rigb.org/assets/uploads/images/induction_ring_1.jpg
[Sah07]	B. Sahu, G.A. Rincon-Mora, An Accurate, Low-Voltage, CMOS Switching Power Supply with Adaptive On-Time Pulse-Frequency Modulation (PFM) Control. IEEE Trans. Circuits Syst. I, Regul. Pap. **54**(2), 312–321 (2007). doi:10.1109/TCSI.2006.887472
[Sam98]	H. Samavati, A. Hajimiri, A.R. Shahani, G.N. Nasserbakht, T.H. Lee, Fractal Capacitors. IEEE J. Solid-State Circuits **33**(12), 2035–2041 (1998). doi:10.1109/4.735545

References

[Sav03] A. Savio, A. Richelli, L. Colalongo, Z.M. Kowacs-Vajna, A Fully-Integrated Self-Tuned Transformer Based Step-Up Converter, in *IEEE Proceedings of the International Symposium on Circuits and Systems*, vol. 1, 2003, pp. 357–360

[Sch64] B.P. Schweitzer, A.B. Rosenstein, Free Running-Switching Mode Power Regulator: Analysis and Design. IEEE Trans. Aerosp. **2**(4), 1171–1180 (1964). doi:10.1109/TA.1964.4319737

[Sch06] G. Schrom, P. Hazucha, F. Paillet, D.S. Gardner, S.T. Moon, T. Karnik, Optimal Design of Monolithic Integrated DC-DC Converters, in *IEEE International Conference on Integrated Circuit Design and Technology ICICDT*, 2006, pp. 1–3. doi:10.1109/ICICDT.2006.220793

[Sed98] A.S. Sedra, K.C. Smith, *Microelectronic Circuits*, 4th edn. (Oxford University Press, New York, 1998). First edition: 1982

[Ser05] B. Serneels, T. Piessens, M. Steyaert, W. Dehaene, A High-Voltage Output Driver in a 2.5-V 0.25 µm CMOS Technology. IEEE J. Solid-State Circuits **40**(3), 576–583 (2005). doi:10.1109/JSSC.2005.843599

[Ser07] B. Serneels, High Voltage Line Drivers for XDSL in Nanometer CMOS, PhD thesis, ESAT-MICAS, K.U. Leuven, Belgium, 2007

[Sev79] R. Severns, A New Current-Fed Converter Topology, in *IEEE Power Electronics Specialists Conference*, 1979, pp. 277–283

[Sta34] Standard Telephones and Cables Limited, Improvements in or Relating to Vibrator Converters for Electric Currents, UK Patent No. 441.394, 1934. Mechanical vibrator fly-back converter, for generating high-voltage out of a battery for vacuum-tube circuits in automobile radios

[Sun02] J. Sun, Small-Signal Modeling of Variable-Frequency Pulsewidth Modulators. IEEE Trans. Aerosp. Electron. Syst. **38**(3), 1104–1108 (2002). doi:10.1109/TAES.2002.1039428

[Tak00] F. Takamura, K. Tanaka, K. Mitsuka, H. Matsushita, Low-Thermal-Resistance Flip-Chip Fine Package for 1-W Voltage Regulator IC, in *IEEE/CPMT International Electronics Manufacturing Technology Symposium*, vol. 26, 2000, pp. 305–310. doi:10.1109/IEMT.2000.910742

[Ter28] F.E. Terman, The Inverted Vacuum Tube, a Voltage-Reducing Power Amplifier. Proc. Inst. Radio Eng. **16**(4), 447–461 (1928)

[Til96] H.A.C. Tilmans, K. Baert, A. Verbist, R. Puers, CMOS Foundry-Based Micromachining. IEEE J. Micromech. Microeng. **6**(1), 122–127 (1996)

[Vee84] H.J.M. Veendrick, Short-Circuit Dissipation of Static CMOS Circuitry and Its Impact on the Design of Buffer Circuits. IEEE J. Solid-State Circuits **19**(4), 468–473 (1984)

[Vil08] G. Villar, E. Alacron, Monolithic Integration of a 3-Level DCM-Operated Low-Floating-Capacitor Buck Converter for DC-DC Step-Down Conversion in Standard CMOS, in *IEEE Power Electronics Specialists Conference*, 2008, pp. 4229–4235

[Wen07] M. Wens, K. Cornelissens, M. Steyaert, A Fully-Integrated 0.18 µm CMOS DC-DC Step-Up Converter, Using a Bondwire Spiral Inductor, in *IEEE Proceedings of the European Solid-State Circuits Conference*, vol. 33, 2007, pp. 268–271

[Wen08a] M. Wens, M. Steyaert, A Fully-Integrated 0.18 µm CMOS DC-DC Step-Down Converter, Using a Bondwire Spiral Inductor, in *IEEE Proceedings of the Custom Integrated Circuits Conference*, vol. 30, 2008, pp. 17–20

[Wen08b] M. Wens, M. Steyaert, A Fully-Integrated 130 nm CMOS DC-DC Step-Down Converter, Regulated by a Constant On/Off-Time Control System, in *IEEE Proceedings of the European Solid-State Circuits Conference*, vol. 34, 2008, pp. 62–65

[Wen09a] M. Wens, J.M. Redoute, T. Blanchaert, N. Bleyaert, M. Steyaert, An Integrated 10 A, 2.2 ns Rise-Time Laser-Diode Driver for Lidar Applications, in *IEEE Proceedings of the European Solid-State Circuits Conference*, vol. 35, 2009, pp. 144–147

[Wen09b] M. Wens, M. Steyaert, An 800 mW Fully-Integrated 130 nm CMOS DC-DC Step-Down Multi-Phase Converter, with On-Chip Spiral Inductors and Capacitors, in *IEEE Energy Conversion Congress and Exposition*, vol. 1, 2009, pp. 3706–3709

[Wes67]   Westinghouse Brake and Signal Company, Voltage Converter Circuits, US Patent No. 3.496.444, 1967. Introduction of multi-phase DC-DC converters, interleaved in time for lower output voltage ripple

[Wes73]   G.W. Wester, R.D. Middlebrook, Low-Frequency Characterization of Switched DC-DC Converters. IEEE Trans. Aerosp. Electron. Syst. **AES-9**(3), 376–385 (1973). doi:10.1109/TAES.1973.309723

[Wib08]   J. Wibben, R. Harjani, A High-Efficiency DC-DC Converter Using 2 nH Integrated Inductors. IEEE J. Solid-State Circuits **43**(4), 844–854 (2008)

[Wik10]   Wikipedia The Free Encyclopedia, Cockcroft-Walton Generator (2010), http://en.wikipedia.org/wiki/Greinacher_multiplier

[Wu98]   T.F. Wu, Y.K. Chen, Modeling PWM DC/DC Converters out of Basic Converter Units. IEEE Trans. Power Electron. **13**(5), 870–881 (1998). doi:10.1109/63.712294

[Wu03]   C.H. Wu, C.C. Tang, S.I. Liu, Analysis of On-Chip Spiral Inductors Using the Distributed Capacitance Model. IEEE J. Solid-State Circuits **38**(6), 1040–1044 (2003). doi:10.1109/JSSC.2003.811965

[Wu09]   J.C. Wu, M.E. Zaghloul, Robust CMOS Micromachined Inductors with Structure Supports for Gilbert Mixer Matching Circuits. IEEE Trans. Circuits Syst. II, Express Briefs **56**(6), 429–433 (2009). doi:10.1109/TCSII.2009.2020925

[Yim02]   S.M. Yim, T. Chen, K.K. O, The Effects of a Ground Shield on the Characteristics and Performance of Spiral Inductors. IEEE J. Solid-State Circuits **37**(2), 237–244 (2002). doi:10.1109/4.982430

[Zar08]   J. Zarebski, K. Gorecki, A SPICE Electrothermal Model of the Selected Class of Monolithic Switching Regulators. IEEE Trans. Power Electron. **23**(2), 1023–1026 (2008). doi:10.1109/TPEL.2008.918377

[Zar10]   J. Zarebski, K. Gorecki, The Electrothermal Large-Signal Model of Power MOS Transistors for SPICE. IEEE Trans. Power Electron. **25**(5), 1265–1274 (2010). doi:10.1109/TPEL.2009.2036850

[Zha06]   F. Zhang, P.R. Kinget, Design of Components and Circuits Underneath Integrated Inductors. IEEE J. Solid-State Circuits **41**(10), 2265–2271 (2006). doi:10.1109/JSSC.2006.881547

Index

A
AC voltage, 104, 106
alternating-current, 4
aluminum, 154, 159
aluminum-oxide ceramic, 227
ampère meter, 224, 228
analog to digital converter, 205
angular frequency, 45
area, 107
asynchronous switching, 118

B
bandwidth, 173
bias current, 202, 207, 227
BiCMOS, 220
binary code, 205
binary counter, 205, 207
bipolar junction transistor, 9, 221, 226
bonding pad, 141, 215, 230, 237, 253
bondwire, 136, 139, 154, 155, 157, 215, 228, 230, 233, 236, 237, 241, 246, 252, 253
bondwire pitch, 230, 236, 237, 252, 253
boundary conduction mode, 77
buffer
    digital tapered, 149, 152, 175, 223
    dynamic losses, 153
    fan-out, 153
    number of stages, 153
    output resistance, 149
    pre-driver, 223
    propagation delay, 153
    scaling factor, 153, 175, 186, 188, 192, 198, 202, 208
    start-up, 209
busy-detector, 184, 187, 190, 197, 201

C
capacitance
    density, 21, 145, 157, 164, 216, 217, 219
    output, 74, 180
capacitor
    aspect ratio, 218
    biasing, 252
    charging, 141
    decouple, 11, 72, 81, 89, 208, 215, 217, 225, 226, 228
    deep-trench, 216
    discharging, 141
    electric series resistance, 142, 145, 180, 216–218, 220
    finger-shaped, 218
    flying, 34, 37, 38, 40, 76
    frequency response, 225
    input decouple, 82, 98, 156, 180, 217, 233, 236, 240, 241, 245, 249, 250, 252, 253
    interleaved wire, 216
    lay-out, 142
    leakage current, 144
    maximum operating voltage, 216
    micro-machined, 216
    MIM, 21, 118, 145, 157, 164, 216, 217, 220, 228, 232, 233, 236, 240, 244, 245, 248–250, 252
    MOM, 21, 216, 217, 220, 252
    MOS, 21, 118, 145, 157, 164, 174, 188, 198, 207, 216, 217, 219, 220, 233, 236, 240, 245, 249, 252
    on-chip, 216
    output, 11, 37, 40, 77, 89, 98, 104, 107, 110, 115, 117–120, 142–144, 152, 162, 184, 215, 217, 220, 230, 233, 240, 241, 245, 249, 250, 252, 253

capacitor (*cont.*)
    parallel resistance, 142, 144
    patterned, 214, 215, 252
    plate, 218
    slots, 252
    trench, 103
causality, 195
ceramic substrate, 227
channel-length modulation, 17
chaos, 171, 185, 197
charge balance, 51, 55, 57, 70, 83, 128, 129, 133, 135, 162
charge-coupling, 175, 207
charging
    capacitor, 33, 34
    capacitor equilibrium, 35
    inductor, 41–43
    inductor-capacitor, 49
chip
    die area, 164
    die temperature, 159, 161
    package, 155, 227, 228, 233
    stacking, 154
clock divider, 207
clock frequency, 205
CMOS
    active area mask, 18
    bulk contact, 221
    contact mask, 18
    deep-submicron, 184, 195, 204, 216, 217
    field oxide, 18
    high-voltage, 205
    introduction, 15
    lay-out, 18, 218, 246
    metal layers, 101, 164, 228, 240, 244, 248, 250
    metal mask, 18
    minimum feature size, 217
    n-well, 221
    n-well mask, 18
    nominal supply voltage, 220
    p-diffusion mask, 18
    polysilicon mask, 18
    scaling laws, 19
    substrate, 214, 215, 227
    substrate contact, 221
    technology, 18
    technology node, 217
    thick-film, 259
    thick-top metal, 228, 232, 233, 236, 240, 244, 245, 249
    variability, 184
    via, 214
Cockcroft-Walton, 5
common emitter, 173
compact fluorescent lamp, 106
comparator, 173, 174, 180, 184, 186, 188, 196, 199, 205, 227
compensation ramp, 171
conduction boundary, 57, 58, 69
conduction losses, 59, 100, 117, 177
conduction mode, 49, 58, 67
constant on/off-time, 181, 188, 194, 203, 207, 232, 236, 240, 247, 257
continuous conduction mode, 49, 67, 71, 73, 75, 78, 80, 85, 87, 92, 94, 96, 100, 101, 103, 104, 108, 113, 118, 119, 130, 135, 149, 151, 162, 171, 172
control strategy, 165
control system
    bang-bang, 175
    boost converter, 255
    COOT, 184, 186, 188, 205, 233, 237, 241
    definition, 11
    efficiency enhancement factor, 257
    F^2SCOOT, 203, 205, 252, 253
    hysteric, 175, 176
    linear series converter, 28
    linear shunt converter, 30
    one-shot, 175
    output voltage ripple, 178
    PFM, 177, 179
    power supply, 170, 209
    PWM, 172, 177, 179, 183, 229, 230, 237, 241
    ripple based, 175
    SCOOT, 196, 199, 245, 246, 249
    stability, 243, 248
    start-up, 210
copper, 154, 159
current
    drain, 226
    feedback, 171
    sensing, 95, 97, 98, 173, 182–184, 195, 204, 229
    source, 207
    unbalance, 108, 222
cutoff-frequency, 197, 198

D
2-D field-solver, 215
D-flipflop, 205, 207
damping
    aperiodically, 44
    critically, 44

Index

damping (cont.)
  damped, 45
  periodically, 44, 47
  undamped, 45
DC voltage, 104
DC-DC converter
  boost, 49, 84, 103, 112, 124, 126, 135, 143, 162, 165, 172, 178, 180, 181, 192, 204, 228, 232
  bridge, 72
  buck, 66, 100, 101, 114, 131, 132, 135, 143, 162, 165, 179, 180, 182, 183, 192, 203, 204, 210, 221, 236, 240, 244, 248, 250
  buck^2, 77
  buck-boost, 91, 104
  capacitive coupled buck, 108
  charge-pump, 32, 210, 257
  control system, 170
  Ćuk, 93
  current-fed bridge, 85
  definition, 2
  fly-back, 11, 104
  forward, 11, 100
  four-phase, single-output buck, 196
  four-phase, two-output SMOC buck, 199
  four-phase boost, 112
  four-phase buck, 114
  full-bridge, 101
  galvanic separated, 99, 116
  half-bridge buck, 102
  inductive, 41
  inverse SEPIC, 95
  inverse Watkins-Johnson, 86
  linear, 27, 209, 210
  linear series, 28, 30, 37, 40, 60, 61, 205, 226, 253, 257
  linear shunt, 29, 202, 208, 211
  measurement, 224
  multi-level, 74
  multi-phase, 107, 171, 193, 194, 216, 257
  multiple-output, 199, 205
  n-phase boost, 110
  n-phase buck, 113
  non-inverting buck-boost, 92
  push-pull boost, 103
  resonant, 105
  rotary, 4
  SEPIC, 94
  series resonant, 104
  series-parallel step-down, 34
  series-parallel step-up, 38
  SIMO, 115, 192
  SIMO boost, 116, 118
  SIMO buck, 117, 119
  single-phase, 107, 194
  single-phase, single-output buck, 184, 186
  single-phase, two-output SIMO boost, 188
  single-phase, two-output SMOC buck, 205
  SMOC, 118, 192, 199, 205, 248
  SMOC boost, 118
  SMOC buck, 119
  step-down, 210
  step-up, 210
  three-level buck, 74
  two-level, 76
  two-output SIMO boost, 118
  two-output SIMO buck, 119
  two-output SMOC boost, 118
  two-output SMOC buck, 119
  two-phase boost, 110, 112
  two-phase buck, 113, 114
  vibratory, 5
  Watkins-Johnson, 79
  zeta, 95
DC-gain, 173
dead-time, 54, 55, 69, 71, 185, 187, 190, 197, 199, 201
design
  parameters, 228
  space, 164
  trade-off's, 166
dielectric, 136, 137
differential equation
  first order, 125
  second-order, 126, 131, 132, 162
digital inverter, 174
digital to analog converter, 205, 207
diode, 146
  forward voltage drop, 152
  laser, 223
  n-MOSFET, 202
  p-MOSFET, 211
  protection, 208
  reverse recovery, 108, 147
  zener, 30
direct-current, 2
discontinuous conduction mode, 53, 68, 71, 73, 78, 80, 86, 88, 94, 96, 127, 129, 133, 140, 149, 151, 156, 162, 171, 172, 176, 178, 181, 184, 186, 188, 192, 195, 199, 204, 205
dual-in-line package, 227
duty-cycle, 49, 105, 171, 174, 186, 188, 191
  definition, 52
dynamic behavior, 171
dynamic losses, 22, 135, 146, 160, 161

E

eddy-currents, 214, 215, 252
effective area, 139
efficiency enhancement factor, 59, 61, 237, 241, 246, 257
electro magnetic interference, 143, 180
energy charging efficiency
    capacitor, 33
    inductor, 43
    inductor-capacitor, 47
energy conversion efficiency
    ideal transformer, 101
equivalent load resistance, 105
equivalent series inductance, 142
equivalent series resistance, 142
equivalent T-circuit, 101
error amplifier, 179
    linear series converter, 28
error-voltage, 171
even harmonics, 144, 180

F

FAIMS, 106
fall-time, 174, 223
FastHenry, 138, 139, 155, 215, 216
feasibility, 164
feed-forward, 170, 203, 205
feed-forward semi-constant on/off-time, 203, 204, 251
feedback, 30, 77, 170, 185, 205, 207, 224, 226, 228, 229, 232, 233, 236, 237, 240, 244, 245, 248, 249, 251, 252
feedback-loop, 179
field-effect transistor, 9
field-solver, 138
figure of merit, 59
finite element, 138
flip-chip, 154
flipflop initialize, 209
flow-chart, 163
form-factor, 110
forward voltage drop, 146
freewheeling diode, 118, 146, 147
freewheeling switch, 146, 147, 152, 182, 183, 192, 195, 204
frequency band, 181
frequency response, 171
frequency spectrum, 144
frequency-domain, 179, 181
fundamental frequency, 109, 172

G

gain, 179
gain-bandwidth, 173, 209
gallium arsenide, 220
galvanic separated, 101
galvanic separation, 11, 99
gate-leakage, 19
gear-box topology, 38, 41
gold, 159
ground-bounce, 223

H

half-bridge driver, 221, 232
heat dissipation, 23
high frequency, 220
high-voltage, 74, 106

I

ideal switching, 3
idle time, 182, 194
impedance transformer, 2
inductance
    bondwire, 136
    density, 21
    input, 230, 241, 246, 253
    magnetizing, 100
    mutual, 4, 100, 137, 214, 216
    negative mutual, 137, 215
    positive mutual, 137
    self, 136, 214
    total, 215
    variation, 232
induction ring, 4
inductor
    bar, 214
    bondwire, 21, 136, 138, 141, 159, 214–216, 228, 230, 236, 237, 252, 255, 257
    core losses, 214
    coupled, 79, 80, 87, 101, 108, 137
    current ripple, 50, 54
    ferro-magnetic core, 214
    hollow-spiral, 21, 138, 214, 216
    input, 104
    lumped model, 136
    metal-track, 21, 136, 138, 139, 141, 142, 159, 164, 214–216, 233, 240, 245, 249, 255, 257
    number of windings, 214
    on-chip, 214
    output, 104
    parasitic series resistance, 136
    parasitic substrate capacitance, 136, 141, 142
    series resistance, 159

Index

inductor (cont.)
    solenoid, 214
    spiral, 137
    substrate capacitance, 215
    thick-film, 214
    windings, 216, 233
input voltage, 170
input voltage ripple, 155, 156
integrator, 172, 179
interconnect, 23, 154–158
interleaving, 195, 201
inversion layer, 16
inverter, 152

J

Joule-losses, 21, 22, 29, 42, 48, 141

L

Laplace-domain, 101
laser imaging detection and ranging, 223
latch, 207
latch-up, 152, 221, 224
lay-out, 108, 154, 158
LC-network, 105, 106
LCC-network, 106
Lenz's law, 45, 55, 152
level-shifter, 173, 175, 185, 187, 191, 202, 207–209, 226
line regulation, 170, 225–227
load regulation, 170, 179, 180, 184, 195, 204, 225–227, 231, 239, 243, 247
low-drop out regulator, 29
low-pass filter, 179, 197, 198, 201, 205, 207

M

magnetic
    core, 103
    coupling factor, 100
    energy, 22
mains power supply, 106
mains-operated, 10, 99
Mathematica, 161
MatLAB, 161
maximal efficiency tracking, 194
maximum operating voltage, 118–120, 157, 164
measurement, 231, 237, 241, 246
metal-tracks, 154, 155, 158
micro-converter, 103
micro-inductors, 103
micro-machining, 103, 214, 216
micro-photograph, 231, 235, 237, 239, 241, 245, 250, 253
micro-transformers, 103

minimum feature size, 147, 153, 164
mismatch, 174, 184, 186, 188, 191, 195, 204
model flow, 162
mono-stable multi-vibrator, 185, 186, 190, 196, 199
Moore's law, 17
MOSFET
    body diode, 228, 233, 236, 245, 249
    bulk, 16, 152, 219, 233, 236, 244, 248
    bulk conduction, 152, 183, 197, 221, 229
    depletion region, 16
    drain, 16, 152, 219
    drain connection, 222
    drain-bulk capacitance, 148, 221
    drain-bulk diode, 152
    drain-source leakage current, 160
    drain/source contact, 219
    early voltage, 149
    enhancement, 16
    fall-time, 149
    finite switching time, 147, 148
    gate, 16, 217, 219
    gate capacitance, 149, 153, 208
    gate contact, 219
    gate leakage, 217
    gate-bulk capacitance, 148
    gate-drain capacitance, 148
    gate-oxide thickness, 217
    gate-source capacitance, 148
    guard ring, 224
    induced channel, 160, 217, 219
    lateral p-DMOSFET, 250
    lay-out, 152, 220, 222, 224
    leakage current, 159
    linear finger, 220
    linear region, 17
    n-type, 16
    n-well, 152
    on-resistance, 146, 147, 159, 164, 210, 257
    oxide thickness, 164
    parasitic capacitance, 147
    rise-time, 149
    saturation region, 16
    saturation voltage, 149
    source, 16, 219
    source connection, 222
    source-bulk capacitance, 148
    stacked, 23, 76, 94, 96, 202, 211, 220, 228, 232, 236, 244, 248, 257
    substrate, 16, 72, 74, 101, 137, 152
    substrate contact, 224
    switch, 146–148, 152, 159, 164
    temperature coefficient, 160

MOSFET (cont.)
  thick-oxide, 23, 76, 220, 257
  threshold voltage, 16
  triode region, 17, 147
  triple-well, 191, 220
  vertical n-DMOSFET, 250
  waffle-shaped, 220, 222
  width, 224
multi-phase, 8
multiplexer, 205, 207

N
negative feedback, 172
Newton-Raphson, 129, 133, 134, 162
noise, 205, 226

O
odd harmonics, 143, 144, 180
off-time, 50, 68
offset, 180
offset-time, 194
on-time, 50, 53, 67, 68
OPAMP, 28, 172, 173, 208, 226
optimization, 216
oscillator, 15
oscilloscope, 225, 227
output filter, 53, 56, 57, 70, 71, 76, 81, 84, 89, 109
output resistance, 105, 246
output stage, 173
output voltage
  boost converter, 129
  buck converter, 130, 134, 135
  control system, 170
  linear series converter, 28
  linear shunt converter, 29
  RMS, 161
  series-parallel step-down, 34
  series-parallel step-up, 39
output voltage ripple, 22
  boost converter, 53, 56, 130
  buck converter, 70, 71, 130, 134, 135
  dependency, 178
  ESR, 142
  F^{2}SCOOT, 203
  input voltage ripple, 157
  load regulation, 204, 231, 243, 247
  metal layers, 164
  multi-phase boost converter, 111
  multi-phase converter, 108
  output power, 179
  SCOOT, 193–195
  small-ripple approximation, 50
  start-up circuit, 211
  subharmonic oscillations, 172
  two-phase boost converter, 110
  two-phase buck, 113
  two-phase buck converter, 113
oxide, 141, 216

P
parallel resonant LC network, 106
parasitic thyristor, 221
pass-transistor, 226
passive integration, 103
passive-die, 154
PCB, 155, 158
PCB track resistance, 224
period, 46
periodically damped, 129
permeability, 139
  relative, 136
permittivity, 141, 217
  relative, 141
phase difference, 109
phase-margin, 209
piecewise linear approximation, 128
positive feedback, 174, 221
positive temperature coefficient, 59, 159
power
  balance, 2, 161, 162
  density, 13, 23, 104, 115, 164, 255, 257, 259
  dissipated, 163
  dissipation, 3, 29, 31
  input, 225
  losses, 160, 161, 220
  maximum output, 255, 256
  output, 30, 107, 231, 237, 241, 246
  supply, 227
power activity probability distribution, 61
power conversion efficiency, 163, 164
  boost converter, 52, 88, 255
  buck converter, 81, 257
  COOT, 184
  DC-DC converter, 225
  definition, 3
  EEF, 59
  F^{2}SCOOT, 203
  galvanic separated converters, 104
  ideal transformer, 100
  linear series converter, 29
  linear shunt converter, 30
  measure, 226
  multi-phase converter, 107
  output power, 194, 231, 237, 241, 246
  SCOOT, 193

Index 279

power conversion efficiency (cont.)
　SEPIC converter, 97
　series-parallel step-down, 37
　series-parallel step-up, 40
　SIMO buck converter, 117
power supply rejection ratio, 207
printed circuit board, 223, 224, 226, 227
probability distribution function, 257
pull-down transistor, 208
pulse frequency modulation, 116, 170, 175, 178, 192, 193, 204
pulse width modulation, 116, 170–172, 178, 179, 192, 194, 195, 204, 228, 257

Q
Q-factor, 108, 138, 164, 215, 252, 255, 257

R
rail-shifter, 208, 209
rail-shifting, 29
real off-time, 54, 69
rectifier
　bridge, 11
　full-bridge, 99, 103–105
　mechanical, 5
　output, 4, 11, 103
reference voltage, 171, 184, 186, 196, 199, 208, 226
resistance
　input, 230
　series, 139, 154, 155, 160
　square, 154, 216, 217, 219
　temperature coefficient, 159
resistive divider, 28, 62, 202
resistive losses, 124, 131, 132, 135, 147
resistivity, 138, 139, 142, 164
resistor
　sense, 225
　variable, 225
resonance frequency, 76, 105
resonant DC-AC converter, 106
rise-time, 174, 223
RLC-circuit, 44
RLC-decoupling, 156
RLC-network, 129

S
schmitt-trigger, 197, 198, 201, 211
self heating, 160, 163
self-starting, 229, 233, 237, 241, 245, 253
semi-constant on/off-time, 193–195, 204, 244, 248, 257
series resonant LC network, 104
series RLC-circuit, 45

settling time, 34
short-circuit, 152–154, 197
signal generator, 226, 227
simulation, 231, 235, 238, 247, 250
sine wave, 109, 226
skin-depth, 139
skin-effect, 139, 154, 155
slots, 215
small-ripple approximation, 49, 50, 52–54, 56, 57, 71, 83, 124
small-signal, 170
SPICE simulations, 66, 71, 73, 76, 78, 80, 83, 84, 86, 88, 91, 93, 94, 96, 147, 160, 163, 171, 180
SPICE-model, 160
square wave, 104, 110, 226
SR-flipflop, 187, 190, 197, 201
stability, 171
start-up, 29, 31, 208
start-up circuit
　boost converter, 229, 233
　buck converter, 187, 237, 241, 245, 248, 253
　concept, 210
　definition, 209
　implementation, 210
　linear series converter, 28
　rail-shifter, 208, 209
　SCOOT, 201
state-space averaging, 171
steady-state model, 124, 126, 132, 135, 136, 177, 204, 228, 231, 235, 238, 243, 247, 250, 257
step-down converter
　battery, 12
　definition, 2
　EEF, 59, 62
　overview, 65
　summary, 81
step-up converter
　battery, 13
　definition, 2
　overview, 82
　summary, 88
step-up/down converter
　definition, 79
　overview, 90
　summary, 97
subharmonic oscillations, 171, 172
superposition, 140
surge-voltage, 223
switch, 146, 229
switch-network, 104, 105, 109

switching frequency, 22, 34, 39, 52, 105, 164
switching losses, 135, 177
switching transients, 149, 152, 220
symmetrical cascoded OTA, 173
symmetrical OTA, 209
synchronous rectification, 118, 184, 186, 188, 199, 205
system-on-chip, 11

T
temperature effects, 177, 231, 237
thermal resistance, 158
time constant
    LC, 45
    RC, 32
    RL, 42
time-delay, 173, 174, 185, 188
time-domain, 107, 195
trade-off, 164
transfer function, 104
transformer, 4, 104, 255
    air-core, 104
    center-tapped, 103, 104
    core losses, 100
    galvanic separation, 99
    ideal, 100
    monolithic, 100
    number of turns, 100
    toroidal, 4
    winding turn ratio, 100
transmission gate, 207
triangular waveform, 110, 171, 173
true RMS, 225
tungsten, 155

U
under-etching, 214

V
vacuum, 141
vacuum-tube, 6
    diode, 7
    inverted, 6
    triode, 6, 7
via, 155
volt meter, 224, 225, 228
volt-second balance, 51, 52, 55, 83, 100
voltage
    multiplier, 5
    reference, 227
    ripple, 208
    source, 224
voltage conversion ratio
    boost converter, 52, 56, 255
    bridge converter, 72
    buck converter, 68, 69, 257
    buck^2 converter, 77
    buck-boost converter, 91
    Ćuk converter, 93
    current-fed bridge converter, 85
    definition, 2
    duty-cycle, 49
    EEF, 59
    fly-back converter, 104
    forward converter, 100
    full-bridge buck converter, 101
    inverse Watkins-Johnson converter, 87
    linear converter, 28
    linear shunt converter, 30, 31
    non-inverting buck-boost converter, 92
    push-pull boost converter, 103
    SEPIC, 94
    series resonant converter, 104
    three-level buck converter, 75
    Watkins-Johnson converter, 79
    zeta converter, 95

CPSIA information can be obtained at www.ICGtesting.com
Printed in the USA
238002LV00002BA/18/P